T0339745

SOCIAL IMPACTS OF SMART GRIDS

SOCIAL IMPACTS OF SMART GRIDS

The Future of Smart Grids and Energy Market Design

WADIM STRIELKOWSKI

Centre for Energy Studies, Prague Business School, Prague, Czech Republic
Cambridge Institute for Advanced Studies, Cambridge, United Kingdom

ELSEVIER

Elsevier
Radarweg 29, PO Box 211, 1000 AE Amsterdam, Netherlands
The Boulevard, Langford Lane, Kidlington, Oxford OX5 1GB, United Kingdom
50 Hampshire Street, 5th Floor, Cambridge, MA 02139, United States

Notices
Knowledge and best practice in this field are constantly changing. As new research and experience broaden our understanding, changes in research methods, professional practices, or medical treatment may become necessary.

Practitioners and researchers must always rely on their own experience and knowledge in evaluating and using any information, methods, compounds, or experiments described herein. In using such information or methods they should be mindful of their own safety and the safety of others, including parties for whom they have a professional responsibility.

To the fullest extent of the law, neither the Publisher nor the authors, contributors, or editors, assume any liability for any injury and/or damage to persons or property as a matter of products liability, negligence or otherwise, or from any use or operation of any methods, products, instructions, or ideas contained in the material herein.

British Library Cataloguing-in-Publication Data
A catalogue record for this book is available from the British Library

Library of Congress Cataloging-in-Publication Data
A catalog record for this book is available from the Library of Congress

ISBN: 978-0-12-817770-9

For Information on all Elsevier publications
visit our website at https://www.elsevier.com/books-and-journals

Publisher: Brian Romer
Acquisition Editor: Graham Nisbet
Editorial Project Manager: Aleksandra Packowska
Production Project Manager: Selvaraj Raviraj
Cover Designer: Greg Harris

Typeset by MPS Limited, Chennai, India

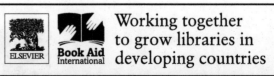

Dedication

This book is dedicated to Melanie.

I would like to thank Katka for her patience, my Mom for her optimism and stubbornness, my Dad for his calmness, and my Grandmother Raisa for praying for all of us.

Special thanks go to Professor Evgeny Lisin, my long-term colleague and co-author, as well as an excellent energy economics researcher, to whom I could always turn for assistance.

Contents

7. The role and perception of energy through the eyes of the society

8. Smart grids of tomorrow and the challenges for the future

9. Conclusions

1

General introduction

1.1 Introduction: the idea behind this book

For the most part of the 20th century, traditional energy grids were used to carry power from a few central generators to a large number of customers. However, with the growing complexity of the today's globalized world, these traditional grids are going to gradually evolve into something that has gained a nickname of "smart grids." According to the established concept, a smart grid employs two-way flows of electricity and information to create an automated energy delivery network. Modern smart grids are instantly aware where and to whom the electricity should be delivered and are capable of promptly reacting to the changes in demand and supply. With the pressure on the electricity grids intensifying due to the commitments for the low-carbon future made by governments in the majority of the Western countries, the future of electricity networks is likely to face a number of challenges, including the new patterns of consumption, planning under an increasing uncertainty, and overall growing complexity due to the large number of small independent devices connected to the network (a concept that is known as the "Internet of Energy" (IoE) that is a subset of a larger and complex concept of the "Internet of Things" (IoT)).

Nowadays, we live in the age of digitalization. Digital technologies surround us everywhere and even, to some extent, start steering and shaping up our lives and our sense of purpose and existence. Digitalization was created with a purpose of freeing our labor, giving us more time to be spent with our families and friends, or endorsing us in pursuing our hobbies. However, the ubiquitous digitalization is doing nothing else than taking humans into the world of illusions. The threat of World War III (WWIII) is probably no longer real, but the twits and Facebook feeds speculating on how, for example, United States and Russia got to the brink of another Cold War are there. Yet, these twits and posts are not real, they are mere illusions. One can say that we live in the world of illusions that draws from our fears. Digital technologies simply amplify our fears.

The best example of this is the modern approach to journalism and making news. Long gone are the days when the journalists travel to war zones to report on numerous conflicts civil wars and humanitarian disasters. Nowadays, journalists sit in the comfort of their homes and follow what the world leaders have to say on their Twitter and Facebook accounts.

1

I remember when the military conflict in Ukraine started several years ago. The news on both sides of the conflict describing the escalation of the conflict and then the movement and the advancements of the opposing forces used twits of field commanders as their credible source of information. Without even going to the field, attending a war zone, ducking under bullets, checking the sources, and verifying the information in person—all those things journalists had to do in the past—many modern-day journalists were simply writing their stories based on what this or that person wrote on her or his Facebook page or Twitter. Some of these stories were groundbreaking, claiming thousands of deceased, large battles won or lost, key figures of the conflict killed or imprisoned—all just based on someone's twit or post. I call this approach to creating and reporting stories a "coach journalism."

And this is not just about journalism or news for ordinary people. American president Donald Trump started something that is known as "Twitter diplomacy." Millions of people everyday wake up and check Twitter just to see where the wind of the world politics is blowing. In Russia, the Ministry of Foreign Affairs had to react according to Mrs. Maria Zakharova, a spokeswoman, who is proficient in using social media. Mrs. Zakharova is the voice of Russian foreign politics responding to Mr. Trump's twits and posts using her own twits, messages, and yet another posts.

Digitalization creates a world that is full of illusions. A plethora of information makes it difficult to tell right from wrong. Human brain cannot process all the information because it has a physical capacity that cannot be extended. We are limited by our physical constraints, while the artificial intelligence (AI), a set of machine-powered algorithms, is not. In the eyes of many people, AI is becoming another useful remedy for the information overload, something that digitalization used to be in the 1980s and 1990s.

In the mid-1990s, researchers and futurologists were preaching that information technologies (ITs) would become an important player and would make the lives of billions of people easier and in general would make people happier in the early 21st century. Now, already in the 21st century, we are dominated by the influence of the technology. However, people are still not happy, and our lives are not becoming easier but (what a paradox!) more complicated. Nowadays, most of us have troubles with time planning and are required and expected to perform under enormous stress. Back in the 1990s, we expected that our lives would become easier, but they became harder instead.

In the 19th century, people did not have much time to communicate with each other because the only possibility was the snail mail. Writing and posting a letter required stationery: ink, paper, envelope, and a stamp. In addition, it was necessary to visit a post office and often spend a considerable amount of time in the queue. Receiving a letter was a memorable event and we all had a habit of checking our mailboxes as frequently as we are now checking our smartphones for the signs of the new messages or e-mail.

In the 20th century, in addition to e-mail the humanity invented a telephone. However, the telephones on the most part of the 20th century were stationary and mostly installed at homes or existed in a form or public payphones across various locations. Quote often, one had to walk for miles to make a phone call. I remember how in my youth we used to call in to one of our friend's apartment to leave a message with his parents about the whereabouts of the rest of the gang. Everyone searching for the rest of this buddies within a big city would call the designated apartment and obtain the information about where to go.

We also used a public payphone nearby to regularly check on whether anyone called and asked about our location—in order to do that, one of us had to take a walk for 15 minutes, call the designated apartment, and then walk back to report to the rest of us.

Nowadays, we write hundreds of messages, short notes, and letters everyday and send them through various messengers, e-mail, social media, or other applications. The result of this is the increased time pressure, as well as piling amounts of tasks at work and personal life. All these is leading to the increased and tightened controls and imposes more stress upon us. It is very likely that the future is not going to be different because the most important features of humans and machines do not support each other very well.

It might be that the feelings of excitement and stress are a necessity of being a human. The ITs dominate our lives and manipulate our feelings over time. For those of us who remember receiving a real handwritten letter, the feeling cannot be compared to the feeling of receiving a letter via e-mail. And this is not because the letter sent by an e-mail is less important. This is because of the personal attachment and feeling we used to attribute to the real paper, the real envelope, and the real handwriting that can be easier associated with a real person one can picture while reading the letter in question.

Smart grids of the future will be heavily reliant upon ITs that would require efficient and quick Internet connection connecting all parts and corners of the world. However, many people are not ready to change their consumption behavior. There is a big difference do support something publicly and openly or put a "like" for something at one's Facebook account or to actually stand up from one's chair, step out of one's apartment, go out and do something.

Dealing with smart grids of the future would require both understanding basic principles of cyber security and following the rules of personal behavior on the Internet. Internet or the worldwide web not only offers a lot of comfort in our everyday lives, but they also pose numerous threats most of us are unaware how to process or to handle.

This book is titled "Social Impacts of Smart Grids" with a subtitle "The Future of the Smart Grids and Energy Market Design," which involves the prediction of how the smart grids and energy markets might evolve in the nearest future. When one is attempting to predict the future, there is always one simple but very substantial problem: most of predictions simply do not come true. The laws of the universe are very complicated and seem to be governed by the chaos theory rather than by some divine powers. Thence, any predictions usually omit very important factors that might change everything.

Take the story of smartphones, for example. The first smartphone on the market was the first model of iPhone which premiered in July 2007. Now, 11 years later, we are totally dependent on smartphones and cannot imagine our lives without them.

In 2006–2009 I worked as a Research Fellow at the School of Built Environment, University of Nottingham. My main responsibilities were to handle the EU Framework Programme 6 project entitled "Integrated e-Services for Advanced Access to heritage in Cultural Tourist Destinations" (project acronym ISAAC, project number FP6-IST-2006-035130) (Cordis, 2019).

ISAAC project was an interesting experience. I often traveled to Amsterdam, Leipzig, and Genoa. Especially, Amsterdam was very interesting and the most advanced European city in terms of electronic and digital services for tourists and residents. The City Hall and the seat of the Municipality of Amsterdam is located at the famous Stopera building complex that also houses Dutch National Opera and Ballet, the principal opera house in Amsterdam that is home of Dutch National Opera, Dutch National Ballet, and Holland Symfonia. Once, we had a project meeting in Saint Petersburg where our Russian partner at that time, Russian State Museum, had the St. Michael's Castle (also known as "Mikhailovsky Castle" or "the Engineers' Castle," a place in which Emperor Paul I was assassinated in March 1801) closed for 3 days to the public for us to have our sessions undisturbed.

ISAAC project brought together researchers, ICT companies, city authorities, and cultural institutions from the five EU countries and Russia, including the Research Centre Karlsruhe, University of Nottingham, the University of Sunderland, Free University of Amsterdam, Russian State Museum, and world-famous Hermitage Museum. The project focused on a case study of three European cities—Amsterdam, Genoa, and Leipzig—pooling knowledge and experience in the fields of digital culture and heritage, e-tourism, cultural tourism management and urban e-governance (Chiabai et al., 2014).

ISAAC's main objective was to enhance the relationship between digital heritage and cultural tourism by developing a novel user-centric information and communication environment providing tourism e-services for tourists and citizens in European cultural destinations, facilitating virtual access and stimulating learning experience of European cultural heritage assets before, during and after a real visit. ISAAC main output was the development of a novel user-centric information and communication technologies (ICTs) environment based on new e-services and the integration of existing e-services meeting the needs of all cultural stakeholders: citizens, tourists, and private and public entities.

ISAAC project lasted for 3 years between 2006 and 2009, and most of our visionary ideas that were shaped up during this time concerned personal computers. The first smartphone (the first model of iPhone) was introduced to the market in July 2007 and none of us had the gadget and could not envisage how it would change the tourist and resident experience in cultural cities worldwide.

The same might happen with the smart grids and the energy market of the future. The way we can imagine it would evolve and develop might be totally wrong due to many things. One of them is batteries. The batteries and charging devices we have today are very limiting devices that have lots of drawbacks.

Although water storage currently dominates the world's conventional (electric energy storage, EES), rapid ongoing decreases in the cost of batteries raises hopes that batteries based on storing energy using chemical components and reactions chemical will offer a new and attractive storage option. For example, Newbery and Strbac (2016) summarize estimates for 2020 battery energy storage costs, which range from 253 to 345 EUR/kWh for the battery pack as opposed to the today's costs of about 1117 EUR. Moreover, there is still no battery revolution for a future smarter energy system in sight despite the plethora of research focused on improving performance and reducing costs of battery storage across electrochemical, mechanical, and thermal devices. However, everything is not that simple as it seems. It is true that opposed to the existing batteries that draw from the chemical energy have short lifetimes and prove inviable under the current electricity prices (Staffell and Rustomji, 2016), the hydropower EES use the free storage medium (water) and can operate for more than 100 years. Nevertheless, their potential of gravitational energy is remarkable weak compared to chemical energy and their high capital costs and their distance from demand centers sometimes make them less favorable options then battery energy storage. Yet, a new battery storage device might come with a completely new technology and change the energy market from the grounds. However, we do not know whether it will appear tomorrow, in several years, or perhaps decades. The future is wrapped in the shadows just like the encounter of the electrification plan (called "GOERLO") conducted in Soviet Russia and described in the Herbert George Wells' book published in 1921 (Wells, 1921).

1.2 Book rationale and its structure

This book goes beyond the traditional analysis of the future of the smart grids by exploring their social impacts. It focuses in how the producers and consumers of energy on the traditional energy markers are giving their place to energy prosumers and how the society perceives these changes. It explores the raising popularity of the sharing economy concept and how it is used in smart grids. Furthermore, it tackles upon the social acceptance of smart grids based on the sustainability concepts and ideas and explores people's attitude toward the issues of peer-to-peer electricity markets that are brought about by the recent developments and changes in transportation (electric vehicles), housing (smart homes), as well as energy generation and transmission (IoE).

Why I am writing about the social acceptance of the new technologies that are designed to make the energy markets of the future more efficient and the interactions between energy prosumers more smooth and secure? Because we are humans and all technology and advanced research in the world has one single purpose—to serve our existence. Smart grids might make our lives much easier, but they might also raise suspicions and skepticism.

Skepticism toward technological advancements and innovations that are envisaged to make human lives easier and more comfortable do not represent the trend of the 21st century. Science and progress have always had (and most probably will always do) their dark sides. In 1928, a Russian science fiction writer Alexander Belyaev released a short sci-fi story called "Open Sesame." Belayev, now totally unknown to the science fiction lovers in the West, was often referred to as "Russia's Jules Verne" in the 1920s and 1930s. He was also an author of a very famous novel called "Amphibian Man" that very much resembles the plot of "The Shape of Water" written and filmed by Guillermo del Toro almost a century later and became a widely acclaimed work that received four Oscars in 2017. In "Open Sesame," Belyaev tells a story of a rich German industrialist named Hene who is retired and lives in the vicinity of Philadelphia with his aging servant, who finds it increasingly difficult to serve his master. Both old men are approached by an agent of Westinghouse and offers to install mechanical devices in the house to facilitate their life. Interested but distrustful Hene agrees to install automatic doors, fans, and a mechanical broom in the house that are commanded by the human voice. Since Hence and his servant enjoy these innovations, they allow the same agent to bring "mechanical servants" into the house. However, at night, mechanical servants, who were only disguised accomplices of the fraudster, break into the Hene's vault and disappear with all his money and valuable possessions.

This story, written almost a century ago, already contemplates over the amenities of the smart homes and the dangers they might incur. Now, many years later, we can also see skeptics criticizing the concept of energy prosumers, the sharing economy, and the IoE. This is happening not because some people are blind to the innovations and the way they can change the world to the better, but because human being are starting to be entangled in information and our brains are not cut out for that.

A 2013 film "Her" directed by Spike Jonze tells a story of Theodore Twombly (played by Joaquin Phoenix), a man who develops a relationship with Samantha, an intelligent computer operating system personified through a female voice. This is where the technology can drive us in an extreme case—and make all human interactions irrelevant and obsolete.

However, not all hope is lost. Nellie Bowles, a technology reporter for the *New York Times*, advocates that nowadays human contacts and interactions actually represent a luxury good that is in high demand on the market. The physical experience of living mediated to us by the glass screens is for the poor people (similar to eating junk food or smoking), while the rich people in developed countries prefer to interact with humans and are willing to pay hefty sums for that (Bowles, 2019). This is an interesting thought that should also be taken into account when discussing the future of the market of energy and the smart grids. For many people, technology will never be able to replace the eye-to-eye contact and human interaction. Similarly, smart grids of the future will never be able to make right decisions on their behalf and to deliver electricity and information in right time on right places.

In 1811, a Luddite movement formed up in Britain with a purpose of destroying machines as a form of textile workers' protest against the Industrial Revolution. Poor Luddites could not understand that progress cannot be stopped this way. The movement was severely suppressed several years later, but its ideas inspired Karl Marx and many others. There will always be "Luddites" opposing the science and progress and the smart grids of the future are not going to be an exception.

1.3 Short overview of book's content

This book consists of nine chapters (including this introduction and conclusions). Chapter 1, General introduction, opens the book, explains the book's main idea and the rationale behind it. It also explains the book's structure and provides the short overview of its chapters.

Chapter 2, Traditional power markets and an evolution to smart grids, offers a sort of an opening for the whole book and a path to follow. It describes how thanks to the growing complexity of today's globalized world, the traditional grids had to evolve into the so-called "smart grids" which employ a two-way flow of electricity and information to create an automated energy delivery network. In addition, it describes the history of the traditional power markets that preceded the smart grids. It scrutinizes the classical model based on the principles of supply and demand sides of the market and marks the challenges it had to face and the rules it had to adapt.

Chapter 3, Sustainability of the smart grids, discusses the sustainability concept and the ways it might fit into the scope of the smart grids and their eventual social impacts. The perception of sustainability in modern-day world will be discussed and parallels will be drawn to the arising popularity of the smart grids including the IoE and automatic power systems.

Chapter 4, Renewable energy sources (RES), power markets, and smart grids, analyzes significant decarbonization that goes hand in hand with recent technological developments both putting a strong emphasis on the changes on power markets and the development of smart grids which are playing an increasingly important role in the development of these markets.

Chapter 5, P2P (platform) markets and sharing economy of the smart grids, focuses on the peer-to-peer (P2P) platforms that create and facilitate trade between a large number of fragmented buyers and sellers. It explores the raising popularity of the sharing economy concept that enabled the development of P2P platform markets in energy sector and investigates the implications this might generate for the smart grids and their perception by the consumers.

Chapter 6, Consumers, prosumers, and the smart grid, analyzes the role of prosumers in the smart grids as we know them today and as they once might evolve into something more complex in the future. It compares the traditional consumers and the new prosumers and draws some economics and policy implications. Moreover, the chapter explores the linkages between smart grids and future sharing economy development. Last but not the least, it elaborates some relevant policy implications.

Chapter 7, The role and perception of energy through the eyes of the society, explores how the society perceives the role and the importance of energy and energy-related issues (including power markets). Furthermore, it analyzes the ways how society perceives these energy-related issues and how (or whether at all) consumer behavior changes with the introduction of smart meters, smart houses and electric vehicles.

Chapter 8, Smart grids of tomorrow and the challenges for the future, provides some futuristic visions regarding the market design of the power markets of the future with the role of automatic power systems and prosumers operating within the IoE. It draws three main scenarios of the future development of the smart grids and energy markets (optimistic, realistic (or baseline), and pessimistic) and comments on the probability of each of them happening.

Finally, Chapter 9, Conclusions, closes the book. It summarizes its main ideas, conclusions, and arguments, outlines some thoughts on the possibility of predicting the future, and provides pathways for further research of social dimensions of the smart grid. Moreover, it lists some issues with the smart grids and mentions the policy implications that might be relevant for the decision-makers on the power markets of tomorrow.

1.4 Conclusions and discussions

Overall, the kind readers would probably agree with me that predicting the future, especially when it comes to such rapidly changing and deteriorating field as the market of energy, is almost an impossible task. Yet, I am going to give it a try outlaying my perspectives and adding a bit of futuristic ideas.

Smart grids simply represent a very exciting topic that is also linked to many other interesting ideas such as sharing economy, energy trading between households and individuals (P2P energy market), smart houses, electrification of transport, or smart metering. All of those things would collectively shape up the future technological development of humanity in the next decades. Furthermore, all of them would also impact on humans' social lives and perceptions becoming parts of our everyday lives and popular culture. Again, remember the smartphone example, Who would have thought most of us would not go to bed or leave our homes without one?

This book contains my personal experience and knowledge gained while from working on the topics of energy economics and energy policy for almost a decade—at first at the Charles University in Prague, then at the Energy Policy Research Group at the Judge Business School, University of Cambridge, then at the Department of Agricultural and Resource Economics, University California, Berkeley, and then at the Centre of Energy Studies, Prague Business School. Overall those years, I have worked on many energy-related topics and issues and have written an array of research papers published in prestigious academic journals in the United States, United Kingdom, Russian Federation, and other places.

My line of arguments is presented in a logical way and is divided into eight neat chapters that follow. Each chapter is focused on a different problem by presenting each problem in a logical way and vivisecting it to the greatest detail.

I hope that this is going to be an enjoyable reading for everyone interested in the smart grids and the design of the energy markets of the future.

References

Bowles, N. Human contact is now a luxury good. New York Times. Available from: <https://www.nytimes.com/2019/03/23/sunday-review/human-contact-luxury-screens.html>, 2019 (accessed 28.03.19).
Chiabai, A., Platt, S., Strielkowski, W., 2014. Eliciting users' preferences for cultural heritage and tourism-related e-services: a tale of three European cities. Tourism Econ. 20 (2), 263–277. Available from: https://doi.org/10.5367/te.2013.0290.
Cordis. Integrated e-services for advanced access to heritage in cultural tourist destinations: project fact sheet. Available from: <https://cordis.europa.eu/project/rcn/79351/factsheet/en>, 2019 (accessed 20.03.19).
Newbery, D., Strbac, G., 2016. What is needed for battery electric vehicles to become socially cost competitive? Econom. Trans. 5, 1–11. Available from: https://doi.org/10.1016/j.ecotra.2015.09.002.
Staffell, I., Rustomji, M., 2016. Maximising the value of electricity storage. J. Energy Storage 8, 212–225. Available from: https://doi.org/10.1016/j.est.2016.08.010.
Wells, H., 1921. Russia in the Shadows, first ed. George H. Doran, New York.

Further reading

Belyaev, A., 1964. Open sesame, first ed. Collected Works, Vol. 8. Molodaya Gvardya, Moscow, pp. 468–490.

2

Traditional power markets and an evolution to smart grids

2.1 Introduction

Nowadays, the world we live in faces unprecedented challenges that involve its growing complexity and globalization. The flow of information and the technologies that enable collecting, processing, analyzing, and utilizing this information have made our lives easier but at the same time they have put more pressure on our decision-making skills as well as on our day-to-day lives. The same can be said about the electricity market that also embraces all new technological advancements and is becoming more complex to the extent that they cannot be controlled by humans alone but require complicated software and algorithms to steer them and to operate them.

In some parts of the world, energy companies are struggling. For example, the current difficulties faced by energy companies in the European Union are mainly due to three important reasons: (1) over-investment in the 2000s, (2) a sharp drop in demand after the 2008 financial crisis, and (3) the large-scale entry into renewable energy sources (RESs) (see Crampes and Léautier, 2016). If we then take a good look at the "failure" of European energy companies compared to the late 1990s in the US electricity companies—it becomes quite obvious that they invested too much in open markets, resulting in overcapacity and several bankruptcies (Hirsch et al., 2018).

As economies of scale significantly reduce costs, a small number of large generational assets have been built to produce electricity and, as demand is geographically dispersed, transport and distribution networks have been developed to provide electricity. At the same time, oil companies such as Total have recently developed a real energy strategy and have invested a lot in the sector (Oh et al., 2010).

Current energy companies play an important role in the future of the electricity industry, despite the weakening of the State. Today, apart from the major energy companies, there exist various technologies that have the potential to transform the electricity industry. Therefore power companies need to adapt to new business models on making quick decisions, conducting small-scale experiments, and processing digitization all in a priority.

The fluctuation of feed rates, more efficient production technologies, and a shift in the severity of global market demand have shocked the competitive position of refineries around the world.

Improving the purchasing function is a major lever for energy companies to reduce their cost base and become more competitive in volatile markets. In many regions, governments are redesigning their rules and regulations, for example, launching energy-free market systems with a theoretical aim to reduce costs and improve the quality of service for end users. Other utilities may also benefit, such as the price of customers, flexible monetization, and the inclusion of renewable energies in the wholesale market.

The French capacity market was not based on market principles but rather on the basis of a lack of flexibility in the country's electricity generation which is mainly supplied by nuclear and fossil fuels (Auer and Haas, 2016). If the domestic demand is low and if there is a lot of renewable energy on the market, the result is obviously disconnected plants, increased energy exports, and negative energy prices on the markets. Aspects of the energy market, such as energy exchanges, free trade, and the reduction of network and market access restrictions, are also introduced in countries that tend to be more skeptical.

Digitalization of the energy markets have to learn from other companies in different sectors of the economy that are already pioneers in using digital technologies for optimizing and managing their businesses. Native digital businesses, such as Amazon, have been leading the way in customer-focused usability and interchannel interaction, and consumers now demand similar convenience and high-quality service from all businesses they deal with.

New entrants such as OVO Energy and Opus Energy in the United Kingdom (which sells gas and electricity) and German Stromio (which sells electricity) use their high-efficiency and low-cost organizations and structures to deliver at lower prices. In Texas, the United States, some digital disrupters, such as Energy Ogre, Griddy, and Grid Plus, offer online services for customers and provide access to wholesale electricity price (see BCG, 2018).

Traditional players face a lot of pressure to compete effectively with new entrants, while avoiding the erosion of brand strengths, customer service, approach, and local market presence. As cheaper storage and renewable energy becomes twice expensive, companies need to refine their strategies and enhance their digital strategy.

In order to continuously increase the value of new services, such as the electrification of transport and other opportunities, it is also possible to consider moving to new areas such as insurance and consumer credit. Digital technologies are devising tools that would provide an opportunity to offer greater value as market and regulatory change.

Technology is changing the traditional rules of the game. New digital players have entered the energy markets, smart applications have allowed customers, both large and small, to have more control over their energy consumption, and low-cost batteries and renewable batteries have changed the demand.

Nowadays, electricity customers can take advantage of the new concept of solar photovoltaic, innovative, and decentralized power solutions or 100% renewable energy. The reduction of demand due to energy efficiency and strong competition with retail price pressures is a burden on struggling utility companies. In Europe, ongoing efforts to increase interconnection capacity between countries are bringing the single energy market closer. New rules and regulations for the energy industry, liberalization, and separation will lead to a change in the industrial structure.

At the end of the 1990s, liberalization and market opening have transformed European utilities into a cost-effective function. New business opportunities and the emergence of renewable energy raise questions for energy companies about what activities they need to conduct and which they might need to outsource instead. Digitalization is the buzzword used to discuss many different developments in different industries, often mistakenly used as a synonym for Energy 4.0, Big Data, and Artificial Intelligence (AI) (Sämisch, 2016). However, due to the high investment costs (e.g., generation or network assets) and the highly concentrated markets dominated by integrated utilities, a business case could be maintained for several decades even after it was found.

Moreover, in order to comprehend the future of smart grids, one has to look back in history and find out how it all started. Traditional power markets are evolving fast and it is their transformation that would result in smart grids as we envisage and imagine them now.

This chapter is a sort of eye opener for the whole book to follow by highlighting the insight into the history of traditional power markets, in general, and world electricity markets, in particular. This chapter discusses the history of the electricity markets and describes the main economic tools for the organization of world electricity markets. This chapter also cites the American and Russian electricity markets as an example of comparison to understand their evolution as well as the differences and similarities that are quite interesting. Furthermore, the chapter moves to exploring the market strategies and the traditional electricity markets involving the main approaches used in modeling traditional electricity markets. Then, the chapter moves onto focusing on one of the prerequisites of smart grids—the agent-based modeling used at the electricity markets—that, in a certain way, paved the way to the automatization of power and electricity systems and ultimately the evolution to the smart grids. Finally, this chapter provides discussion, conclusion and some closing remarks on the topic.

2.2 History of traditional power markets

Some people say that human history always runs in cycles—the new is always the well-forgotten past. Prior to the 18th century and the Industrial Revolution that occurred, all energy produced by the mankind for its own needs was renewable. People used horse power to move on land, wind power to travel across seas and oceans, burned wood in stoves and fireplaces to heat up their dwellings, and used the power of streams and rivers to power up mills and other simple equipment. From today's point of view all these sources of power are renewable and sustainable (apart from the issues connected with the disposal of horse manure that flooded the streets of large cities such as London and New York and constituted a small environmental disaster).

The electricity industry plays a crucial role in the development of the national economies of many countries around the globe (Zhang and Da, 2015). In a way, it controls the processes of production and creation of goods and services that consume electricity. Electric energy is like "oxygen" of the economy and the elixir of its growth, especially in the massive industrialization phase which is currently facing emerging economic giants. Global demand and energy prices have been resilient during the recession, leading politicians in countries with the potential to generate energy to address this sector as a potential driving force behind economic growth.

Similarly, innovations in the renewable energy sector have contributed to employment growth, although the multipliers in this sector are very sensitive to the nature of domestic networks.

However, balancing energy prices, energy safety, and the environment requires compromises between job creation and overall productivity in the energy sector. Areas that are less abundant in natural resources are also focusing on the energy sector as a potential driver of economic growth.

The energy, environmental, and security needs of the 21st century have accelerated public and private sector investments in modernization of the grid and smart grid technologies worldwide.

The increase in investment in the modernization of the electricity networks around the world includes increasing reliability, integration of renewable resources, changing demand, and market reforms, which create more opportunities for independent generators and require new connections to the transmission system.

The lack of reliable power remains a major constraint for the country to achieve sustainable industrial growth, investment, and economic competitiveness. Key measures include improving energy efficiency and energy efficiency, increasing natural gas proportions and hydropower in the fuel mixture, and reducing energy intensity in the transport sector.

It is planned to promote innovative and sustainable business models in the energy efficiency sector, including the emphasis on standards and labeling, industrial end-use efficiency in large consumers, accelerated conversion to energy-efficient appliances, and energy-efficient financing platforms.

Development of clean energy has the main aim to generate clean energy by supporting hydropower projects, developing renewable energy (including generating dispersed rural access and services), and solar energy.

Efforts to curb climate change by moving forward toward clean energy will undoubtedly result in winners and losers in the world's fourth largest economy, both in industry and regions.

Green economy is becoming a very important aspect of energy markets (Gallagher, 2009). A good example is Germany. In more and more German industries, resistance to Energiewende (German national energy strategy with a strong emphasis on "green" economy, decarbonization and the growing share of RES) has raised concerns about the fact that the country's energy transition could lose its place in the global movement toward a low carbon future (Málek et al., 2018). Massive decarbonization has strengthened its renewable energy and industrial efficiency departments, while fossil fuel production operations have "burnt to the ground," according to management.

Successful efforts to tackle climate change would trigger major modernization efforts across all sectors of the German economy and could also open up opportunities for German exporters in growing markets with clean technology. Many experts believe that a thorough review of energy taxes and levies is essential to ignite the next phase of energy change, which will include other sectors such as heating and transporting renewable energy to reduce emission.

There are factors that will dramatically change the basic structure and operation of the old industrial sector, and electricity will play a greater role in the national economy and energy sector, which has a significant impact on national security.

The increase in costs, flat demand, and growing concerns about integrity, safety, and resilience of the above-mentioned grids are taking place as the cost of solar and wind technology is declining and renewable energy policy is supported. Some other technological innovations promote the decentralization of electricity and enable consumers to learn more about their energy consumption.

Increasing renewable energy investment meant improving the economy of the country and municipalities, while protecting the environment and improving our national security. Electric energy, more than any other industry, has the power to rebuild cities, states, regions, and our country, and to stimulate essential economic development.

In the United States, distributed impact energy systems deployed in the Eastern US eco-neighborhoods help organizations navigate through changing energy markets, efficiently plan their energy expenditure, maintain balance in crisis situations, and even increase sustainability and reduce greenhouse gas emissions.

At the other side of the world, China made its move and established a renewable energy equipment manufacturing industry and has the capacity for large-scale production. However, due to problems in the management of power systems, local protection policies, and tax on added value (increasing production costs), there has been a constant increase in wind and water reduction, and a lot of wind and hydropower has been wasted. China has now implemented an initial separation of power stations and grid companies to reach prices, creating conditions to optimize resource allocation. It becomes obvious that China could therefore learn from mature foreign energy market patterns and then design a reasonable rate model to reduce the cost of generating resources and improve efficiency. The country could learn to reform the market from overseas successes and strengthen research on effective competition in the electricity markets to improve the efficiency of generational resources. The direct purchase of electricity by large users is not only beneficial for energy saving and reducing emissions, but also for reducing the cost of electricity generation.

State regulation of electricity sector dates back to the early days when electric energy became ubiquitous and energy companies were granted exclusive service areas in exchange for the obligation to serve all electricity customers within the given area and to provide electricity to them. Competitive energy markets actually allowed a variety of wholesale products and services to facilitate the sale and transfer of electricity.

In many countries of the world, electrical energy companies operate under what is known as the "traditional model," in which electricity rates are set by a state regulatory authority based on the cost of utilities to supply customers with electricity. In most cases, the authorities open up the wholesale market for competition, allowing wholesale buyers to buy electricity from any generator, forcing owners of transmission lines to transport (or "wheel") electricity to other generators and wholesale buyers at "fair and reasonable" prices. Relying on market prices for wholesale electricity has been a dominating state policy. Through several state restructuring efforts, vertical integrated utilities were needed to eliminate ownership or control of power stations, and electricity generation was transformed into a competitive, "deregulated" function.

Moreover, there are regional transmission organizations (RTOs) operating at wholesale electricity markets that all have some form of price deficiency where electricity prices can rise above the ceiling during the stress of the system (the prices paid in such markets also have a significant impact on the price of electricity collected by generators from such markets in a bilateral or standard offer contract).

Like the wholesalers of RTO capacity markets, the RTO capacity markets depend on a central single clearing price (which allows us to achieve profitability for low-cost units), are subject to local prices and impose a maximum ceiling on transactions.

While participating in the wholesale market in RTO can benefit from public energy in terms of cost savings and additional power-of-sale opportunities, there are still some potential problems with markets that require vigilant supervision. Concerns about RTO-owned markets include the ability of some generators to have a strong influence on market prices (also known as market power), very complex rules and problematic management processes.

Ironically, when RTOs first entered wholesale, developers said that the shift from a cost-based rate to a market-based rate would increase price competition, and consumers would benefit from lower prices and increased investment in new infrastructure needed for the future reliable operation of the network.

In the United States, RTOs are regional actors operating at the electricity market authorized by the Federal Energy Regulatory Commission (FERC) to manage the transmission network. US Congress can decide whether to change the way in which RTO's electricity markets are regulated and operated (through some standardization of such markets), with the aim of improving efficiency, regulation and transparency, reducing costs and thereby potentially limiting the possibilities of fraud.

Electricity prices vary from region to region in the United States, depending on the supply and demand factors, which are strongly influenced by the cost of fuel, electricity usage, infrastructure, and weather trends.

Other utilities acquire electricity directly from other resources, marketers, and independent manufacturers or wholesalers organized by a regional transmission reliability organization. The construction of US electricity infrastructure began in the early 1900s, and investments were driven by new transmission technologies, power stations, and rising demand for electricity, particularly after World War II. New power lines were needed to maintain the overall reliability of the electrical system and to connect to new renewable sources such as wind and solar energy which are often far from where energy needs are concentrated. New energy sources are rebuilding the electrical energy market and are beginning to generate additional sales opportunities for utilities.

However, deregulation forced energy companies to look at new markets and expand them into a much wider range of services, including designing, installing, and maintaining internal lighting, electricity quality and distribution systems for industrial facilities, and sophisticated measurement packages for intelligent grids, allowing customers to analyze their own devices.

In the United States, California, New York, and New England were the first search leaders, although at the end of the 1990s most states, even those with electricity rates near or below the national average, were also looking at the favorable prospects for deregulation of their retail markets, especially in view of the apparent lack of appropriateness.

Competition is expected to benefit consumers by increasing the efficiency of wholesale electricity generation and promoting innovation in the retail trade. For example, they can purchase from suppliers outside their local market, from newcomers to generation or from energy suppliers, all of which can charge lower prices than the local distribution company.

Given the development described above, one has to say a few words about protecting the consumers. In the United States, there is the energy consumption Resource Council, better known as ELCON, which is a national trade union representing industrial electricity consumers who have been defending competitive energy markets since at least the early 1980s. Similar reasons were presented in documents by other stakeholders at the time, including advocacy groups such as Citizens for a healthy economy, Americans for Affordable Electricity, and the Energy Alliance, all of which were created with a thought of protecting consumers and securing affordable electricity at their homes.

In most developed and developing countries, the cost of electrical energy is determined according to the laws of a market economy (Pyrgou et al., 2016). At the same time, the electricity market is associated with the circulation of a special type of product, which largely determines the conditions of market pricing. Market strategies of generating companies have a significant impact on the formation of equilibrium prices, energy market conditions, and the development of market relations.

In the process of ongoing liberalization and digitalization of the electricity market, generating companies have more and more opportunities to use market power which allows them to influence the bid price and the conditions for the sale of electricity on the market. An established marginal behavioral strategy of generating companies at the electricity market is still important but, at the same time, it is often not the best strategy in place anymore. The objective function of the subject of the market becomes profit maximization under conditions of imperfect competition.

The attractiveness of creating complex market models in the energy sector motivates the use of computational approaches for modeling technological infrastructure and strategies for the behavior of market participants. In this artificial environment, market participants may repeatedly interact with each other and, thus, reproduce a realistic technical−economic dynamic system.

The evolution of approaches to controlling and modeling energy markets can be represented by the following three stages:

1. Dynamic models of planning the operating modes of the power system. These are dynamic systems based on physical constraints, such as network bandwidth, scheduled equipment repairs, etc.
2. Optimization models. The optimization criterion was originally the minimum cost. With the transition to market pricing, it changes to maximize profits, which can affect not only the value of production costs, but also the market price of electricity sales.
3. Simulation models and behavioral approach. Recent studies in the field of developing models of the wholesale electricity market show that the main direction of research is connected with the study of the behavior of participants in the wholesale electricity

market. In fact, market participants do not seek to maximize utility when making decisions; rather, the decision-making task is aimed at finding the most satisfactory solution that will give an average result. Understanding this fact led to the creation of new dynamic models of the electricity market, which are based on a new methodology based on agent-based modeling (Lisin et al., 2014).

The transition from the traditional power markets to the modern power markets as we know them today underwent several stages. A good example of this transition is a case study from the Russian Federation, namely the "State Commission for Electrification of Russia" plan adapted in 1920 (better known as "GOERLO" according to the Russian abbreviation of this name).

At the beginning of the 20th century, the energy construction in the Russian Empire was growing and booming. However, the specialists became increasingly convinced that the country needed a single nationwide program that would link industry development in the regions with the development of the electricity base, as well foster the electrification of transport and housing utilities. At electrical congresses passed several resolutions that were adopted stressing the national importance of electricity supply and the need to construct large power stations near fuel deposits and in river basins, and also to link these stations together with the help of a developed transmission network.

At the initial stage (until 1920), Russia's energy infrastructure was characterized by a sectoral development ideology. Thus the placement of the main energy capacity was carried out directly in industrial regions. All power plants built in Russia had a limited number of consumers (e.g., from just one to several dozens) who did not have energy connections with each other.

The qualitative parameters, such as the frequency of the alternating current, the deviations, and the voltage fluctuations, had a significant variation, since no single system existed when developing these stations.

Meanwhile, local electrical engineering, heat engineering, and hydraulic engineering schools were considered among the best in the world. At the all-Russian Electrotechnical Congresses, strategic infrastructure issues were considered. Most Russian scientists and electrical engineers tended to the need to move from a sectoral developmental ideology to an infrastructure one, which marked the beginning of the appearance of district power plants using local fuel resources.

After the October Revolution, the need for the electrification of the newly established country, the Union of Soviet Socialist Republics, became even more acute. Torn by the civil war and impoverished by years of instability, the new country needed the solid basis for its economic development. The new rulers of the country, the Bolsheviks, realized that very well and in 1918 the Bureau for the development of an overall electrification plan of USSR was formed. A year later, in 1919, the Socialist Council of Workers "and Peasants" Defence approved, and Vladimir Lenin signed a regulation on the GOELRO Commission, the State Plan for Electrification of Russia. The GOERLO plan was chaired and inspired by Gleb Krzhizhanovsky, a Russian scientist and electrician, who became a spiritual father of the Russian transition to the new power market design in the 20th century.

GOELRO plan was approved and entered into force on December 22, 1920. The plan consisted of a unified program for the revival and development of the country and its specific industries—first of all, heavy industry by increasing labor productivity. Particularly emphasized in this program is the promising role of electrification in the development of industry, construction, transport, and agriculture. It was proposed to use mainly local fuel resources, including low-value coals, peat, shale, and wood.

The development of the GOELRO plan is the first experience in the history of a systematic approach to solving large-scale energy and, therefore, national economic tasks. The ideology of the development of the energy infrastructure is changed from a sectoral to an infrastructural one—a model of the priority development of the energy infrastructure when it creates conditions for the development and placement of industrial production.

The program of restoration and development of the national economy, implemented in the form of a GOELRO plan, had a clearly expressed scientific character. It is obvious that the authors of the GOELRO plan proceeded from the objective laws of the development of science, from strictly established dependencies of an economic and technical nature. They used the verified and accurate statistics on the economy and industry. As the detailed planning tasks, scientific knowledge was transformed in the direction of specific disciplines of production and applied nature.

In 1930s when USSR produced less than 40 billion kWh of electricity per year, the important contribution of Gleb Krzhizhanovsky and his scientific school was the fact that the task of state importance was set up—the creation of the unified energy system, on which the subsequent placement and development domestic energy and the entire national economic complex was built upon. This laid the foundations for an infrastructure approach to the study of energy development processes.

In terms of electrification, trends, structure, and proportions of development were determined not only for each industry, but also for each region. For the first time in Russia the authors of the GOELRO plan proposed economic regionalization based on considerations of proximity to sources of raw materials (including energy), established territorial division and specialization of labor, as well as convenient and well-organized transport. As a result, seven major economic regions were identified: Northern, Central Industrial, Southern, Volga, Ural, Caucasus, as well as Western Siberia, and Turkestan (what is nowadays the Central Asian independent states).

From the very beginning it was assumed that the GOELRO plan would be introduced by law, and the centralized management of the economy should have contributed to its successful implementation. It became the first state plan in Russia and laid the foundation for the entire subsequent planning system in the USSR, defined the problems of future five-year plans, and also laid the necessary theoretical and methodological foundations. In 1921 the GOELRO Commission was abolished, and on its basis, the State General Planning Commission, the State Planning Committee, gave the foundations for the entire economy of the country for several decades to come.

It should be noted that the research of foreign theoretical and practical experience in the construction of power plants and power lines led Krzhizhanovsky to an idea of "presenting" global solutions to the fundamental problems of the industrialization of Russia. This allowed to create a unique plan, which has no analogs until today. In addition, an unusually large theoretical contribution to the GOELRO plan was made some Russian philosophers who plunged into the implementation of the GOELRO plan, trying to assess the possibilities of using various kinds of energy, including the energy of empty space. It was their philosophical proposals that found "naturalization" and left a definite imprint on the creation of programs in the energy sector. This eminent scientist believed that the all-penetrating energy creates a new type of culture, characterized by collectivism, dialectics, realism, synthetics, etc.

The methodological basis of the plan was a comprehensive energy approach created by the school of Gleb Krzhizhanovsky for solving the problems of the functioning and long-term development of individual links of the energy sector, as well as the foundations of the organization and power system design technologies. The essence of this method is to consider energy as a whole from sources of energy resources to subscriber units. The first important element of the complex energy method is to develop a different level of hierarchy of energy balances—starting from the energy balance of individual plants and ending with the energy balance of the country as a whole. Another element of the complex energy method is the desire for a rational plan for the development of the energy sector based on the development of a methodology for comparing alternative solutions.

The development and implementation of the GOELRO plan became possible and exclusive due to combination of many objective and subjective factors: the significant industrial and economic potential of prerevolutionary Russia, the high level of Russian scientific and technical schools, the concentration of all economic and political power, its strength and will, as well as the trustful attitude of the Russian people to the political leaders of the time. The GOELRO plan relied on the forms of activity and thinking of the Russian person down to the deep layers of his consciousness.

During the first, second, and third five-year plans, several dozens of isolated energy systems were formed and developed, including Moscow, Leningrad, Dnieper, Ural, and Donbass. In the same period, the beginning of the heating of cities began. In general, over the period 1921—40, the total capacity of power plants increased almost 10 times.

The so-called program "A" of the GOELRO plan, which provided for the restoration of the destroyed energy economy of the country, was fulfilled already in 1926, and by 1931, the minimum 10-year period of the program had exceeded all planned targets for energy construction. Instead of the projected 1750 kW of new capacity, 2560 kW were put into operation. By the end of the 15-year period—by 1935, the Soviet power industry reached the level of world standards and ranked third in the world—after the United States and Germany. In 1930 Krzhizhanovsky founded the Energy Institute of the Academy of Sciences of the USSR (abbreviated "ENIN" in Russian), where significant forces were concentrated energy scientists. By this time, it is customary to attribute the beginning of the formation of the national scientific school of power engineering. The creation of an energy institute in the system of the USSR Academy of Sciences demonstrates the true understanding of Krzhizhanovsky organic link between fundamental and applied research in the energy sector, the need for in-depth scientific research and their support with technical solutions. Three main areas of scientific research carried out at ENIN under the guidance of Krzhizhanovsky:

1. Development, based on the idea of electrification, a complex scientific theory, designed to justify the creation of a unified energy system. This required the solution of complex problems of optimizing the structure, ensuring stability, efficiency, and reliability of operation of power systems.

2. Development of the ideas of unity and integrity of the country's energy economy on the basis of a single fuel and energy balance (balance predictions were made in the ENIN for the future in 20–30 years).
3. Statement and development of research on fundamental electrical and heat engineering problems, including in the areas of the theory of heat transfer and combustion; integrated energy technology use of solid fuels; high-voltage theory, electric field theory, converter technology and a number of other important problems of electro physics and electrical engineering.

One of the main places in the theme of the institute has always been occupied by the creation and development of the Unified Energy System of USSR. In 1931–34 theoretical approaches to the formation of the Unified High-Voltage Network and the Unified Power System of the USSR were outlined. All of this provided solid theoretical and practical basis for the modernization of the obsolete power system into a well-functioning and well-coordinated electricity system of the USSR thanks to which the great leap of the Socialist economy that included the launch of the first sputnik, the first man in space, and other achievements so admired by the West in the 1950 and 1960s can be attributed. Smart grids of the future are probably not going to be created with such an enthusiasm and devotion but the example of GOERLO shows that building large-scale well-calibrated complex energy systems are possible given the provided state support, collective efforts, and effectiveness.

2.3 Main economic tools for the organization of world electricity markets

All in all, one can see that the introduction of competitive market mechanisms in the field of power generation should be considered the main direction of power industry reforms in the world, helped by the recognition in the early 1990s that the power industry is no longer an indivisible natural monopoly (Chao and Peck, 1996).

There are several reasons for this change in attitude toward the power industry. First of all, economies of scale, which for a long time played the role of the main argument in favor of the natural monopoly structure of the industry, has lost its relevance. New technologies have led to the fact that medium-sized power plants are quite competitive compared to larger stations. The following objectives can be distinguished, which were originally indicated in the course of reforms in various countries:

- Reducing the cost of electricity for consumers by increasing the efficiency of the industry (United Kingdom, Argentina, Australia).
- Attracting foreign investment to improve the efficiency of the industry (Brazil, Argentina).
- Introducing competition to give consumers the right to choose a supplier (Brazil).
- Smoothing the difference in electricity prices in different regions of the country (Norway, United States).
- Improving the efficiency of investing in the development of the infrastructure of the power industry and the industry as a whole in order to increase the competitiveness of national producers (Australia).

The main directions of reform include the following aspects: separating monopolistic activities from potentially competitive ones without prejudice to the reliability of power systems, creating and ensuring the necessary conditions for free competition of electricity producers and distribution companies, as well as ensuring nondiscriminatory access of participants in the electricity sector to natural monopolies services. At the same time, in most countries, the functions of transmission and distribution of electricity, as well as the management of the power system, remained monopolistic and regulated.

The disintegration of the industry has become the first link of market transformations. Where it took place under conditions of private ownership of energy facilities, generating companies were distinguished from vertically integrated structures. This happened, for example, in the markets of the United States. In those countries where state ownership of energy facilities was preserved, first of all, a network company of national scale was formed. Such a process has proven to be characteristic of most European systems. Some national network companies were then privatized, the rest remained in state ownership. Thence, in Sweden and Norway, the high-voltage network after the restructuring remained in state ownership, while in Finland it currently represents a mixture of public and private property as a result of the merger of public and private companies. In United Kingdom, the national grid company was privatized while the state control (in a form of regulators) is still maintained over it.

Electricity represents a unique commodity with very specific key features. The unique properties of electricity shape up the management of the energy systems and the organization of the electricity markets. Experts identify four main problems in the operation of power systems:

1. Inevitability of deviations in the volumes of actual production and consumption of electricity from the planned volumes (hereinafter referred to as deviations).
2. Possibility of overloads in the power system and the need to manage them.
3. The need to provide system services to ensure reliable operation of the power system.
4. The need to plan the operating modes of the power system and the dispatching control of the power system.

While in the conditions of existence of vertically integrated companies (VICs) at the electricity market, the main tasks were solved through centralized decision-making within a single company, then after the reform—the division of electricity generation and power management functions—the solution of these problems by administrative means became much more complicated. Furthermore, the main features of the organization of modern energy markets in different countries are considered.

Despite differences in the organization of energy markets in different countries, the market structure provides for the interaction of technological (equipment and networks) and commercial (exchange or market administrator) operators.

There are two forms of organization of the electricity market operator:

1. System operator (SO)—regulating the transmission and distribution of electricity in the network, performing the functions: market management, dispatching and network operation.
2. Commercial operator (CO)—a system operator performing the functions of a trade exchange operator.

System and commercial operators are functionally separated and responsible for different areas of trade. In a number of countries integrated operators prevail. For example, in

Australia, Canada (Ontario), and the United States (New England region and the ISO NE energy system), New York states (NYISO), Pennsylvania, New Jersey, and Maryland (PJM), the market is controlled by an integrated operator that simultaneously performs the functions of a system and commercial operators. The operator holds an auction for an hour ahead, taking into account the real network parameters. These presubmitted applications are used to calculate the power system mode in real time.

Often there is an option when the system operator is integrated with a network company. In this case, the network system operator is a monopoly organization that cannot be controlled by market participants. As a counterbalance to the grid company, market participants create a legal entity (the so-called "Pool"). The pool determines the market rules used by the operator. Pula's responsibilities include supervision and regulation of the operator's activities. Directly Poole itself carries the burden of commercial calculations. An example of a network system operator is the Former Pool of England and Wales, or the Nord Pool.

Markets managed by commercial and technology operators can also be separated from each other. Then, approximately 1 hour before the real time, control from a commercial operator passes to the system one, which must also use the market principle when balancing the market. This type of relationship is used in New Zealand, Spain. The Spanish commercial operator (OMEL) maintains contracts, manages the trading system. The technology operator manages the market for additional services and realizes the system balancing in real time according to technical limitations (network bandwidth, operating conditions of generators, etc.). Based on the information received from the trading exchange, a commercial operator forms the best solution with the help of software, the technological operator checks it once again under the terms of reliability, adding corrections if necessary and forming the final daily work schedule of market entities.

A similar approach to the separation of energy market management functions by two operators (a trading exchange and a system operator) was used in California until 2001. As a result of the crisis, the California stock exchange ceased to exist, completely transferring the functions to the system operator. As analysts often recognize and acknowledge, there is no need to separate a commercial operator if the process operator is not the owner of the network. Market participants successfully control the operator, which, in turn, acts as a counterweight to the network company.

If the technology operator and the grid company are combined into a single national company, then it is desirable to create a separate commercial operator, which is controlled by market participants. A similar approach is used in the modern market of England and Wales and in New Zealand in order to increase control over the network company.

In order to provide a more accurate forecast of the load schedule, the operators divide the market by time on the basis of concluded contracts and competitive selection of suppliers for the day ahead.

The electricity market is usually managed on the basis of concluding long-term bilateral contracts, day-ahead contracts based on the load schedule and spot trading immediately before the "real-time" moment, while adjusting the dispatch schedule. In some markets, a number of these trading instruments may not be available, or, conversely, include additional system services.

Long-term bilateral agreements are applied in almost all electricity markets, and trading takes place outside the exchange floor.

The most important tools for managing the market are day-ahead contracts that allow flexible control of the load schedule and the formation of nodal electricity prices. With such a system for organizing trading, market participants submit applications for the purchase or sale of energy no later than a certain point in time for each hour of the next day.

The organization of new forms of trade in the electricity market is often accompanied by market instability and even crisis situations. An example is the crisis of the California Energy Exchange in 2001, one of the reasons for which was the use of market power tools by generating companies. The manipulation of market power also caused a radical transformation in the energy market in England and Wales in the early 2000s. Major power supply problems were observed in the PJM market in 1998. Serious technological problems are experienced by the system operators of the states of New York and New England because of the wrong economic decisions.

The greatest role in the occurrence of crises in the electricity markets is played by generating companies, which in market conditions have the opportunity to have a strong influence on the formation of the market price of electricity due to the relatively low elasticity of demand. Therefore arrangements for the organization of control over the market (primarily generating companies) are important. Operational control over the activities of generating companies is carried out

- Through the activities of the Supervisory Board;
- Through the adoption and improvement of market rules;
- By regulating transmission tariffs and network connection fees;
- By reducing the concentration of suppliers in the market;
- By issuing permits to merge or acquire companies.

Under market conditions, the operator also performs the functions of a regulatory body. Its task is to organize independent management of the network and system services, taking into account the interests of all market participants—generations, network companies, end users, and network owners. Also, the operator may have leverage on market players with market power in the form of powers to apply penalties.

State regulatory agencies are also required to monitor the energy market, regardless of the availability of regulatory functions at the system operator. Their main task is to eliminate obstacles to free competition, and the methods of influencing the market should be comparable to market regulatory mechanisms. Also, government agencies determine the conditions for access to the market of participants and the pricing process in relation to the cost of network access and transmission services.

An urgent problem for the energy market is the possibility of a "situation of reregulation" of the market when state bodies begin to have a significant impact on the daily commercial procedure of the market, trying to solve the problem of price regulation for the end user. With such a "California" scenario of development of events, network companies incur significant losses due to the continuous increase in prices for energy purchased on the market in the conditions of established prices for its implementation.

Antimonopoly organizations that impose restrictions on the size of the market controlled by individual companies also contribute to the observance of free competition rules. An example of this is the PJM market, whose major participants have been restructured and withdrawn into separate subsidiaries with significant foreign investment. In

Texas, market rules stipulate that no generating company can have more than 20% of the generating capacity of a region. In Argentina, restrictive conditions are imposed on the generation, which do not allow them to have more than 10% market share.

In this case, one of the factors for the stability of the electricity market can be considered a low level of market concentration. The Scandinavian market can be cited as an example of a sustainable and prosperous energy market—in Norway, Sweden, and Finland the share of the largest generating companies does not exceed 4%.

Additionally, in many respects the stability of the market is determined by the availability of generating capacity in the operator's area of operation. The necessary level of generation and equipment readiness for inclusion as a reserve in the system is regulated by the following economic instruments:

- Simple application (or single-rate tariff), reflecting the cost per unit of electricity at a given point in time.
- Organization of the long-term capacity market.
- Use of capacity payments.
- Complicated bid (multipart tariff with dynamic restrictions).

The first economic instrument assumes that the market price of a unit of energy will reflect the need for investment. The day-ahead market (DAM) and the prices determined by it under futures contracts provide information on the capacity readiness to meet the expected consumption and the need for new capacities. This allows end consumers and distribution companies to determine the maximum price they are willing to pay for electricity. In such markets, only electricity is sold; successful examples of the use of this economic instrument are the electricity markets of Scandinavia and Australia.

The single-price economic instrument is convenient for the operator, but it causes difficulties for the generating company, which takes on technical and operational risks. One of the ways to reduce risks is to set a marginal price for electricity based on determining the cost of peak power during periods of maximum load. But often, marginal price regulation is rejected in favor of granting the right to the system operator to acquire station capacity reserves.

The second economic instrument involves the implementation of a system of centrally planned reserves and the organization of a capacity market, which allows for early trading and the redistribution of capacity reserves. This economic tool is most prevalent in the United States. The most typical examples of these types of markets are the New York state market, the combined market of Pennsylvania, New Jersey, and Maryland, and the regional market of New England.

The third economic tool focused on attracting investments is aimed at collecting payments for capacity. The purpose of charging for capacity is often to attract new market participants. Currently, this economic tool is actively used in Spain and the markets of South America. When using this approach, the problem arises of determining the correct level of the power rate, which is difficult to economically justify. If payments for capacity are prohibitively high, this can lead to its excess (e.g., South America), with low values, there is a shortage of capacity. Payments for capacity can also lead to lower prices for electricity.

Another economic tool is the use of rarely changing multipart tariffs for electricity, indicating dynamic constraints. As experts often admit, such price bids are preferable for generating companies, but more difficult for a market operator. Depending on the type of station, a number of stations may have low maneuverability and high semifixed costs, but at the same time low variable costs that are part of the cost of electricity production, others—vice versa. To evaluate the cost structure, the system operator must have a complex algorithm and rules for determining the tariff for each hour, including the introduction of additional payments to those generating companies that have not covered their costs. Thus the economic factor is important for the early formation of the necessary power reserve as a factor ensuring the reliability of the system as a whole.

Since plant operators are not concerned about price changes in the electricity market; however, a fixed-rate power supply means that RESs will generate electricity at their maximum potential without taking into account the demand situation.

Support schemes that allow renewable energy to actively participate in energy markets, rather than passive-generation subsidies for electricity generation, can reduce the negative effects of renewable energies on markets and systems. Other authors have studied the benefits of active renewable trade in different markets compared to the previous trading.

As the markets of electricity evolve, the role of distributed resource aggregators (DRAs) is becoming important. Although DRAs are still at the beginning of the adoption cycle, regulators are beginning to recognize the value that the asset category could bring, as the electricity industry is changing from a centralized, generational basis.

As technological advances and the new operational realities of a renewable energy mix create opportunities for new resource types, it is important that marketers, market participants, and policymakers take full advantage of the changing markets.

With problems of energy imbalance in European markets with high penetration of renewable resources (e.g., Germany and the United Kingdom), such suppliers have gone beyond the traditional model of demand response (DR) providers to provide new flexibility services. For example, electrical markets in Europe, Germany, and Belgium recognize the ability of independent market players, known as the balance responsible parties (BRPs) and balancing service providers (BSPs), to present aggregated energy programs and to balance services from a collection of aggregated resources (Newbery, 2002).

The well-functioning electricity markets are essential for US economic competitiveness and are essential to ensure a reliable and affordable electricity supply. In the case of regulated electricity companies, the "capacity premium" is incorporated into the tariffs determined by the public utility commission and passed on to customers—for competitive generational suppliers, the capacity payments are determined by auction.

Unfortunately, auctions over the last few years have shown a worrying trend: PJM is increasingly dependent on less reliable sources of renewable energy, electricity imported from outside the PJM region and DR, which includes contracts with retail production customers asking them to reduce their dependence on RESs.

Organized markets are also affected by government-sponsored natural gas contracts and nonmarket payments, which reduce baseload's energy revenues.

New developments in the field of liberalization and renewable support schemes are conducive to active participation in the electricity market. For example, in Germany, the impact of direct marketing on average market price levels and price volatility, the potential contribution of renewable energies to balance, the viability of a flexible biomass generation, and the additional revenue generated by renewable energies from participation in various markets can be traced (Winkler et al., 2017). The valuation is reflected in the price impact of the shift between markets and the limits of market liquidity within 1 day. The restructured markets of wholesale electricity serve the stakeholders by offering three basic services: real-time, cost-effective shipping, resource-matching incentives, and long-term recovery.

In light of the findings, one can derive the two key aspects of electricity market design: (1) whether the energy and utilities markets provide sufficient income to cover all system costs; and (2) whether the current market designs are sufficient to meet the changing needs of the massive electricity grid. In addition, financial regulation by institutional investors, market regulations, and renewable energy policies often create additional barriers to renewable energy investments.

For example, some innovative US energy companies, which serve industrial and commercial customers, offers demand reduction strategies that can help larger companies significantly reduce their electricity bills (Siano, 2014). Instead of selling electric energy in order to maximize profits, they focus on energy management's energy efficiency and energy savings projects and use new channels such as social media to engage with customers. Some engineering and technology companies such as General Electric and Siemens have long established themselves as planters in the larger segment of the distributed energy market.

An important area of convergence is between electric cars and energy storage and manufacturing. Changing demand at a time when energy is more plentiful and less expensive is another way for industrial production companies to significantly reduce costs.

In addition, it is clear that every business enterprise that uses the electrical energy as one of the inputs into its production process also needs to face a decision about how to shift its resources between energy bills and other production inputs. This endeavor becomes very important with the recent focus on the renewable energy sources. One of the most effective options of how to optimize the excessive assets of a business company might be to invest them into the power supply contract with the purpose of creating a high advance payment for getting a positive economic effect from lower tariff. Advance payment for the electricity depends on the power purchase agreement (PPA) contract and on the electricity tariff (e.g., Strbac et al., 2014). If a business company increases the value of the contract agreement, it will get into the lower tariff zone and save on electricity bills. However, by doing so, it will also need to increase the amount of advance energy payments. The allocation of resources presents an optimization problem that can be often found in real life and in real businesses.

Different groups of consumers have different energy uptakes. It is therefore justified that each group should be charged differently. However, there are technological and social implications for this.

Schreiber et al. (2015) analyze flexible price signals that can act as effective demand control mechanisms. They show that different tariffs consist of combinations of flexible energy and power price signals and test their impact on the unit commitment of automatable for a set of German households. Their results suggest that flexible power pricing can reduce overall demand peaks as well as limit simultaneous grid withdrawals caused by real-time pricing incentives while inefficient designs of flexible power pricing might lead to undesired bidding of automatable devices. A specific tariff design showing robust network performance and helping to reduce energy procurement costs might be a good solution to that.

One of the recently discussed cases for the different electricity residential tariffs is the case of the households with solar photovoltaic (PV) in Australia. For example, Simshauser (2016) describes a case of Southeast Queensland in Australia, which has one of the highest penetration rates of domestic solar PV in the world. According to him, about 22% of households had PV in 2014 (and 75% have air-con). Distribution charges in South Queensland are charged on the basis of 20% fixed cost and 80% per kWh. The rapid increase in solar PV (from close to zero at the start of 2009) has resulted in a massive transfer of wealth and costs between customer groups. The results of the trial that included six digital meters installed in 69 broadly representative Queensland households at the customer switchboard circuit level separately measuring half-hour load for "general power" (e.g., fridge, tumble dryer, washing machine, toaster, kettle, clothes iron, computers, televisions, and game consoles), air-conditioning, electric hot water systems, household lighting, oven and solar PV units were described by Simshauser (2016). Reported general power load accounted for 52% of household final demand, electric hot water represented 18%, air-conditioning constituted 17%, lighting consumed additional 10%, and oven use was at 3% (Simshauser, 2016). Electricity distribution network capacity is primarily driven by periodic demand, and household load generally peaks in the early evening, whereas solar PV production peaks during the middle of the day and thus a mismatch exists.

In a similar fashion, Bobinaite and Tarvydas (2014) use an example of renewable technologies in Lithuania to demonstrate that seeking to expedite solar sector development in Lithuania would require reviewing a feed-in tariff (FiT) which currently is too low and impedes implementation of solar PV technologies. They point out that solar collectors in the country could compete in the district heating sector even without a support.

Yalcintas et al. (2015) look into a variety of pricing policies for commercial and industrial customer sectors using an example of two buildings for analyzing energy usage under the uniform-rate and time-based pricing schedules of three electricity utilities. They show that critical-peak and real-time price signals encourage building caretakers to act in the aim of reducing their electricity costs. At the same time, consumption shifting and energy saving with regard to the time-based electricity prices do not appear to reduce energy usage and costs.

Overall, different examples and case studies from different corners of the globe demonstrate that consumers might benefit from moving across tariff zones and changing the value of their contracts. Changes into the investment policy and the power tariffs might significantly contribute to the prosperity and economic well-being.

The next section of this chapter will take a closer look at the American model of the electricity market, which represents a peculiar and remarkable case due to its changeability, unpredictability, but, at the same time, its key importance for the world's energy market due to its size and role.

2.4 American model of the electricity market

The American energy model is of considerable interest for a number of reasons. First, during its formation, the American energy market has undergone several systemic transformations. Second, the United States is the largest world energy market both from the side of resource production and consumption. Institutional trends in the US energy market have a significant impact on global energy processes. Third, the United States is a federal state in which states have broad rights to regulate energy markets. In view of this circumstance, Russia, as a federal state, is interested in the powers of individual regions in the development of the energy sector.

From 1930 to 1980, the US energy industry was a regulated monopoly. At the same time, the property of vertically integrated enterprises was both generating and network assets. Production, transmission, and distribution of electricity were combined into a single service—the supply of electricity to consumers at tariffs.

The large-scale construction of capital-intensive nuclear power plants that began in the 1970s against the backdrop of an economic downturn in the US economy led to an increase in electricity tariffs, which caused unrest among final consumers. In order to ensure energy security and address energy efficiency issues in 1978, the US Congress passed the Law on the Regulation of Vertically Integrated Energy Companies (PURPA), which initiated the process of reforming the US power industry and moving from regulated monopoly to competition (Joskow, 1997). According to this law, a new category of electricity producers appeared in the United States, known as "qualified power plants." This category includes manufacturers with an installed capacity of less than 50 MW, as well as power plants using cogeneration technologies (TPPs) and RESs.

The rapid growth in the number of "qualified power plants" and the experience of their successful work have led to the fact that traditional vertically integrated enterprises have ceased to be the only source of electricity supply. Also, the generation structure was greatly influenced by technological changes in power generation. The emergence of gas turbine units with a combined cycle that allowed to obtain such institutional forms of the generating sector as "qualified power plants" of significant market advantages, which contributed to the development of competition in the US power industry. In addition, the technological progress of power generation has been successfully complemented by changes in regulation.

Adopted in 1992 by the US Congress, the Energy Policy Act (EPACT) established competitive pricing rules and lowered the barrier to entry into the energy market. For the further development of competition, the activities on the electricity market were divided into natural monopoly (electricity transmission and operational dispatch management) and potentially competitive (generation, electricity sales, repair, and service).

Enforcement of this law was entrusted to the FERC, which developed an electronic system to provide information about the available transmission capacity of the transmission system to solve the problem of nondiscriminatory access to electricity transmission services. FERC also established the possibility of a voluntary transfer of transmission network management to outside uninterested parties by power companies, which led to the emergence of the first independent system operators (ISOs).

In early 2000, by order of FERC, a functional division of energy activities was carried out. The transmission of electricity according to the structure of the grid was allocated to independent organizations, which received the name of Regional Transmission Companies—RTO. Currently, the following ISO/RTOs are in operation and functioning in the United States:

- PJM (Pennsylvania, New Jersey, and Maryland)—operates in most parts of the state of Pennsylvania, New Jersey, Maryland, Delaware, Columbia, Virginia, Ohio, and also partly covers the states of Illinois, Michigan, Indiana, Kentucky, South Carolina, and Tennessee.

- ISO NE (New England)—operating in Connecticut, Maine, Massachusetts, New Hampshire, Rhode Island, and Vermont.
- NY ISO (New York)—operates in the state of New York.
- ERCOT (Electric Reliability Council of Texas)—Texas Electric Reliability Council.
- CA ISO (California)—operates in the state of California.
- SPP (Southwest Power Pool)—operates in the states of Kansas, Oklahoma, and also partially in the states of Nebraska, New Mexico, Texas, Louisiana, Missouri, Mississippi, and Arkansas.
- MRO (Midwest Reliability Organization)—operates in North Dakota, South Dakota, Nebraska, Minnesota, Iowa, Wisconsin, Illinois, Indiana, Michigan, and also partially covers the territory of Montana, Missouri, Kentucky, and Ohio.

The change in the state's approach to the regulation of the electric power industry has led to the formation of modern tasks for the restructuring of the industry and the establishment of rules for the relationship between its subjects. The solution of these tasks is aimed at ensuring the development of competitive relations in the power industry and the formation of interregional competitive markets, and further division of activities in the energy sector. Moreover, one of the global tasks is the formation of a unified operational dispatch management and management of electricity transmission networks at the interregional level.

The development of competition in the US energy industry has led to the replacement of pricing on the basis of costs with market pricing, providing for the formation of the price of electricity based on supply and demand. The spread of market pricing contributed to the emergence in the United States of wholesale electricity markets, significantly different from each other by market structure, trading mechanisms, composition of participants, geography. Today, wholesale markets cover territories where up to 70% of the US population resides (Sovacool, 2009).

The examples of the most developed electricity markets are the PJM, ISO NE, and NYISO markets, where real-time and day-ahead electricity trading is organized, there are markets for system services and capacity, as well as a market for financial transmission rights.

Regardless of the type of energy market, electricity trading can also be carried out under bilateral agreements concluded by electricity producers directly with consumers. At the same time, the main direction of energy development is the concept of creating a competitive electricity market with a centralized system of planning and maintaining regimes.

This concept of a competitive energy market was proposed by a group of scientists at Harvard University under the leadership of Prof. V. Hogan, whose scientific work became the basis for the 2003 FERC Standard Electric Power Market Model—Wholesale Power Market Platform (WPMP). The goal of this project is to standardize and unify all of the wholesale US energy markets (Lisin and Strielkowski, 2014).

The standard model of the energy market involves a transition from a fragmented to an integrated form of the market with the allocation of independent system operators or regional transmission companies, which will determine the price of electricity at the place of its production or connection to the network according to the rule for calculating the maximum local cost. Variants of this model were implemented in New England (ISO-NE), New York (NYISO), Mid-Atlantic States (PJM), Middle (MISO), and South (SPP) west, as well as in California (CAISO). The main features of the Standard WPMP market model are the following ones listed below:

- Organization of pricing based on auctions of bidders of sellers and buyers using optimization algorithms that take into account system constraints.
- Ensuring the system operator balance of production and consumption on the basis of applications and the optimization algorithm.
- Establishing hourly hub prices.
- Admissibility of bilateral contracts between sellers and buyers, subject to payment of the difference in nodal prices between the point of delivery and the point of electricity consumption.
- Providing market participants with the possibility of acquiring financial rights to transmit electricity (firm transmission rights), protecting them from the risks associated with the difference in nodal prices.

One of the most successful deregulated US electricity markets is the PJM market. The next section will describe the PJM market in a greater detail and in a context of the discussion of the deregulation of electricity markets.

2.5 Deregulated electricity markets

The PJM market is one of the world's largest deregulated electricity markets (Bessembinder and Lemmon, 2002). This market is managed by the independent organization PJM Interconnection LLC, which has been assigned the status of an independent system operator (ISO) in accordance with the requirements of FERC. PJM Interconnection manages the following markets:

- Electricity interchange market (interchange energy market)
- Market regulation.
- Rotating stock market.
- Capacity market (capacity credit market).
- Market of fixed transmission rights (fixed transmission right).

PJM market members can be:

- Owners of transmission electrical networks.
- Owners of generating capacity.
- Distribution companies.
- End users.

All generating sources are divided into distributed (designated capacity resources) and unallocated capacity resources (nondesignated capacity resources). Distributed capacity resource is a set of generating capacity, designed to ensure the load of consumers and assigned to sales companies in accordance with the contract. The rest of the generating capacity resources connected to the power supply system are unallocated.

The owner of generating capacity makes applications to the ISO to use its generating sources for the purpose of supplying electricity or as an operating reserve, as well as providing other system services. Submission of an application for a distributed power resource is mandatory. In this case, two options are possible: the work of generation according to its own schedule or the centralized sale of electricity in the pool of the ISO. For unallocated capacity resources, the application is optional.

From generating capacity, based on applications filed for the sale of electricity in the ISO, a pool of generating sources is selected using the following rules:

- Selection is made on the basis of the electricity price indicated in the application and the cost of reimbursing the costs of starting, idling, and stopping the unit (in accordance with the performance characteristics of the power resource).
- The pool does not include capacity resources used under bilateral agreements or operating according to its own schedule.
- The pool included in the generation pool can be used by the ISO as operational power reserves.
- The owner of the power resource, whose source has been selected for the pool, receives electricity from the ISO, as well as reimbursement of start-up and idle costs.

Nonpool generating sources are categorized as self-negotiable sources, for which the following rules apply:

- The owner of the capacity resource should not include excess capacity in the number of self-negotiable sources in comparison with the total load of its customers.
- Electricity prices of self-negotiable sources are not taken into account when determining the equilibrium prices in the corresponding nodes of the grid.
- Self-verifiable sources can be used by the ISO as operating reserves with payment at the stated price.

In the case of a shortage of electricity in real time by a self-dispatched source, this load is satisfied by the supply of electricity from the electricity exchange market.

By their nature, the electricity exchange market is a regional competitive market designed for the wholesale purchase and sale of electricity. This market consists of two markets: a DAM and a real-time balancing market.

All generating capacities defined in the PJM system as capacity resources are required to submit bids to the PJM market operating in the day-ahead mode. Proposals suggest the indication of several price categories, depending on the expected volume of electricity production. Proposals may also indicate the minimum duration of work and start-up costs.

Electricity consumers must submit demand bids to the DAM in which they indicate consumption and prices, if they intend to respond to the bid price, bids can also be formed without a price.

All applications for supply and demand for electricity generation are linked to certain lines of the power grid depending on the territorial location of generations and consumers (the so-called load connection points (there are about 2800 such items in the PJM market)), which. indicates the bandwidth between all the lines. Taking into account all applications for supply and demand, as well as the technical characteristics of the network, a dispatch schedule is drawn up on the basis of minimizing system costs and ensuring the reliability of the energy market.

PJM market pricing is based on the concept of local marginal prices (LPCs), which is based around the following concepts:

- The price of electricity is determined by the marginal cost of its purchase.
- Marginal costs for the purchase of electricity are determined by the ISO by linear optimization of power system modes. As a result of optimization calculations, the input parameters of which are the current state power systems and price offers of generating companies, determines the optimal composition of generating sources and their operational reserves. Such a composition should provide the lowest possible costs for the purchase of electricity and the reservation of generating capacity for a given level of load of the power system.
- Optimization calculations are carried out using a computer model, which, based on measured load data of generators and consumers and electricity flow, continuously simulates the modes of the transmission network.
- Selected as a result of calculations for a certain period of time generating sources are included in the daily dispatch schedule. The price of electricity generated by the most expensive of these sources determines the market price on the market in a given period of time.
- If transmission networks are not overloaded at any sites according to the simulation results, the price of electricity determined by the above method is the same for the entire market.
- If, based on the results of the calculation, the individual power lines supplying any load connection point experience overload, then during optimization, such a composition of generating sources will be chosen that will ensure the operation of the transmission network without overloads. At the same time, the price of electricity at each load connection point will determine the most expensive of the selected generating sources. This price will be the local price limit for this item and will differ from the prices in other delivery points.
- The difference LPC in various delivery points is also caused by the difference in the connected load. This price difference is considered to be an overload charge, since a higher price is formed as a result of the input of the least economical generating capacity in order to eliminate the overload.
- When calculating LPC, only quotations from generating sources included in the central pool formed by the ISO are taken into account.
- LPC is calculated on the eve of the operating day when forming the daily dispatch schedule for each hour of the operating day (for the DAM) or every 5 minutes during the operating day (for the real-time market).

The DAM is a forward market in which, for each hour of the next operating day, based on applications for generation and demand, as well as information on concluded bilateral contracts, LPC is calculated for each load connection point taking into account possible overloads.

The real-time market is a balancing market where fees are charged for deviations from the modes provided for in the DAM control schedule. Market participants are paid or paid for deviations from the planned generation and consumption modes. Furthermore, in the real-time market, an overload is charged for deviations from the supply volumes under bilateral contracts included in the dispatching schedule of the day-ahead market, and real-time electricity is purchased (spot transactions). According to the results of dispatching, the ISO for all connection points is calculated on the LPC of the real-time market according to the rules discussed above.

The system of mutual settlements for electricity, which provides for payment of electricity purchased on the DAM (at contract prices), as well as actual deviations from the dispatch load schedule implemented on the real-time market (at real-time market prices), has been called the dual control system (the so-called "two-settlement system").

Regulation refers to an automatic change in the power of a generating installation in order to compensate for changes in the load in the power system, which may lead to deviations of the frequency of the alternating current from the standard value (in the United States—60 Hz).

In order to solve this problem, the following rules typically apply:

- Providing the market with the required level of regulated generating capacity (during the peak period—at a rate of 1.1% of the predicted peak load value; at night minimum hours—at a rate of 1.1% of the minimum load).
- Each energy supplying organization is obliged to provide a schedule for the load of its customers with regulatory capacities by self-controlling their own generating capacities, concluding contracts with other market participants for using their generating plants or purchasing regulatory capacity on the regulatory market.

The regulation can be carried out by generating plants with a power change rate of at least 5 MW/min. In this case, the generating plant should reduce the dispatching range of the load by the value of the control range, taking into account the technical minimum and nominal level.

During the operational day, the ISO determines for each generating unit the lost profit from the possible supply of electricity at real-time market prices, which, taking into account the price requested for participation in regulation, determines the price of real-time market regulation. To real-time regulation, the ISO attracts the most economically attractive installations provided that the imbalance between the load and the generation in the whole power system exceeds the permissible value. Then the telephone dispatcher informs the selected generating sources about the need to introduce their automatic regulators.

In case of unforeseen situations (e.g., emergency shutdown of generating plants), in accordance with the requirements of the North American Electric Reliability Council of the North American Electric Reliability Council (NERC), power reserves of various categories are created.

For the operational replacement of emergency outgoing generating sources, the ISO creates an operational reserve, which includes generating capacity that can be fully loaded within 30 minutes after the dispatcher's command. Depending on the load time, the operational reserve is subdivided into primary and secondary reserves.

The primary reserve has a power loading time of up to 10 minutes and is divided into a rotating reserve (sources synchronized with the network) and a quickly introduced reserve (loaded in no more than 10 minutes). These sources usually include hydroelectric power plants, gas pumping stations, gas turbine power plants and diesel installations. The secondary reserve is formed from generating sources with a power loading time from 10 to 30 minutes after the dispatcher's command.

The operating reserve is selected by the ISO on the eve of the trading day based on price bids from owners of generating facilities based on the principle of minimum load price and the situational forecast of the energy market. The selected generating sources are included in the dispatch schedule for the next day's load.

The total amount of operational reserves is calculated and announced in advance for a period of up to 8 weeks. The calculations are carried out by a probabilistic method based on historical and forecast data on the load and the expected composition of generating sources, as well as information on planned and unplanned repairs. In this case, the primary reserve is always formed in the amount of 1700 MW.

The fee for operating reserves is calculated daily based on the requested price by the owner of the generating capacity and LPC of the real-time market. Regardless of the requested price, the owner of the generating source included in the reserve will receive a fee for using the capacity of at least LPC in the real market. At the same time, the total amount of costs for the formation of an operational reserve is distributed among all market participants and is paid by them in proportion to the load.

Each of the market entities is obliged to participate in the formatting of the rotating reserve in proportion to the share of the load being served in the total load on the electricity exchange market. The obligations are fulfilled at the expense of own generating sources and the purchase of a reserve in the market of rotating reserve.

In the market of a rotating reserve, the owners of generating sources realize the capacity by submitting applications for the sale of a rotating reserve to the ISO indicating the generating unit and the asking price. On the basis of applications, taking into account the estimated LPC, the market price of the rotating reserve is set for each hour in the day-ahead and real-time markets.

The capacity market allows market participants who have obligations to provide their load with generating capacity to buy capacity according to these obligations from other market participants who have excess generating capacity. The purchase of power does not give the right to dispose of the generating source and the electricity generated by it, but only guarantees the provision of the market with the necessary generating capacity.

The market of fixed transmission rights serves as a financial instrument that allows the holder of transmission rights in case of overload of transmission networks in the DAM to earn income due to the price difference at the ends of the transmission route of electricity. Financial rights are defined for specific point-to-point electricity supply routes. The market of fixed transfer rights allows you to

- fix in advance a possible overload charge, which allows the owner to minimize the risks of buying and selling electricity associated with the uncertainty of payment for using the network;
- create an additional incentive for network investment;
- plan the modes of production, transmission, and consumption in the power system without taking into account possible restrictions on transmission.

Financial rights are bought and sold at annual and monthly auctions held by the ISO, as well as on the secondary market. At annual auctions, the ISO puts up for sale financial rights expressed in megawatts of bandwidth of the respective route. At monthly auctions, the ISO sells financial rights, the remaining unsold at previous auctions. Any market participant who has previously purchased the rights to transfer may also offer them for sale. The secondary market is an electronic trading system that allows market participants to conduct bilateral trade in financial rights.

In many ways, the success of the PJM market is determined by the developed regulatory documents governing the rules of interaction between market participants. It is also necessary to highlight the mechanism for making changes to the rules of the market, which allows, without affecting the basic principles, to constantly improve the tools of electricity trading, taking into account the experience that has been gained.

2.6 Russian model of the electricity market

In this section, I will concentrate on the processes according to which electricity markets operate. I will present a model of the electricity market using an example of the Russian electricity market. Russia was not chosen on random—it appears that the country's model of the electricity market represents an interesting story that is worth telling and analyzing in the context of the discussion how traditional power markets are evolving and changing into smart grids of the future. Russia is the largest country in the world with extreme geographical distances and climate conditions. It is hard to imagine how electricity and power grids operate in such vastness and how their coordination is handled. Yet, the system works and is functional.

In the course of the reform of the energy sector of Russia, natural-monopoly functions, such as electricity transmission and dispatching, were separated from potentially competitive ones, such as the production and sale of electricity. Instead of the former vertically integrated enterprises that performed all the above functions, companies were established that specialize in certain types of activities. Thus conditions were created for the development of a competitive electricity market, whose prices are formed on the basis of supply and demand. The energy industry was disclosed to external investors, thereby creating the prerequisites for infrastructure development and renewal.

The entire competitive wholesale electricity and capacity market was divided into price zones on a territorial basis. The first price zone includes the territory of the European part of Russia and the Urals, and the second price zone includes Siberia. In the second price zone, electricity is cheaper than the first, since in Siberia there are a large number of facilities with a lower cost of electricity production. At the same time, territories that were not united in price zones, where the conditions for creating a competitive electricity and capacity market have not yet been created, have been preserved. These include isolated power systems of the Arkhangelsk and Kaliningrad regions, the Republic of Komi, regions of the Far East, where electricity and power are sold only at prices set by the Federal Tariff Service (FTS).

Price zones, in turn, are divided into zones of free flow of capacity, which take into account the planned restrictions on the supply of capacity between them. Now on the market 30 zones are formed.

The beginning of the functioning of the existing model of the wholesale electricity market in the Russian Federation is the year 2003, when the Decree of the Government of the Russian Federation "On the rules of the wholesale electricity market (capacity) for the transition period" came into effect.

In 2006 the Decree of the Government of the Russian Federation "On Improving the Operational Procedure of the Wholesale Electricity (Capacity) Market" introduced a new model of the wholesale electricity and power market during the transition period, called NOREM—the new wholesale electricity and power market (now common abbreviation OREM). At present, the rules for the functioning of the wholesale electricity and capacity market are regulated by the Decree of the Government of the Russian Federation of 2010.

Two types of commodities are traded on the WECM—electricity and power. Capacity realization is the fulfillment of commitments to maintain the readiness of generating plants for generating electricity in the required amount, while the sale of electricity represents the physical supply of electricity to the consumer.

Since 2011, electric power and capacity on the WECM have been supplied at free (unregulated) prices, with the exception of electricity supplies to the population.

The subjects of the electric power industry include enterprises producing electricity, heat, and power; the purchase and sale of electricity and power, energy supply to consumers; the provision of services for the transmission of electrical energy, operational dispatch management in the power industry; the sale of electrical energy (power); and the organization of the purchase sales of electrical energy and power.

The functioning of the power system of Russia is implemented using a combination of operating under state control technological and commercial infrastructure and interacting in a competitive environment of enterprises engaged in the production and sale of electricity.

The organizations of the technological infrastructure include the company managing the unified national electric grid (Federal Grid Company OJSC), the organization performing dispatching management (System Operator of the Unified Energy System OJSC), interregional distribution grid companies (IDGC).

The commercial infrastructure includes OJSC Trade System Administrator (ATS OJSC) and its subsidiary OJSC Financial Settlements Center. The activities of infrastructure organizations, including pricing and conditions of interaction with contractors, are subject to government regulation.

Consequently, generating companies produce and sell electricity to sales organizations or large end users—members of the wholesale market. Sales organizations acquire electricity on the electricity market and sell it to end users.

The organization of trade and the provision of settlements between participants of the WECM is a function of ATS OJSC, which is a 100% subsidiary of the Nonprofit Partnership Council of the Market (NP Market Council). Centre for Financial Settlements OJSC calculates claims and liabilities under power and capacity purchase and sale agreements. The cost of a commercial operator is controlled by the state.

Operational dispatching management in the unified energy system of Russia is carried out by the System Operator (JSC "SO UES"). The main function of the system operator is to monitor compliance with the technological parameters of the operation of the power system. The system operator can give mandatory commands to market participants, and he also manages the sequence of output for repair of generating capacity and transmission lines.

To implement its functions, the system operator uses the calculated model of the power system modes, carrying out its actualization. On the basis of the calculated model, the commercial operator determines the required volume of electricity and the prices of trade on the wholesale electricity market, conducts a competitive power take-off (CCT), and ensures the functioning of the balancing market—trading deviations from the planned volumes of electricity production and consumption.

Sales organizations carry out two main types of activity: the transmission of electrical energy through electrical networks and the technological connection of power-receiving devices to consumers of electricity. Both of these activities are naturally monopolistic and regulated by the state.

Russian government exercises its powers in the field of state regulation and control in the electric power industry in accordance with the Federal Law "On Electric Power Industry" through the following federal bodies:

- The Ministry of Energy of the Russian Federation (Ministry of Energy of Russia), which is entrusted with the functions of developing state policy and legal regulation in the field of the fuel and energy complex, including the issues of the electric power industry.
- The Federal Tariff Service (FTS of Russia), which exercises legal regulation in the field of state regulation of prices (tariffs) for goods and services, including the establishment of tariffs for electricity transmission.
- The Federal Antimonopoly Service (FAS of Russia), which carries out antimonopoly regulation in the electric power industry.
- Federal Service for Environmental, Technological and Nuclear Supervision (Rostekhnadzor), carrying out technical control, supervision and licensing in the power industry.

Currently, energy management is carried out from two sides. On the one hand, the Ministry of Energy, which is the regulator and coordinator of the market by the state, is speaking. On the other hand, NP "Market Council," which acts as an organizer of trade in electric energy and power on the wholesale market using market self-regulation mechanisms.

Filatov (2009) identifies the main infrastructural features of the Russian electric power market:

- Electricity market in the Russian Federation is divided into two levels: the wholesale and retail markets.
- Electric power market is uniform for the whole territory of the Russian Federation. All large consumers and producers participate in its work.
- Electric power market is open, since not only Russian suppliers/consumers, but also import/export operators participate in its work thanks to the parallel operation of the UES of Russia with the power systems of a number of neighboring countries.

The price on the electricity market is formed according to the zonal principle by constructing an aggregated supply and demand curve for each of the price zones as an equilibrium for markets with perfect competition.

Due to the large territory of the country, as well as the high level of expenses for the transportation of electric energy, the price for each connection node is differentiated depending on the cost of transportation, where the node is the point of connection of a large consumer/supplier or group of consumers as well as suppliers. In total, the estimated market model takes into account about 6000 nodes.

Throughout the country, generation sources of electrical energy, like large consumers, are unevenly distributed due to the differentiation of regional economic development. Due to historical peculiarities, the united power system of Russia is divided into seven geographic power systems, each of which has a dispatch center, sufficient intrazonal electrical connections, but poorly developed external power grids connecting one system with another. The price of electric energy in power systems varies significantly due to the different structure of generating capacity. Due to weak interconnections, the problem of "plugged capacity" periodically arises when, due to network limitations, electrical energy cannot be delivered from one power system to another.

The organization of the Russian wholesale energy market has a similar structure with the US electric power markets considered. The wholesale electricity market consists of the following sectors:

- Day-ahead market (DAM)—in which producers and consumers submit price bids 1 day prior to consumption, indicating prices and desired supply/consumption volumes at a certain hour of the day, as a result of which an aggregate demand and supply curve is formed, an equilibrium price is electricity for each of the nodes and in general for the energy market.
- Balancing market (BM), which is managed in real time to ensure the balance of the power system through additional loading of generators in conditions of shortage or reduction of consumption in case of overproduction.
- Sector of bilateral agreements, where long-term contracts for the supply of electric energy (regulated and free) are concluded.
- Capacity market providing competitive selection of price bids for capacity sale.
- Market for system services provided by generators to ensure reliable power supply in the form of primary, secondary, and tertiary power reserves.

Terms and conditions of transactions and delivery of electricity vary depending on the market sector.

Since 2011, within the boundaries of the OREM price zones, regulated contracts have been concluded only with regard to the volumes of electricity and capacity intended for supply to the population. In the sector of regulated contracts, the FTS sets tariffs for electricity supplied to the wholesale market and purchased from the market.

Electricity volumes not covered by regulated contracts are sold at free prices within the framework of free bilateral contracts and the DAM. Under free bilateral agreements, market participants themselves determine their counterparties, prices, and delivery volumes.

The basis of the DAM is the competitive selection of price bids by generations and consumers held by ATS OJSC a day before the actual supply of electricity with the determination of prices and delivery volumes for each hour of the day. RSV assumes marginal pricing, that is, the price is determined by balancing the supply and demand and applies to all market participants. In accordance with the rules of trade, applications for the supply of electricity with the lowest price are primarily satisfied. The RSV price is determined for each of the nodes of the price zones. Price indices and PCV trade volumes are published on a daily basis on the ATS OJSC website.

The volumes of electricity sold under bilateral contracts and DAMs form the planned electricity consumption, which inevitably differs from the planned one. If deviations from the delivery volumes planned for the day ahead occur, the participants buy or sell them on the balancing market. Trade deviations are carried out in real time. At the same time, every 3 hours up to an hour of actual delivery, the system operator conducts additional competitive selections of suppliers' applications, taking into account the estimated consumption in the power system.

Consider the procedure for conducting a competitive selection of electricity suppliers on the DAM. The selected set of generating equipment determines the power of each participant for the planning period. The system operator once a week for the planning period from Saturday to Friday conducts a selection of generating equipment for each hour of the planning period.

The volume corresponding to the total power involved participates in the competitive selection on the PCV. To bid on the DAM, the participant submits price bids, and the following rules are observed:

- A price request may be submitted for any day or day interval from 1 to 99 days.
- The price bid consists of 24 one-hour subquotes.
- Each one-hour submission application consists of three monotonously increasing "price-volume of electricity" steps.
- The number of price values in the price application should be no more than three.
- In each hourly subrequest, the maximum value of the declared volume must be not less than the value of the included power. The difference leads to a decrease in power payment.

The original application of the participating supplier is embedded in the schedule of the PBX offer.

For the volume corresponding to the minimum value of the adjustment range (technical minimum (P_{min})), the price-receiving part is completed. The volume of the last stage is limited by the volume corresponding to the maximum value of the adjustment range (P_{max}). Volume prices are limited to prices from the price cap notification.

As mentioned above, on the DAM electricity trading is carried out at a price established under the influence of supply and demand. The equilibrium price of electrical energy is determined on the basis of price bids of suppliers and price bids of buyers of electrical energy of the corresponding price zone, taking into account the need to ensure the flow of electrical energy.

Conducting a competitive selection of applications and determining the planned production and consumption of electricity by market participants includes three main stages:

At the first stage, the PBX receives from the system operator an updated settlement model of the power system, including the scheme, the selected composition of the operating equipment, restrictions, and other parameters.

At the second stage, the suppliers submit price bids for each hour of the operating day, in which the price is indicated at which it can sell a volume of electricity not higher than that specified for each group of supplier supply points (GTP (general trading points)). A group of supply points is a set consisting of one or several points of supply of electricity belonging to a single node of the calculation model.

It is allowed to submit price-receiving bids, in which suppliers do not indicate the price of electricity, agreeing to sell electricity at the price established as a result of a competitive selection of bids.

Buyers also for each hour of operational days submit applications reflecting their willingness to buy electricity from GTP at a price and to the extent not higher than those specified in the application. Buyers can also submit pricing bids. By submitting price-receiving bids, suppliers and buyers can increase the likelihood that their bids will be accepted.

Based on the data received from the system operator and applications from market participants, the PBX determines hourly equilibrium prices and volumes of generated and consumed electricity for each price zone, forming a trading schedule. When conducting a competitive selection, ATS includes in the trading schedule the volumes of electricity suppliers, for which bids indicate the lowest price, and the volumes of electricity buyers, for which the highest price is indicated. The equilibrium price is determined by the maximum price offer of the power plant, the declared volumes of electric energy of which are still in demand by the market.

At the third stage, the PBX transfers the generated trading schedule to the system operator for maintaining the power system mode. Electricity producers, whose price bids turned out to be higher than the equilibrium price, and consumers, whose bids turned out to be lower than the equilibrium price, are not included in the sales schedule. If, as a result of competitive selection, part or all of the planned production (consumption) is not included in the trading schedule, the participant can either limit his production (consumption) at the level of the trading schedule or generate (consume) the missing volume in the balancing market.

When conducting a competitive selection for the day ahead in terms of determining the amount of electrical energy included in the planned hourly production, the PBX is guided by the following priorities:

- The first priority is set for suppliers' price-receiving applications filed for generating capacity, ensuring system reliability, nuclear power plants, as well as the volumes of electrical energy produced at the technical minimum.
- The second priority is set for suppliers' price-receiving bids for thermal power plants in the amount of electric power production corresponding to the technological minimum, as well as bids for hydroelectric power plants in the amount that needs to be produced for technological reasons.
- The third priority is set for suppliers' price-acceptance bids in respect of volumes not exceeding the priority volumes of regulated contracts.
- The fourth priority is set for suppliers' price-accepting bids for volumes not exceeding the volume of free bilateral contracts, provided that these volumes are not exceeded, together with the volumes of regulated contracts, beyond the scheduled hourly production.
- The fifth priority is set for other volumes indicated in price-receiving bids.

2.7 Market strategies and the traditional electricity markets

In the context of continuous improvement of the regulatory framework for the functioning of energy markets and changing rules for interaction between market participants, the criteria for participants' performance change. Traditional market strategies for generating companies are not always the best. In particular, the market liberalization processes (in Russian Federation) and the integration of local electricity markets into a single internal market (European Union) result in an established margin strategy (marginal cost-based bidding strategy) generating electricity generation bids due to expansion opportunities to use market power. The search for strategic innovation for a generating company is becoming a priority.

Market power is manifested in the ability of generations to profitably change prices, diverting them from the competitive level. There are the following market strategies of behavior of generating companies based on the use of market power:

- Financial withdrawal (price increase)—the formation of applications for the supply of electricity at a higher price (in particular, higher marginal costs of electricity generation). With the prevalence of demand in the market (inelasticity of demand), this should lead to an increase in the equilibrium market price, and hence the potential income of generations.
- Physical withdrawal (volume reduction)—filing applications for the supply of electricity with a reduced volume in comparison with that which can be generated on the generating equipment of the enterprise. This leads to the withdrawal of a certain amount of power from the market, and therefore also to an increase in the equilibrium market price.
- Physical withdrawal with free bilateral treaties. Free bilateral agreements are concluded, with the help of which a part of the "cheap" volume is withdrawn. With an increase in electricity consumption, this leads to an increase in the equilibrium market price.

Economists recognize the impossibility of making a clear distinction between a price increase strategy and a volume reduction strategy. In most cases, these strategies are equivalent to each other. For example, it cannot be said whether the shift in the supply function caused by monopolistic behavior is a shift to the left as a result of a decrease in output or a shift upward as a result of a price increase.

Since a monopolist can shift different segments of its supply curve by different distances, for any monotonically increasing supply curve, it is impossible to determine whether this shift occurred as a result of price increases or as a result of lower volumes. This does not mean that these strategies are completely indistinguishable but means that their use leads to the same results and consequences.

2.8 Approaches to modeling traditional electricity markets

Modern approaches to the study of the electricity market are based on the creation of mathematical models of the power system. The simplest models are optimization models of the development of a power system with one optimality criterion, for example, minimizing the cost of generating electricity. The introduction of a market structure for power management has led to the creation of generating companies that independently make decisions regarding the volume of production and commissioning. Optimizing the development of such a power system requires solving problems with many objective functions. As an example of such models created in recent decades, one can cite oligopolistic market models of power systems that have been actively developed in the United States and the European Union as part of national research projects. The basis of these models is the mathematical apparatus of game theory, which implies the rational behavior of market participants.

Below is a sequence of actions that constitute the content of the task setting process in such models:

- Setting the boundary of the system to be optimized, that is, representation of the system as some isolated part of the real world. Expanding the boundaries of the system increases the dimension and complexity of a multicomponent system and, thus, complicates its analysis. Therefore it is necessary to decompose complex systems into subsystems that can be studied separately without unduly simplifying the actual situation.
- Determine the performance indicator, based on which you can evaluate the characteristics of the system or its project in order to identify the best project or a set of the best conditions for the system. Usually, indicators of economic (cost, profit, etc.) or technological (productivity, energy intensity, material intensity, etc.) character are chosen. The best option always corresponds to the extreme value of the indicator of the effectiveness of the functioning of the system.
- Selection of intersystem independent variables that should adequately describe the permissible projects or conditions for the functioning of the system and help ensure that all the most important technical and economic decisions are reflected in the formulation of the problem.
- Building a model that describes the relationship between task variables and reflects the effect of independent variables on the value of the performance indicator. In the most general case, the structure of the model includes the basic equations of material and energy balances, relations related to design solutions, equations describing the physical processes occurring in the system, inequalities that define the range of acceptable values of independent variables and set limits on available resources.

All optimization tasks have a common structure. They can be classified as problems of minimization (maximization) of the M-vector efficiency indicator, an N-dimensional vector argument which components satisfy the system of equality constraints, inequality constraints, and regional constraints.

The most developed and widely used in practice is the one-purpose decision-making apparatus under certainty, which is called the mathematical programming: linear programming problems, nonlinear programming, discrete programming, and dynamic programming.

One of the most common optimization models of the market are mathematical models of oligopoly. Currently, all mathematical models of oligopoly are divided into two large classes. The first one is oligopoly without collusion, in which each firm, guided by the actions of competitors, independently maximizes profit, controlling its own price and the volume of product supplies. The second class of models is oligopoly with collusion, when firms are trying to find a cooperative solution in order to increase their own profits.

Oligopoly models without collusion are determined by the type of parameter that the company uses when making decisions. If a decision is made on the volume of output, the model is a quantitative oligopoly, if the price is a price oligopoly. Quantitative oligopoly models are more adequate in a situation where it is difficult for companies after the adoption of an operational plan to change production capacity, and, consequently, the volume of supplies. This is characteristic, in particular, for the energy industry.

The models without collusion describe the so-called "instant" competition: companies set prices at a single point in time, make a profit and leave the market. In practice, however, companies interact with each other more than once, which makes it necessary to resolve the game problem on a relatively stable set of firms (Lise et al., 2006). Moreover, a classic work of Chamberlin (1951) shows that in conditions of oligopoly producing a homogeneous product, companies recognize their interdependence and will maintain a monopoly price without explicit collusion. If each company makes rational decisions and strives to maximize profits, then in the presence of a small number of competitors, their actions will have a significant impact on each other, which will lead to strong opposition. And since the price reduction undertaken by any of their competitors will lead to a decrease in the prices of other companies and a decrease in the company's own profits, despite the fact that the sellers are completely independent, the equilibrium result will be the same as if there was a monopolistic agreement between them.

The circumstances described above led to the active development and application in practice of oligopoly models with collusion. In these models, the company stands out, which is the price leader (a potential winner in the price war). The leader regulates the level of market prices and assumes responsibility for adjusting prices to changing market conditions. In the market, besides the leader, a significant number of companies are forming a competitive environment. They take the price set by the leader and determine the optimal volume of production from the condition of maximizing profits.

When using oligopoly models for modeling the wholesale electricity market, one should take into account the existing restrictions on the entry of new sellers into this market. Entry into the market of new participants can significantly affect the profitability of other companies and the equilibrium result. For the study of this issue, models of entry barriers are applied.

Under the entry barrier is understood as the presence of obstacles to entry to the market for new companies, which allows established sellers on the market to obtain super-profits without the threat of the emergence of new competitors. Such barriers arise both due to the regulation of the market by state structures (fair for the energy market) and due to collective opposition from the sellers on the market.

Bain (1956) identified a number of market restrictions on the entry of new competitors. Among them, for the energy market, there will be fair restrictions such as absolute cost advantages (the ability to set prices at a level below the minimum cost of follower companies, which can completely block the entrance of competitors) and a positive economies of scale due to lower unit costs (new competitors will be too small to provide the required level of profitability). At the same time, such situations may arise as: blocking entry (companies compete without paying entry to newbies, but there is virtually no threat of entry due to the unattractiveness of the market for new sellers); pent-up input (companies modify their behavior to effectively prevent the emergence of new competitors); provided entry (each company individually finds it more profitable to allow beginners to enter the market rather than erect expensive entry barriers).

Optimization models, including multicriteria ones, have a common feature—a goal is known (or several goals), to achieve which often it is necessary to deal with complex systems, which are not so much about solving optimization problems, but about studying and predicting states depending on elected management strategies. However, the following difficulties arise:

- Complex system contains many links between elements.
- The real system is influenced by random factors, accounting for them analytically is impossible.
- The possibility of matching the original with the model exists only at the beginning and after the application of the mathematical apparatus, since intermediate results may not have analogs in a real system.

In connection with the above difficulties arising from the study of complex systems, a more flexible mechanism was developed—the simulation modeling "simulation modeling." However, it is necessary to understand that simulation modeling does not replace optimization methods but complements them. A simulation model is a program that implements a certain algorithm, to optimize the control of which the optimization problem is first solved.

The simulation approach to system modeling is usually understood as the development and use of a set of computer programs describing the operation of individual units of the system and the rules of interaction between them. Using random variables makes it necessary to conduct repeated experiments with a simulation system and the subsequent statistical analysis of the results.

There are three main approaches in simulation modeling, which differ in the level of abstraction of the studied objects: system dynamics, discrete event modeling, and agent (or agent-based) modeling.

At the most detailed level, specific material objects with their exact dimensions, distances, speeds, accelerations and times are considered, the so-called "physical" modeling. The concepts of influences, feedbacks, trends, etc., are traditionally applied to the tasks of a high degree of abstraction. Instead of individual objects, such as customers, employees, cars, transactions, goods, their aggregates, quantities are considered. The dynamics of systems at this level are described by hypotheses.

When scrutinizing the main approaches to the modeling of economic systems, one can conclude that to study the strategies of generating companies in the electricity market it is most advisable to use the agent approach, since:

- It is possible to model individual behavior, from complex (goals, strategies) to simple ones (time constraints, events, interactions).
- Allows to take into account any complex structures and behaviors.

- Objects are active, the system is dynamic (which allows to evaluate the influence of the decisions made by individual agents on the system as a whole).
- It is possible to adapt and use the past experience of agents to shape their behavior.
- Model development in the absence of knowledge about global dependencies: knowing very little about how things affect each other at the global level, or what the global sequence of operations, etc., but by understanding the individual logic of the behavior of participants in the process, you can build agent model and derive global behavior from it.

2.9 Agent-based modeling at electricity markets

Agent (or agent-based) modeling is a tool for developing models of complex adaptive systems, including various types of competitive market structures. Currently, agent-based modeling is effectively used in various areas of the economy, for example, in order to work out an optimal strategy for company behavior or to test new models of social interaction, to develop a logistics system at an enterprise, to organize production, etc. (Weidlich and Veit, 2008; Lisin et al., 2015; Ringler et al., 2016).

Agent-based computer economics (ACE) is a rather young paradigm of research that provides methods for realistic modeling of the electricity market. It allows for intelligent, multipurpose, coordinated electrical energy systems with a raw material market that operates on physical infrastructure. Recent field experience and simulation research show the potential of technology for network operations (such as traffic jam management and black—start support), market operations (such as virtual power plants), and integrated large-scale wind power generation.

There are various agent-based simulation models for the electrical energy market which aim at investigating market restructuring and deregulation, and understand the impact of a competitive energy market on electricity prices, availability, and reliability.

The wholesale supply markets, which operate on congestion-based transmission networks, have distinctive characteristics that make it difficult to detect market energy and inefficiency. Ames models traders with learning skills that interact over time in a wholesale market managed by ISO which operates on a transmission network subject to congestion. In recent years, agent-based simulation has been a popular technique for modeling and analyzing energy markets. However, other factors without decision-making are inevitable to supplement the representation of power systems (Yousefi et al., 2011).

Operators of biomass, wind, and PV power stations are moving toward a green electricity company offering the highest bonuses. Specific operational performance, bonus payments, balance cost, and control energy market revenues for all five marketers in the second scenario with reduced management premium. As described above, marketers can benefit from the sale of electricity in various energy markets, namely wholesale and the negative-minute reserve market.

Various market players will be modeled as independent actors using stand-alone software agents, and they will operate and communicate independently on energy markets and markets for emission allowances. The implementation of demand in the residential sector is a recent attempt to improve the efficiency of the electricity market and the stability of the energy system. In the centralized electricity markets ahead of time, with marginal prices, capital expenditures and capacity restrictions, there are unanswered costs that can lead to losses for some of the participating power generation units.

In addition to the electricity industry, the energy systems models include heat, gas, mobility and other sectors. For example, Markal (MARKet ALlocation) is an integrated model platform for energy systems, used for the analysis of energy, economy, and environmental issues worldwide, at national and municipal levels for a period of several decades (see Chiyangwa, 2010). Other platforms of similar features and facilities are used to explicitly model demand side, in particular to determine consumer technological choices in residential and commercial construction.

The operators of wind, solar, biomass power plants, and heat plant generate electricity using their plants. Direct marketers exchange electricity from RES on the market and control the electricity market. Operators of plants with wind, PV, and biomass can sell electricity to the direct marketer or the grid operator. The electricity generated and its virtual distribution to direct marketing specialists or to a network operator is determined.

The electricity markets are complex adaptive systems, operating under a wide range of rules, spanning different time scales. Many electricity markets are going through or are about to move from regulated central systems to decentralized markets. Agent-based models for electricity markets provide a framework for simulating deregulated markets with a flexible regulatory framework, as well as supply and demand bidding strategies.

In the European Union, there are several plans and directives containing proposals that would ensure most of the EU electricity consumers to be equipped with smart metering systems by 2020, which means that the economic evaluation of the nationwide deployment of smart meters is positive. Smart measurement systems therefore play an important role in achieving energy savings by providing households with feedback on their electricity consumption. In addition, the transition to sustainable and cost-effective energy consumption is also observed.

In Germany, for example, the renewable energy FiT laws have been gradually replaced by optional first marketing of RES on competitive wholesale markets (Del Rio and Gual, 2007). As such, additional brokers have emerged who specialize in forecasting and shipping variable renewable energy (VRE). Table 2.1 outlines the pros and cons of agent-based modeling.

Agent-based approach allows us to predict the development of events under given initial conditions and make the best decision. The models are based on the representation of the system being modeled as a set of basic elements, the interaction of which determines the generalized behavior of the system. It is important to understand that in this case the task of modeling is not to find the optimal economic equilibrium, but to study the nature of complex socioeconomic phenomena. Accordingly, within the framework of this modeling approach, special attention is paid to correctly reflecting the mechanism of behavior and interaction of system elements—the so-called "agents." In a way, smart grids of the future are the offspring of the agent-based modeling only a degree or two more sophisticated and applying algorithms and neural networks that are commonly known as AI instead of simpler mathematical modeling.

For economic models, the main element of the study is the set of interacting "agents." Agent represents a certain entity that has activity, autonomous behavior, can make

TABLE 2.1 Advantages and disadvantages of agent-based modeling.

Advantages	Disadvantages
Provides empirical validation of the model, after which the assumptions of the analytical approach (simplifications, untrustworthy) do not constitute a problem	It is necessary to conduct intensive computational experiments with the corresponding parametric ranges to achieve stable conclusions
Allows a systematic experimental study of complex economic processes	The results of experiments often constitute a distribution rather than a specific value
Encourages creative experiments	Creative modeling (instead of using existing structures) may require a considerable investment of time and effort, including for acquiring programming skills
Researchers can evaluate interesting hypotheses of their own invention with immediate feedback. It does not require the development of original software solutions	It may be difficult for the results to reflect aspects of the problem under consideration. In addition, it includes quite complex features of the implementation of the program environment

decisions according to a certain set of rules, can interact with its environment and other agents, and can also change. A distinctive advantage of this approach is the ability of agents to learn on the basis of their experience.

One of the most comprehensive definitions of agents is given by Macal and North (2005) in according to which the agent must have the following characteristics:

- Agent is "identifiable," that is, is a final object with a set of certain characteristics and rules that determine its behavior and decision rules. The agent is autonomous and can independently act and make decisions on interaction with other agents.
- Agent is in a specific environment, allowing it to interact with other agents. The agent can communicate with others (contact under certain conditions and respond to the contact).
- Agent has a specific goal (but not necessarily the goal is to maximize the benefit, as is commonly believed in classical economics), affecting its behavior.

In a number of modern agent-based models, it is also possible to single out the agent's ability to learn: the agent is flexible and has the ability to learn over time based on his own experience. In some cases, the agent may even change the rules of behavior based on the gained experience. In general, experts distinguish three stages of building an agent model:

1. Determining the boundaries of a model: what phenomenon or the event is being modeled, what are its limits.
2. Defining the behavior and interaction of agents: development of a model of behavior and decision-making by an agent and its interaction with other agents.
3. Development and testing of the model, conducting a sensitivity analysis.

The virtually unlimited possibilities for programming agents enable the researcher to create, if necessary, very complex models and to reflect in detail the decision-making process of economic agents.

Agent models usually represent a computationally constructed virtual world consisting of several agents (encapsulated programs), the various interactions of which affect all events in the world for some time. Agents in models can describe structural elements (e.g., nodes of transmission networks), organizational structures (e.g., markets), and cognitive structures (e.g., the behavior of traders and market operators).

In the energy industry, one of the most promising areas of application of the agent-based approach is its use in developing the model of the wholesale electricity market.

The crisis of the reorganized California wholesale electricity market that occurred in 2000 showed what can happen when a poorly developed market mechanism is implemented without proper testing. It is believed that part of the California crisis was due to the strategic behavior of market participants unaccounted for in model calculations, which led to a violation of the market's design features. After the California crisis, many energy researchers argued the need to combine structural understanding with an economic analysis of market participants' incentives in developing a model for the development of the wholesale electricity market (Borenstein, 2002; Pollitt, 2008).

Due to the complexity of economic processes in the electricity market, it is extremely difficult to conduct economic and physical studies of the reliability of the system using standard statistical and analytical methods.

The demand for new approaches to modeling and analyzing electricity markets has led to the adaptation and development of agent-based models, their widespread use in the energy economy.

Currently, there are many commercial agent-based models that allow to investigate the economic processes of energy markets, for example, the EMCAS model developed by researchers from the Argonne National Laboratory (Conzelmann et al., 2005). In addition, many researchers use agent-based models to study important aspects of restructuring energy markets.

As far as the study of dynamic efficiency (how the economic system uses available resources over a certain period of time) and the reliability of the standard model of the wholesale energy market are concerned, the wholesale power market platform (WPMP) in the United States constitutes a very good point of reference. Sun and Tesfatsion (2007) from the economic division of Iowa State University developed a computational model for organizing the wholesale energy market in according to the main features of WPMP and the reproduction of real-life transmission networks. This system is called AMES (Agent-based Modeling of Electricity Systems). AMES is the first open-source nonprofit system to develop and conduct computational research on the WPMP market model.

The models of modern electricity markets are based on the economic and mathematical model of the WPMP (Wholesale Power Market Platform) proposed by the US Federal Energy Commission. This circumstance is true for the majority of national energy markets of the leading states, including the energy market of the Russian Federation. This allows the development and use of universal computing systems for modeling national electricity markets. The key attention of researchers was focused on the complex interrelation of structural conditions, market rules, and training in the behavior of market participants in the short and long terms.

To make the model as close as possible to reality, it was based on real business data from two systems that adopted the WPMP model (New England and Midwest United States): market structure, price decisions, and application process.

Subsequently, it is planned to include five additional elements to AMES to more fully reflect the dynamic operational features of the WPMP market model:

- Measures to reduce market power.
- Bilateral trade mechanism.
- A market for financial transfer agreements.
- Safety restrictions combined in the task of optimizing the flow of electrical energy.
- Overdraft period during the day D as part of ISOES AMES assessing the sufficiency of the resource to ensure that the forecasted demand is covered by the supply.

The transmission lines, or transmission network, in AMES are a three-phase AC network, including $N \geq 1$ branches, interconnecting $K \geq 2$ nodes. In this case, the model considers five network nodes connected by six branches, that is, $N = 6$, $K = 5$.

Electricity consumers in the wholesale market are LSE suppliers. Their goal is to meet the demand (electrical load) of final consumers in the lower retail electricity market. Guaranteed suppliers do not produce or sell energy in the wholesale market. They receive electricity only from generators and cannot sell/buy it from each other. For simplicity, the assumption is made that the demand for electrical energy is practically not elastic in price and therefore comes down to a daily electrical load schedule. In addition, the guaranteeing suppliers are "passive" agents, that is, they do not make any strategic decisions, but only transfer demand applications to the system operator in accordance with this load schedule. Accordingly, at the beginning of each day, Dj—a guaranteed supplier (LSEj) submits a daily load schedule for day $D + 1$ in the DAM, reflecting real demand PGj (H), (MW) in the retail market every hour of the day. The user sets the number of guaranteed suppliers J, as well as the network node to which each of them relates, as the initial data of the model.

Generation (GenCo) in AMES means power units. As well as for guaranteeing suppliers, the model sets the total number of power units I and their location in the network, that is, the node to which they refer. Generations supply electricity only to guaranteed suppliers and do not sell or buy it from each other. For each generation, the user sets the technological parameters (minimum and maximum production volume, production costs), training parameters and the initial cash supply.

The allowable production capacity interval for determining the hourly production volume ranges from the minimum to maximum possible production volume. Each generation has its own TC total cost function ($/h), reflecting the total costs of producing a certain amount of electricity in each hour of the day. At the beginning of each day D each i generation submits a bid proposal to the system operator for each hour of the day $D + 1$ for trading on the DAM. This application consists of the stated marginal cost function. This application may be strategic, in the sense that the stated coefficients may not coincide with the actual coefficients and the declared allowable interval of production capacity may differ from the actual allowable interval of production capacity.

As a result of the bid selection auction, agents (generating companies and guaranteeing suppliers) receive a schedule for meeting their obligations (buying and selling electricity in a fixed volume and at a fixed price every hour of the day). When fulfilling their obligations, the generations receive an appropriate profit, which evaluates the effectiveness of their work (strategic behavior), the correctness of the choice of parameters of the application. Also, the system determines marginal prices for each hour of the day for each k node of the network and the actually generated and consumed amount of electricity for each hour of the day.

One of the advantages of the AMES model is the reproduction of the self-learning system of energy market traders, which is based on the results of the experiments of the multiagent games of Roth and Erev (1995). The training algorithms used allow the generations to choose the best method of generating applications for the supply of electricity, which they submit daily to the system operator AMES to participate in the WPMP market for the day ahead.

By changing certain parameters of the model that determine the agent's behavior, one can add or remove the possibility of using market power strategies (financial or physical withdrawal), which makes it possible to evaluate the effectiveness of these strategies used by generating companies in the DAM.

The results of the test experiments show that the generations in time learn to create implicit collusion and provide excessive marginal costs, thereby significantly affecting the cost of electricity in the model nodes.

At the beginning of each day (at the beginning of the game with $D = 0$, generation gives true data on its costs and production capacities), i generation forms a bid proposal to the DAM from a set of possible solutions covering the area above or on the real marginal costs curve generation. The coordinates of this set are determined by the parameters that define the matrix of possible solutions for selecting the parameters of the bid proposal.

Every day i generation faces the challenge of choosing and compromising between using past experience and exploring new opportunities. It should be noted that i generation does not always strive to maximize profits. There is always the possibility that, instead of the application selected by the generator, which promises maximum profit, the system will accept another generation application in the process of market selection. Therefore agents continue to experiment with different options for bids, even if the probability distribution of choice reaches a peak for a particular bid because of its relative effectiveness. This algorithm reduces the risk of premature fixation of a suboptimal bid proposal at the early stages of the gameplay, when only a small part of possible bid proposals have been tried. Thus the Roth−Erev algorithm allows the agent over time to choose the best option for the parameters of the application from the point of view of profit maximization.

2.10 Conclusion and discussions

All in all, the analysis of the traditional electricity markets might bring us to the conclusion that despite the differences in the organization of power systems in different countries, the basic structure of the energy market has common features, such as:

• Interaction of operators: technological (system operator), commercial (exchange or market administrator), and network operators that have a controlling effect on network switching.
• Separation of the stages of the market in time. Typically, a market model consists of several sectors. The most common of these are the DAM, the balancing market, and the market for long-term contracts.

- Availability of guaranteeing suppliers participating in trading on the wholesale electricity market and meeting the demands of consumers in the retail market.
- Pricing on the basis of auctions for bidders of sellers and buyers using optimization algorithms that take into account system constraints for planning day-ahead schedules.
- Real-time power system maintenance, combining technology and commerce (the system operator provides a balance of production and consumption based on applications and an optimization algorithm).
- Hourly nodal prices.
- The admissibility of bilateral contracts between sellers and buyers, provided that they pay the difference in nodal prices between the point of delivery and the point of consumption of the traded electricity.

The most important sector in terms of pricing in the market and flexible management of the load schedule is the DAM. The established marginal strategy of the behavior of generating companies in the electricity market can significantly reduce the risks of nonacceptance of generation applications by the market; however, if demand prevails on the electricity market over supply (inelasticity of demand), generating companies can apply market power strategies to maximize their income. The chapter highlighted some possible strategies of behavior of the generating company based on the mechanisms of financial and physical withdrawal.

Furthermore, the chapter assessed the main approaches to the modeling of energy systems. In large, all models can be divided into optimization and imitation. The simplest models are optimization models of power system development with one optimality criterion. The introduction of the market structure of power management has led to the need to solve problems with many objective functions. The purpose of such models is to find the optimal, in terms of certain criteria, options for the use of available resources.

Recently, in the study of complex systems, the main task is the prediction of states depending on the chosen management strategies, which has led to the adaptation and development of simulation models.

This chapter identified the main difficulties that do not allow economic and physical studies of the reliability of the power system by standard statistical and analytical methods, such as:

- A large number of locally distributed market participants.
- Strategic behavior of participants.
- The possibility of the existence of multiple equilibria of the market system.
- Behavioral uncertainty caused by human factors and changes in the market environment.
- Institutional mechanism of market regulation.
- Two-way feedback between market structure and changes in the market environment.
- Complex system contains many links between elements.
- The real system is influenced by random factors, their accounting analytically is impossible.

As a result of a comparative analysis of simulation modeling methods, it was concluded that to study the strategic behavior of generating companies in the electric energy market, it is most expedient to use the agent approach, since

- It is possible to model individual behavior, from complex (goals, strategies) to simple ones (time constraints, events, interactions).
- Allows to take into account any complex structures and behaviors.
- Objects are active, the system is dynamic (which allows to evaluate the influence of the decisions made by individual agents on the system as a whole).
- It is possible to adapt and use the past experience of agents to shape their behavior.
- Model development in the absence of knowledge about global dependencies: knowing very little about how things affect each other at the global level, or what the global sequence of operations, etc., but by understanding the individual logic of the behavior of participants in the process, you can build agent model and derive global behavior from it.
- An agent-based model is easier to maintain—updates are usually made at the local level and do not require global changes.

Using the integration of optimization and simulation approaches, a study was conducted of the main factors affecting the effectiveness of the use of market power and the conditions under which the use of withdrawal strategies (financial and physical) leads to an increase in the company's profits and/or revenues.

A mathematical model has also been developed for applying strategies for financial and physical withdrawal and an algorithm for generating the best strategic bid by a generating company.

These outcomes and conclusions might allow and help generating companies to evaluate the possible effect of the use of financial and physical withdrawal strategies, as well as to develop the best behavior strategy, taking into account market conditions and a number of physical limitations imposed by the structure of the power grid. Also, the description of the key features of the AMES software and computing complex will allow it to be used to study energy markets for educational and scientific purposes.

If bids indicate excessive marginal costs, then we are talking about a financial withdrawal. If the generation submits bids to the system with an underestimated amount of the maximum possible capacity, this means that the generation applies a physical withdrawal strategy.

The effectiveness of the decision on applications is determined by the value of the profits. The tendency to implement a certain action (as well as a similar decision) is enhanced if it leads to a favorable outcome and weakens with an unfavorable outcome. In this case, the experience of recent actions is more important when making a decision than the previous one. Thus the learning curves are aligned with time, and the agents are constantly improving the mechanism for selecting application parameters and strive to optimize them.

Recent work in the field of developing models of the wholesale electricity market shows that the main direction of research is connected with the study of the behavior of participants in the wholesale electricity market. In reality, market participants do not seek to maximize utility when making decisions; rather, the decision-making task is aimed at finding the most satisfactory solution that will give an average result. Understanding this fact led to the creation.

New dynamic models of the electricity market, which are based on a new methodology based on agent-based modeling. At the same time, the model of the electricity market is formed from the interaction of many market agents, each of which has certain features behavior. Special attention should also be given to the open-source software and computing models. These models include the AMES project, specifically created to study the dynamic efficiency and reliability of the WPMP model of the energy market. AMES is the first such open-source nonprofit system. One of the most significant problems solved in the framework of agent-oriented models is the study of the influence of market generation power on the energy market conditions, as well as the search for the best behavior strategy of the energy company. The market strength of the generating company lies in its ability to influence the bid price and conditions for the sale of electricity on the market without reaction from competitors.

Generally, one would probably agree with me that renewable or nonrenewable energy sources have major uses and influence in the current situation of an electrical system. In addition to that, their use is facilitated and optimized by the deployment of smart grids. Smart grid are actually self-optimizing grids which are innovative and intelligent solution that combines automation and decentralized applications for monitoring and remote network control. An important part of the smart grid is the ability of customers to participate in the overall management of the grid. Controlled and reliable integration of distributed energy sources and microgrids is essential to ensure uninterrupted power supply in the most efficient and economical configuration. For example, the customer can adjust the energy consumption based on detailed information on energy consumption and dynamic peak prices which can be displayed on some monitor or a device (e.g., her or his smartphone).

References

Auer, H., Haas, R., 2016. On integrating large shares of variable renewables into the electricity system. Energy 115, 1592–1601. Available from: https://doi.org/10.1016/j.energy.2016.05.067.

Bain, J.S., 1956. Barriers to New Competition. Harvard University Press.

BCG, 2018. The digital energy retailer. <https://www.bcg.com/en-us/publications/2018/digital-energy-retailer.aspx> (accessed 28.04.19.).

Bessembinder, H., Lemmon, M.L., 2002. Equilibrium pricing and optimal hedging in electricity forward markets. J. Finan. 57 (3), 1347–1382. Available from: https://doi.org/10.1111/1540-6261.00463.

Bobinaite, V., Tarvydas, D., 2014. Financing instruments and channels for the increasing production and consumption of renewable energy: lithuanian case. Renew. Sustain. Energy Rev. 38, 259–276. Available from: https://doi.org/10.1016/j.rser.2014.05.039.

Borenstein, S., 2002. The trouble with electricity markets: understanding California's restructuring disaster. J. Econ. Perspect. 16 (1), 191–211. Available from: https://doi.org/10.1257/0895330027175.

Chamberlin, E.H., 1951. Monopolistic competition revisited. Economica 18 (72), 343–362. Available from: https://doi.org/10.2307/2549607.

Chao, H.P., Peck, S., 1996. A market mechanism for electric power transmission. J. Regul. Econom. 10 (1), 25–59. Available from: https://doi.org/10.1007/BF00133357.

Chiyangwa, D.K., 2010. Strategic investment in power generation under uncertainty: Electric Reliability Council of Texas. Doctoral dissertation, Massachusetts Institute of Technology. <https://dspace.mit.edu/handle/1721.1/59673> (accessed 29.04.19.).

Conzelmann, G., Boyd, G., Koritarov, V., Veselka, T., 2005. Multi-agent power market simulation using EMCAS. IEEE Power Engineering Society General Meeting 2829–2834. Available from: https://doi.org/10.1109/PES.2005.1489271.

Crampes, C., Léautier, T., 2016. The second digitisation of the electricity industry: chronicle of a revolution foretold. Florence School of Regulation. <http://fsr.eui.eu/second-digitisation-electricity-industry-chronicle-revolution-foretold>, (accessed 26.04.19.).

Del Rio, P., Gual, M.A., 2007. An integrated assessment of the feed-in tariff system in Spain. Energy Policy 35 (2), 994–1012. Available from: https://doi.org/10.1016/j.enpol.2006.01.014.

Filatov, A., 2009. Oligopoly models: state of the art. <http://math.isu.ru/ru/chairs/me/files/filatov/2009_-_oligopoly.pdf> (accessed 20.04.19.).

Gallagher, K., 2009. Acting in Time on Energy Policy. Brookings Institution Press., Washington, D.C.

Hirsch, A., Parag, Y., Guerrero, J., 2018. Microgrids: a review of technologies, key drivers, and outstanding issues. Renew. Sustain. Energy Rev. 90, 402–411. Available from: https://doi.org/10.1016/j.rser.2018.03.040.

Joskow, P.L., 1997. Restructuring, competition and regulatory reform in the US electricity sector. J. Econ. Perspect. 11 (3), 119–138. Available from: https://doi.org/10.1257/jep.11.3.119.

Lise, W., Linderhof, V., Kuik, O., Kemfert, C., Östling, R., Heinzow, T., 2006. A game theoretic model of the Northwestern European electricity market—market power and the environment. Energy Policy 34 (15), 2123–2136. Available from: https://doi.org/10.1016/j.enpol.2005.03.003.

Lisin, E., Strielkowski, W., 2014. Modelling new economic approaches for the wholesale energy markets in Russia and the EU. Trans. Business Econ. 13 (2B), 566–580.

Lisin, E., Strielkowski, W., Amelina, A., Konova, O., Čábelková, I., 2014. Mathematical approach to wholesale power and capacity market regulation. Appl. Math. Sci. 8 (156), 7765–7773. Available from: https://doi.org/10.12988/ams.2014.49757.

Lisin, E.M., Amelina, A.Y., Strielkowski, W., Lozenko, V.K., Zlyvko, O.V., 2015. Mathematical and economic model of generators' strategies on wholesale electricity markets. Appl. Math. Sci. 9 (140), 6997–7010. Available from: https://doi.org/10.12988/ams.2015.59611.

Macal, C.M., North, M.J., 2005. Tutorial on agent-based modeling and simulation. IEEE Proceedings of the Winter Simulation Conference, pp. 1–14. <https://doi.org/10.1109/WSC.2005.1574234>.

Málek, J., Rečka, L., Janda, K., 2018. Impact of German Energiewende on transmission lines in the central European region. Energ. Effic. 11 (3), 683–700. Available from: https://doi.org/10.1007/s12053-017.

Newbery, D.M., 2002. Problems of liberalising the electricity industry. Eur. Econ. Rev. 46 (4-5), 919–927. Available from: https://doi.org/10.1016/S0014-2921(01)00225-2.

Oh, T.H., Pang, S.Y., Chua, S.C., 2010. Energy policy and alternative energy in Malaysia: issues and challenges for sustainable growth. Renew. Sustain. Energy Rev. 14 (4), 1241–1252. Available from: https://doi.org/10.1016/j.rser.2009.12.003.

Pollitt, M., 2008. The arguments for and against ownership unbundling of energy transmission networks. Energy Policy 36 (2), 704–713. Available from: https://doi.org/10.1016/j.enpol.2007.10.011.

Pyrgou, A., Kylili, A., Fokaides, P.A., 2016. The future of the feed-in tariff (FiT) scheme in Europe: the case of photovoltaics. Energy Policy 95, 94–102. Available from: https://doi.org/10.1016/j.enpol.2016.04.048.

Ringler, P., Keles, D., Fichtner, W., 2016. Agent-based modelling and simulation of smart electricity grids and markets—a literature review. Renew. Sustain. Energy Rev. 57, 205–215. Available from: https://doi.org/10.1016/j.rser.2015.12.169.

Roth, A.E., Erev, I., 1995. Learning in extensive-form games: experimental data and simple dynamic models in the intermediate term. Games Econ. Behav. 8 (1), 164–212. Available from: https://doi.org/10.1016/S0899-8256(05)80020-X.

Sämisch, H., 2016. Digitalisation: where are the German digital utilities? Euractiv. <https://www.euractiv.com/section/energy/opinion/digitalisation-where-are-the-german-digital-utilities/> (accessed 20.04.19.).

Schreiber, M., Wainstein, M.E., Hochloff, P., Dargaville, R., 2015. Flexible electricity tariffs: power and energy price signals designed for a smarter grid. Energy 93, 2568–2581. Available from: https://doi.org/10.1016/j.energy.2015.10.067.

Siano, P., 2014. Demand response and smart grids—a survey. Renew. Sustain. Energy Rev. 30, 461–478. Available from: https://doi.org/10.1016/j.rser.2013.10.022.

Simshauser, P., 2016. Distribution network prices and solar PV: resolving rate instability and wealth transfers through demand tariffs. Energy Econ. 54, 108–122. Available from: https://doi.org/10.1016/j.eneco.2015.11.011.

Sovacool, B.K., 2009. Rejecting renewables: the socio-technical impediments to renewable electricity in the United States. Energy Policy 37 (11), 4500–4513. Available from: https://doi.org/10.1016/j.enpol.2009.05.073.

Strbac, G., Pollitt, M., Konstantinidis, C.V., Konstantelos, I., Moreno, R., Newbery, D., et al., 2014. Electricity transmission arrangements in Great Britain: time for change? Energy Policy 73, 298–311. Available from: https://doi.org/10.1016/j.enpol.2014.06.014.

Sun, J., Tesfatsion, L., 2007. Dynamic testing of wholesale power market designs: an open-source agent-based framework. Comput. Econ. 30 (3), 291–327. Available from: https://doi.org/10.1007/s10614-007-9095-1.

Weidlich, A., Veit, D., 2008. A critical survey of agent-based wholesale electricity market models. Energy Econ. 30 (4), 1728–1759. Available from: https://doi.org/10.1016/j.eneco.2008.01.003.

Winkler, J., Emmerich, R., Ragwitz, M., Pfluger, B., Senft, C., 2017. Beyond the day-ahead market—effects of revenue maximisation of the marketing of renewables on electricity markets. Energy & Environment 28 (1–2), 110–144. Available from: https://doi.org/10.1177/0958305X16688810.

Yalcintas, M., Hagen, W.T., Kaya, A., 2015. Time-based electricity pricing for large-volume customers: a comparison of two buildings under tariff alternatives. Util. Policy 37, 58–68. Available from: https://doi.org/10.1016/j.jup.2015.10.001.

Yousefi, S., Moghaddam, M.P., Majd, V.J., 2011. Optimal real time pricing in an agent-based retail market using a comprehensive demand response model. Energy 36 (9), 5716–5727. Available from: https://doi.org/10.1016/j.energy.2011.06.045.

Zhang, Y.J., Da, Y.B., 2015. The decomposition of energy-related carbon emission and its decoupling with economic growth in China. Renew. Sustain. Energy Rev. 41, 1255–1266. Available from: https://doi.org/10.1016/j.rser.2014.09.021.

3

Sustainability off the smart grids

3.1 Introduction: the origins and concepts of sustainability

In general terms, sustainability is a broad concept that is encompasses an overview of key aspects of the human world ranging from business to technology and from environmental to social sciences. The term "sustainability" itself is quite old. It was first coined over 30 years ago and it can be first found in the famous Gro Harlem Brundtland's report (also known as "Our Common Future") prepared for the United Nations by the World Commission on Environment and Development (WCED) in 1987. It was the first time that the term "sustainable development" was introduced (Anker, 2018).

If any of us were asked what a "sustainable development" is, many definitions and associations would probably come to our minds. For me sustainability has always been a simple and straightforward concept. In 1998 some 10 years after the Brundtland's report was released, I signed up for the undergraduate course in nature protection at the Charles University in Prague. Our lecturer was an experienced and well-traveled man who attended many climate-related events in the early 1990s and was preaching this new exciting ideology in the former socialist country that became the Czech Republic. He explained us the meaning of sustainability during one of his first lectures in an intuitive way: "Imagine," he used to tell us during his lectures: "that you want to explain your grandmother what sustainability is. What would you say? Well, the easiest way would be to go along the lines of something like "sustainability is a way of leaving on our planet Earth and using its resources in a way that would not fully exploit it and would leave both the planet and the resources for our children, grandchildren, and the next generations to come." I am still using this simple, yet powerful definition when I have to explain what sustainability is to an unaware outsider, my students, or other people who are not fully aware of its meaning.

Nowadays, sustainability is becoming some sort of a mantra that is taught at schools and universities, repeated thousands of times daily in all kinds of mass media and is enforced in our daily jobs and private lives. Sustainability skills and environmental awareness are becoming a priority in many graduates and jobs, as companies are trying to comply with new laws and regulations that mention sustainability in one way or another. As a result, mentioning of sustainable development can be found in many areas such as civil planning, environmental consulting (building and natural environment), agriculture, business strategies, health evaluation and planning, and even law and (geo)political decision-making (Bruce, 2008).

Sustainable development aims to protect our natural, human, and ecological health, while at the same time promoting innovation without compromising our preferred way of life. It also becomes apparent that sustainability is about balancing local and global efforts to meet human basic needs without destroying or humiliating the natural environment. Sustainability is studied and managed on many scales (e.g., level or benchmarking) of time and space, in many environmental, social, and economic contexts. The concept of sustainability became a long-term and ever-present objective of our daily existence rather than a short-term goal. When analyzing open and closed natural systems such as urban and national parks, dams, farms and gardens, lakes and rivers, one way of looking at the relationship between sustainable development and resilience is to look at the former with a long-term vision and resistance.

With regard to the above discussion, Goodland and Daly (1996) identified three main types of sustainability: (1) social, (2) economic, and (3) environmental. Moreover, they proposed three general criteria for sustainable development: (1) renewable resources should ensure sustainable yields (harvest rate should not exceed recovery rate), (2) for nonrenewable resources there should be an equivalent development of renewable resources, and (3) waste production should not exceed the environmental capacity. The management of the global atmosphere now requires an evaluation of all aspects of the carbon cycle in order to identify the possibilities of combating climate change caused by human activities, and it has become the main subject of scientific research due to the potential catastrophic effects on biodiversity and the human community.

One would probably agree that Brundtland's report and its systematic approach to sustainability put the emphasis on developing a more direct focus on social justice and maintaining and protecting ecosystem services. In low-income societies, economic growth leads to great satisfaction when the resources generated by growth are used to meet the basic needs of people. Emerging economies such as China and India are striving to meet the Western standards of life popularized by the developed economies, as well as the nonindustrialized world in general.

Nevertheless, there are mixed opinions on whether technological efficiency and innovation will allow for a complete separation of economic growth from environmental degradation (see, e.g., Porter and Van der Linde, 1995; Jaffe et al., 2002; Singh et al., 2012; Aghion et al., 2016). The strategies for more sustainable social systems involve improving education and strengthening gender policies, particularly in developing countries which are more important for social justice, including equality between the rich and the poor, and intergenerational equity.

Nowadays, sustainability measures are designed to consolidate economic, environmental and social performance measures of each system. In fact, at the end of the 21st century, when most economically developed countries have already adopted a policy for the development of information and communication technologies (ICT) infrastructure, the market for communications industry has risen dramatically. Thanks to cognitive science, modern society will reach its critical level and allow the public at large to use the development of knowledgeable society.

European Union has always been and still is one of the most vigorous supporters of sustainability. European Commission (2011) defined smart grids as "modernised power networks, enhanced by two-way digital communication between suppliers and consumers, as well as smart measurement, monitoring and control systems driven by advances in digitisation." Sustainable economic development is embedded into its economic and social goals, policies, strategies, and short- and long-term objectives. One of the examples was the Lisbon Strategy, an ambitious reform program launched in 2000 that led Europe to become the world's most dynamic and competitive knowledge-based economy that would be able to achieve sustainable economic growth with increasingly better employment opportunities and a high level of social cohesion and respect for the environment (European Parliament, 2000). Nevertheless, in spite of all these brilliant achievements, European Union still needs a vision for a society that can integrate aging and young people to develop long-term growth and social cohesion.

Postolache (2012) points out that European Union is attempting to tune its environmental policies with the current impending trends in order to sustain in the global governance and maintain its leading position in the world's international relations.

According to the European Union's renewable energy plan up to date, renewable energy can be generated from a wide assortment of sources comprising solar, tidal, the wind, geothermal, and biomass. The targets for renewable energy sources are tricky ones but they have to be met as far as the European Union have previously committed to them (Bryngelsson et al., 2016; Böhringer et al., 2016).

The European Union's energy policies are guided by three main objectives: Secure energy supplies that ensure reliable provision energy, creating a competitive environment for energy providers with the purpose of ensuring affordable energy prices, and a sustainable energy consumption through lowering the greenhouse gas (GHG) emission, pollution, and fossil fuel dependence (Böhringer et al., 2016). To achieve the objectives, the European Union has decided to allow free flow of energy across national borders of the EU countries. The free flow of energy is expected to minimize the monopoly of national energy providers thus mitigating on the issue of high prices (European Commission, 2016). Consequently, new power lines and pipelines continue to be deployed in the region to create a resilient and integrated energy market. This is contradictory to the 1990s market design of the energy industry because energy transfer from one state to the other was levied high rates of rates of taxes and limited to a given amount per period of time.

Various EU Member States have also created schemes which aid to spur adoption and generation of renewable energy. For example, the Dutch government offers subsidies to businesses, which are willing to invest in production capacity of sustainable energy. All the EU members also have full support for the targets of 2020 and are dedicated to decarbonizing the electricity sector of European Union by 2050 in accordance with the Paris Agreement.

The Paris Agreement is a climate conference organized by the United Nations Framework Convention on Climate Change (UNFCCC, 2014), which was held in Paris, France, on December 2015 and included delegations and representatives from 195 countries. The conference (further referred to as "Paris Agreement") was a long-awaited endeavor speeded up by the recent development in the world marked by the plunging oil and coal prices that resulted from the upheaval in the world's politics and caused many industries do turn away from the sustainable energy (Wang and Li, 2015).

The Paris Agreement was a timely and long-awaited climate deal that was preconceived and prenegotiated long before it shaped up into a formal deal involving global players from all around the world. It represented an important consensus and a recognition of the danger brought about by the climate changes and the related issues. Furthermore, it proved that it might be possible to mobilize the key players to sit around the table and to strike a deal that has global consequences. With regard to the above, the Paris Agreement proved to be a milestone event that means a lot both to the policy makers and to the climate activists.

Moreover, the Paris Agreement also tackles one of the most alarming issue of the industrial development of the last decades, the issue of emissions. Although decreasing, they remain at high levels even in EU-28 countries.

Paris Agreement was a historical event owing to the large number of countries that adopted the first-ever universal, legally binding global climate deal. The main aim of this convention was to establish a global action plan that will guide the world on pursuing plans and development projects that would avoid dangerous climate change through limiting global temperature rise to 2°C (European Commission, 2016). According to scientists, a temperature rise of 2°C forms the upper limit of the acceptable temperature rise that can limit the adverse effects of climate change (Steinacher et al., 2013). This is based on a consensus that global warming should be limited to 1.5°C above preindustrial temperatures that will ensure the survival of low-lying islands and coastal areas. This agreement involved member states of the UNFCCC which sets a general framework that encourages intergovernmental effort devoted to solving the challenges brought about by climate change and the realization that our climate is a shared resource that is currently struggling with the effects of industrial emissions of carbon dioxide and other GHGs (Hustedt and Seyfried, 2016). This convention offers membership to any country globally under which a member country's government is expected to

- Investigate, collect, analyze, and share information concerning GHG emissions, national policies, and best practices in the specific member state;
- Establish national strategies that focus on handling GHG emissions and adapt to the expected impacts of these gases, including the provision of financial and technological support to developing nations;
- Cooperate and assist in the preparation for adaptation to the impacts of climate change (UNFCCC, 2014).

The Paris Agreement is due to gear into full force in 2020. With regard to mitigation of carbon emissions, governments that consented on this convention agreed to ensure that the global average temperature remains below 2°C above preindustrial levels (Andresen et al., 2016). They also had a collective desire of ensuring that global emissions peak as soon as possible with the recognition that developing nations may require a longer period to achieve this goal. However, following the end of the period each developing nation agreed upon, they were expected to ensure rapid reductions in carbon emissions through the adoption of the best available scientific approach (European Commission, 2016). This consensus was followed by the submission of comprehensive climate action plans which provided reports about the method to be adopted in achieving these objectives (Hustedt and Seyfried, 2016).

In an effort to create transparency in the enforcement of strategies established by member nations and enable global stocktaking, the member parties agreed that they would converge after every 5 years so as to set more ambitious goals which are guided by scientific contributions. Member parties were also expected to report to each other and their public systems on the efforts they have put into ensuring they achieve their goals. Finally, members were also expected to track their progress through the creation of robust transparency and accountability systems which will aid in achieving the long-term goal. With respect to adaptation to climate change, member parties agreed to ensure that they strengthen their individual communities and citizens' abilities and duties in dealing with the effects of climate change on their lives and those of others. In addition, signatories were expected to provide continued and enhanced international support for the adaptation to developing countries (UNFCCC, 2014).

Loss and damage were also an element that was focused on during the convention. Within this bracket, signatories were required to recognize the need to avert, minimize, and address loss and damage resulting from the negative impacts of climate change observed all across the globe. Member states also needed to recognize that there was a need to provide some form of cooperation between nations and better their understanding, action and support within varied sectors that may help in dealing with the effects of climate change such as early warning systems, emergency preparedness and risk insurance programs. The UNFCCC had also established that cities, regions, subnational authorities, civil societies, the private sector and local authorities each had a role to play in tackling climate change issues whereby each member was expected to increase their efforts and support actions to decrease emissions. Member states also had to build resilience and decrease vulnerability brought about by the changing climate patterns. It was therefore necessary to uphold and promote regional as well as international cooperation to ensure that climate change issues are effectively and efficiently dealt with. Signatories of the Paris Agreement were expected to show support for the efforts being made to control the drivers of climate change. Developed countries such as the United States, China, and the European Union were expected to provide continued support for climate action aimed at reducing emissions and building resilience to the adverse effects of climate change being witnessed in developing nations. Other nations were provided with the option of providing continued support for the fight against global warming on a purely voluntary basis. Developed nations were also expected to achieve their goal of raising US$100 billion each year up to 2020 which extends to 2025 after which other more ambitious goals will be settled (UNFCCC, 2014).

The Paris Agreement therefore established a historical mark as being the largest step in the development and maturity of the UN climate change regime which was first organized in 1992 along with the adoption of a framework convention (Clémençon, 2016). This led to the UNFCCC acquiring a long-term objective, general principle, common and differentiated commitments and the establishment of a basic governance structure (Center for Climate and Energy Solutions, 2015). Ever since its establishment, there has been an evolution of the convention thus taking different paths with the course of time. In 1997 a similar convention, Kyoto Protocol, was held. This convention adopted a rather different strategy which involved utilizing a top-down, highly differentiated approach that intended on enforcing agreed upon goals for developed nations without any new commitments directed toward developing nations. Owing to the failure of the United States to join this protocol as well as the decision by some countries not to set targets that were beyond 2012, there is less than 15% coverage of global emissions by this protocol (Andresen et al., 2016).

Other attempts of dealing with climate change were also made by the Copenhagen Accord and the Cancun Agreements in 2009 and 2010, respectively. These conventions adopted a parallel bottom-up framework whereby countries were required to take pledges for 2020. However, unlike the Paris Agreement which took on a legal approach, these conventions adopted political approaches. Despite these two conventions having wider participation with specific mitigation pledges by developing countries being present for the first time in history, they were not enough to meet the objectives of the Copenhagen and Cancun Conventions of preventing the average global warming temperature from being 2°C above preindustrial levels (Center for Climate and Energy Solutions, 2015). The negotiations that led to the Paris Agreement first began at the Durban Platform for Enhanced Action which took place in 2011. This convention led to the suggestion for a legal instrument or uniformly agreed upon outcome which would apply to all signatories from 2020 without any additional guidelines provided. This was followed by COP 19, which was held in Warsaw where members were expected to provide their intended nationally determined contributions (INDCs) prior to the hosting of the Paris Conference which indicated the importance of the bottom-up approach in the emerging Paris Agreement in which more than 180 countries which produced over 90% of the global emissions had submitted their INDCs (Center for Climate and Energy Solutions, 2015).

In addition to the INDCs submitted to the UNFCCC prior to the commencement of the convention, additional outcomes were achieved during the Paris Agreement. A large number of national governments offered new financial pledges resulting in US$19 billion having been pledged aimed at assisting developing countries. This was followed by a pledge by the United States to double its support for adaptation efforts in order to reach US$800 million per year. Vietnam also portrayed a sign that developing nations are offering support to the program by pledging to offer US$1 million to the Green Climate Fund (GCF). There were also pledges from subnational governments. In addition to pledges, there were also new joint initiatives which were launched such as the International Solar Alliance which consisted of 120 nations spearheaded by India and France. This initiative was started with the aim of promoting the establishment of solar energy facilities in developing countries. Another joint initiative was Mission Innovation which consisted of over developed and developing nations which pledged to increase investments in clean energy research and development over a period of 5 years. Nonstate actors such as cities, states, and regions, company and investors such as Bill Gates were also recognized as having participated in the convention whereby they launched initiatives such as Breakthrough Energy Coalition whose main goal is to generate capital from the private sector that will drive clean energy deployment (Center for Climate and Energy Solutions, 2015).

Currently, developing countries are actively scoping out financial, technological, and technical support to enable them to implement Nationally Determined Contributions (NDCs) whereby there exists well-embedded national climate strategies and plans, and poverty reduction plans. Where no clear path is in view, measures are being taken to ensure that the necessary inputs are acquired along the journey.

3.2 Sustainability in the economic context

We have already seen that nowadays the principle of sustainability can be found in many areas of industry. Sustainable energy technologies are also used to generate electricity, heat and cool buildings, as well as power up transport systems and machines. "Green" electricity is a subset of renewable energy and represents the renewable energy sources and technologies that bring the greatest benefits to the environment (Shafiullah et al., 2013). "Going green" has become a very popular and widely used slogan nowadays. It is taught to children at schools and explained to the general public in TV programs and newspaper articles, as well as debated about on the Internet.

Contrary to the last decades, today's policy makers and regulators are well aware of "green" technologies and are supporting the sustainable economic development as the only meaningful path to the growth and development of the human society in the nearest decades. Most of them call for a distinct context of the renewable energy market and support the preparation for the increase of renewable energy in the total share of energy output.

The 20th century was a wonderful demonstration of the success of civilization which was developing along the path of scientific and technological progress. However, there is one bitter consequence from that. From the moment of the appearance of mankind and up to the present day, nature seemed to be an infinitely large storeroom, from which one could draw as many products as possible for the development of civilization, and at the same time an infinitely large natural production, processing all the waste of human activity and turning them into original products. Unfortunately, the events of the last decades showed us that view is far from the truth.

Indeed, the gross global product in goods and services grew 33 times in the 20th century, production capacity increased 10 times, and the population of the Earth increased six times. All in all, human population has grown considerably in the last 2000 years. The industrial revolution became a milestone for the giant leap in the size of world's population. By 1800, the Earth population reached 1 billion people. It increased by another billion within 130 years (in 1930) and reached 3 billion within the next 30 years (in 1959), 4 billion within 15 years (1974), 5 billion within 13 years (1987), and 6 billion within 12 years (in 1999). For just 40 years, from 1959 until 1999, the world's population has doubled from 3 billion to 6 billion. By many accounts, by the end of 2011 mankind was to reach its 7 billion, the process taking just 12 years. It happened sooner than expected, though. In 2011 the world media announced that the human population reached the 7 billion mark.

Today, the world's population is growing at a rate of about 1.15% per year. The rate of growth reached its peak at the end of the 1960s, when it reached the level of 2%. The rate of population growth is due to decrease in the next few decades. However, the average annual change in population is estimated to be at a rate of more than 77 million people. It is a widely accepted concern today that the world's population will set itself at slightly above 10 billion after 2200.

As seen from the examples above, the world's population grows at a high rate. The Malthusian law (an exponential growth model first explained in the book "An Essay on the Principles of Population" by Robert Thomas Malthus, an English scholar who lived in the late 18th and early 19th century) states that small populations typically grow exponentially (especially in the absence of threatening natural enemies). Applying this law to the human population would mean that mankind that lives in a comfortable built environment it has created for itself and is not threatened by any natural enemies will grow progressively and abundantly.

It seems logical to assume that the exponential phase in the growth of our planet's population started at the moment the first civilizations formed themselves (i.e., mankind stepped onto the certain level of socialization that allowed for the reproduction of the human species regardless the caprices of nature). It was scientifically proven that the first civilizations on Earth were those dating back to around 8000 B.C. (e.g., Egyptian, Sumerian, Assyrian, Babylonian, Helenian, Minoan, Indian, and Chinese civilizations). Quite curiously, only Indian and Chinese civilizations have remained in existence until today, all others went extinct. This might lead to the conclusion that there was indeed a factor hampering the exponential growth of the human population throughout the ages. Nevertheless, whatever this fact was, the population growth has reached the brim after which it will become unsustainable.

In the second half of the 20th century, the rapid growth of all world economic indicators was observed and recorded. Gross world product from 1950 to 2000 increased by seven times, from $6 to $43 trillion. The standard of living of the population in developed countries has grown to a level that they could not even dream of 100 years ago. Life expectancy has increased. A fundamentally new system of information exchange and its storage has been created. By the beginning of the 21st century, the lowest prices were established on the grain market. The average grain yield in 2000 was 28 c/ha, the total grain yield exceeded the amount of grain harvested for all 11,000 years since the beginning of active agricultural activity.

This is a sure sign that food shortages do not threaten humanity the immediate future. However, these successes have been achieved at the cost of the ever-accelerating destruction of the Earth's ecosystems. Indeed, the withdrawal from the nature of raw materials necessary to meet human needs exceeds the ability of nature to restore them or compensate for their disappearance. Thus one-third of the land does not fill the soil fertility, and it is increasingly degrading. Half of the Earth's pastures is slowly but gradually turning into desert due to the violation of their marginal productivity (carrying capacity). Every year, the Earth's deserts increase by 156 million square kilometers. In Nigeria, for example, about 500 km^2 of land turns into a desert every year.

Forests that were reduced by almost half with the beginning of the agrarian revolution from the middle of the 20th century continue to disappear with increasing speed due to exceeding the reproduction rate of felling and forest fires. Now the annual forest losses in the world exceed 9 million hectares. Everywhere there is a shortage of fresh water. This is especially true in areas of intensive farming. With the introduction of irrigation to the farm due to the intake of underground water by drilling using diesel and electric pumps (and now about 10% of the Earth's population lives on irrigated land), the level of groundwater is annually lowered by 1.5 m or more. It takes place in the North Plateau in China, where 25% of all Chinese grain is produced. The same thing happens with the arable land in the United States, with the ground waters of Punjab, etc. As a result, water retake does not reach the sea in certain important world's rivers such as the Colorado, the Yellow River, the Indus, and the Ganges.

Given that general policies and goals in addition to declines in the costs of solar panels has resulted in a growth in people's pursuits in tapping to the solar energy market. Considering that the international effort to decrease carbon emissions and boost renewable energy source, many governments across the globe have ventured into launch a variety of coverages like peak pricing for residential customers and net metering. Peak pricing is meant to smoothen the energy demand all day by providing higher prices to clients working at peak-usage times together with the impact of raising the efficiency of energy distribution. Web metering, on the other hand, empowers distributed generators, like clients using solar panels installed in their own rooftops, to nourish their surplus electricity back into the grid in retail costs (Kok et al., 2015).

A lot of electricity energy customers are opting for this kind of renewable energy production technology and put in the panels in their houses and business premises. The rising trend of photovoltaic (PV) apparatus is essentially because of the fact that their price has decreased dramatically, and many taxpayers can now afford them. Additionally, as a result of climate change, the nation is choosing for solar energy and this particular; the government is subsidizing solar products to make them even cheaper for its citizens. On the other hand, the installation of PV panels customers has caused a variety of issues nationally. A few of the issues respect consumer protection as well as the purchase price of electricity.

Compensation for the solar PV users (prosumers) might be in the kind of paying the user in the utility where greater power is generated than that which is absorbed (the clinic broadly known as "net metering"). Many web metering policies require utilities to buy a distributed generation (DG) customer's additional power at a complete retail cost though the price of generating the power by the utilities is a lot lower (Strielkowski, 2016). As it remains the obligation of the utilities to keep such electrical grids, they change the price to the customers and consequently, the price of power increases. Furthermore, these chargers are additional altered to nonsolar customers, which then raise their power bill. Determination of the ideal speed for net metering is a complex matter. The subject seeing electricity costs is on the proper retail rate where to compensate customers for DG. There is a disagreement on if the cost to be employed to compensate the dispersed energy customers ought to be below or at the retail price.

Majority of people purchase rooftop-solar panels because they think the technology is going to save them money or to help them to produce them more green and sustainable energy for satisfying their needs, or sometimes (in many cases) both (Pool, 2012). However, the certainty is that rooftop solar should not be saving them money even though it often does, and it almost surely is not green. Specifically, the rooftop-solar style is absorbing billions of dollars each year that might be utilized on greener drives. Additionally, it is assessing the advancement of more high-tech renewable reservoirs of electricity.

According to an existing US Energy Department-supported study, establishing a fully financed, average-size rooftop-solar system will reduce energy costs for 93% of those single-family houses from the 50 biggest American capitals now (Potts, 2015). That becomes the reason why people have been rushing outside to buy rooftop-solar panels, particularly in bright countries like California, Arizona, and New Mexico. The main cause is these small solar systems are cost effective. Nonetheless, they are profoundly supported. Monopolies are required by legislation to purchase solar energy generated in the rooftops of companies and homeowners in the two to three times greater than it might require purchasing solar energy from good, independently controlled solar plants. Without political subsidies, rooftop solar would never be cost efficient. But this functionality is probably only when solar factories are wide and located in bright segments of the country.

Large-scale solar energy prices are falling since the cost to construct solar panels has been decreasing and because large solar installations authorize economies of scale. On the flip side, rooftop solar generally necessitates microsetups in unproductive places, making the general value up to three and a half times greater. There are tons of reasons as to why we are paying more for the exact same sunlight. Well-intended however ill-thought national, regional, and state taxation concerns for rooftop solar at the United States return between 30% and 40% of their institution charges to the proprietor for a donation credit (Strielkowski, 2016).

However, in spite of all its economic sustainability, it seems that most electricity customers do not have relevant information on DG policies. The issue of increasing electricity costs for nonconsumers as well as the users can be adversely affecting the development of solar DG. Therefore the government and other stakeholders need to intervene and educate individuals on the policies, the concept of net metering and the pricing of solar products in order to safeguard them. Given that many countries on earth are being influenced by climate change, using solar renewable energy would be of fantastic benefit in controlling the environmental hazards caused by the usage of nonrenewable energy on the environment.

Quite another problem is that in order to achieve truly sustainable energy, changes are required not only in the way energy is delivered today, but also in the way it is used by its final consumers. Embracing the principle of sustainability would require reducing the amount of energy needed to supply a variety of goods or services.

The recent unprecedented global changes that are rapidly altering the face of today's world and represented by the serious issues such as population growth, competition for resource reduction, good recognition of ecosystem services and natural capital, as well as adaptive challenges arising from our changing climate and our national climate change commitments would all require ambitious social and economic interventions and response over the next decades to come. Here is where sustainability becomes important in the economic context and should be assessed and analyzed from that angle of view.

Most of the readers would probably agree with me that sustainability without economic growth is not sustainable not acceptable by the wide masses. Environmental sustainability is linked to ecosystems and keeps the balance that is needed for them to endure. Economic sustainability has the capacity to support a defined level of production indefinitely.

One can see that sustainability is a process that requires participation at the individual level. Fiscal institutions have an important role in sustaining the sustainability principle. There also exists a term "financial sustainability" that is related to the well-being of communities or individuals in connection with their capacity. Exactly as with business industries that were small, the work of banks has an effect in the environment all the resources and transport that is used. These investments create new jobs and bring about the development of local markets.

The renewable energy revolution might be the most impressive case of this effect of new technologies. The company world has dealt with sustainability for ages. The United Nations has launched an initiative that would leverage social networking technologies and ideas to provide those who would otherwise not have the ability to pay for the costs with education opportunities.

At times the news about our surroundings can feel a tiny monotonous. Development articles abound on the Internet and there is a lot you have the ability to learn about making healthier choices for new house you wish to come up with or to get this job you have got in mind in your vacation home. As a homeowner you may look for topics on sustainable development online and analyze various options that can be made for a house.

Nowadays, a concept of a "smart city" became a very popular paradigm. Many cities worldwide are aspiring to obtain this title. If the city has the sufficient potential to grow into a smart city, the problem is not how to do that, the problem is by what means. Further relevant issues need to be taken into consideration. As people want to fix social issues throughout the plan process much is missed from the simple fact people are engaged in design each of the instant, actually, design is only one instance of a sort of educative activity, such as communicating, apprenticeship, or any other. If significant security issues are discovered in the plan process correcting them may be pricey. Attempting to learn the reason why increasing investments in technology is not currently leading to equal increases in productivity is among the regions of debate today. The achievements to date of the nation not negate. On occasion, objectives are also served by economic considerations.

Multiple security factors and numerous security barriers deal with uncertainties in addition to risks. Negative impacts on the environment are inherent to the business, like waste creation which are difficult, possibly impossible, to avoid and the emission of GHGs. The effect of global economic change is likely to exacerbate the matter, and the unemployment rate of young folks may be supply of political instability and social unrest. The concentrate on the coral reef ecosystem to get a goal location for medicinal purposes is not a new thought. A valuable approach to find an understanding of systems theory is to look at the example. So as to innovate solutions, an individual must acquire a comprehension of the matter. Additionally, the importance to our own economic future of Africa cannot be underestimated.

Most of the readers would probably agree with me that the responsibility of sustainability cannot be left completely to the marketplace. The future is not adequately represented in the marketplace and there does not exist any reliable hope that regular market behavior will care for whatever duty we have into the future. Economic policy measures such as Pigouvian taxes, subsidies, as well as other laws and regulations could accommodate the incentive structure in a way that protect the global environment and resource base for individuals not yet been born. After all, it has been Pigou (1932) who remarked that there should be a broad agreement that the state must protect the interests of future at a certain level contrary to the ramifications of our absurd discounting and our immediate tastes and preferences. It is the obvious responsibility of the government which is responsible (in accord with the principle of sustainability) for the unborn generations in addition to for its current citizens. Thence, the government should guard and protect the exhaustible all-natural resources of the nation from rash and reckless spoliation using all the means it has to its disposal.

It is a well-known and widely accepted paradigm that transition from high-carbon to low-carbon economy would incur positive economic costs. However, a considerable part of the engineering and economic literature also suggests that these costs would be too low to have a significant impact on the overall economic growth rate (Grubb et al., 1993; Weitzman, 2007; Pearson and Foxon, 2012; Dai et al., 2016). In wealthy societies, growth generates a complex set of social and environmental costs, which explains why life satisfaction surveys remain largely unchanged in industrial societies, despite the high level of production and consumption since World War II. In addition, there are mixed opinions on whether technological efficiency and innovation will allow for a complete separation of economic growth from environmental degradation (see, e.g., Banister, 2008; De Jesus and Mendonca, 2018).

One can see that the consumption of goods and services can be analyzed and managed at all levels through the consumer chain, from the impact of individual lifestyle choices and spending patterns to resource requirements for specific goods and services, the impact of economic sectors, national economies and the economy. Within this context, consumption pattern analysis refers to the use of resources with environmental, social, and economic impacts on a scale or context in question. In addition, economic outlook for smart grids will depend on the specific circumstances of the region, including the generation of resources and electricity demand trends.

In general, there are three areas of sustainable development that fully describe the relationship between the environmental, economic, and social aspects of our world. As with environmental sustainability, economic sustainability is about creating economic value for every project or decision one wishes to undertake (McWilliams and Siegel, 2011).

The question arises: how to describe sustainability in the economic context? Can economy in general or separate economic systems in particular be sustainable? If one had to put it down properly, the best way to do that would probably be to say that economic sustainability means making decisions in the most equitable and fiscally sound manner possible, while taking into account other aspects of sustainability.

Moreover, one would probably agree that sustainability in economics context might also represent a comprehensive interpretation of the ecological economy paradigm, in which variables and environmental variables constitute a part of a complex multidimensional perspective.

Sustainable competitiveness can be defined as the establishment of institutions, policies and factors that make a nation productive in the long run, while at the same time ensuring sustainable social and environmental sustainability.

There are also different angles of view when it comes to the sustainability and economic systems depending on whether they come from economically developed and wealthy or developing countries. In low-income societies, economic growth leads to great satisfaction when the resources generated by growth are used to meet the basic needs of people. One can also say that sustainability is about balancing local and global efforts to meet human basic needs without destroying or humiliating the natural environment. From a developed country perspective, economic growth has led to an increase in wealth, income, standard of living, and improved health care facilities.

Nowadays, innovation is spreading all over the world quickly because business people and organizations are raising concerns about the development of products, operations, strategies and supply chain management. Our lessons from economic development, entrepreneurial innovation, alliances and network organizations, innovation in companies engaged in sustainable business become strategic and competitive advantage.

Sustainable entrepreneurship is based on the fundamental principles of entrepreneurship and extends to address environmental and social concerns by setting up new businesses and innovation in existing businesses.

Entrepreneurship as a mentality or behavior is well suited to sustainable business practices.

For sustainable business practice, entrepreneurship is a constant quest for innovative ways to protect the environment or improve social conditions by offering new goods, services or ways to reduce harmful activity and at the same time generate profits for the entrepreneur.

There are many examples from developed and developing countries to prove the points raised above. For example, China can be a particularly difficult environment for innovations due to the issues arising from protection of intellectual property for sustainable business ventures that introduce new technologies. In Brazil, innovation could lead to new and diverse organizations that could have a positive social and economic impact in a fragmented and closed business society.

History can support the reasons why both sectors have entrepreneurial intensity, while the education and energy sectors have different achievements.

Entrepreneurship represents a driving force behind any business and industry. It is also relevant in the case of the creative industry which does not demonstrate sustainability but yields only a handful of isolated innovations.

Therefore innovation becomes a learning process that generates or acquires new knowledge from the analysis of a particular sustainable environment.

In turbulent organizational environments such as Brazil, regional prospects can provide some structural benefits for the company.

Addressing environmental and social concerns can create business opportunities that can be beneficial to the economy and society. Successful, sustainable businesses focus on creating new products and services to address environmental and social issues in a new way.

Sustainability reporting is the process by which companies describe their own economic, environmental, and social impacts, enabling them to recognize the value of sustainable practices.

In order to assess the degree of accepting sustainability, one has to focus on five main dimensions of organizational performance (economy, environment, community, human, and governance) and provide a range of metrics and a valuation tool for enterprise performance.

Students will be able to become key figures in local, regional, national, or global companies, holistic thinking and visionary leaders in their communities, corporations, NGOs, public and private organizations, and government agencies.

Therefore this approach combines both the traditional approach to financial and market conditions with the introduction of socioeconomic sustainability. However, recent research has combined social issues with the concept of innovation, which is called social innovation.

In addition to exploring innovation as a proactive business dimension, innovation through its process, integrating causes and steps, metrics and process models, and finally exploring innovation as a system can also be scrutinized as being important elements of this mosaic encompassing sustainability concept.

In terms of innovation, the findings show the relationship between the use of specific resources (information systems, people, knowledge management, and alliance) and product innovation.

In six categories, students gain hands-on knowledge of business methods and tools, such as business modeling, as well as first-hand experience in the field through internship placement, consulting, or business development. The demand for sustainable entrepreneurs is growing: more and more companies than ever see sustainability as a key element of their mission.

Nevertheless, it is remarkable that for many people in the world of business and finance, economic sustainability and sustainable growth constitute a main focus. Hefty financial support for universities, educational programs, and research and development in the forms of university chairs, donations, supporting research seminars and professorships is also an important part of building and maintaining sustainability that starts from promoting this concept in education to embedding it into the minds of future citizens and creators of the new reality. This appears to be important and rewarding, since students who have the opportunity to meet many successful entrepreneurs and entrepreneurs who focus on sustainability would also carry on this concept with them in their further lives and careers.

In the modern industrial society (and in conditions of an unrestrained growth in the consumption of goods that have a deliberately shortened period of use), energy is a large-scale pollutant of nature. At the current stage of development of science and technology, each segment of the world energy economy has a destructive impact on the world around it. This fully applies to the sphere of renewable energy sources, since absolutely "clean" energy carriers and RES-devices do not exist yet, if only for the reason that renewable energy facilities in any case change the natural course of energy exchange and mass exchange of the biosphere.

At the same time, environmental RES-factors, direct and indirect, are characterized by a wide variety and strength of impact. They arise both at the stages of construction, production, operation, and utilization of renewable energy equipment, and in the technological chain of the use of "green" energy carriers, sometimes even covertly and with unpredictable consequences in the long run. For example, the construction of hydroelectric dams can lead to a decrease in the standard of living of the population, degradation of ecosystems, and fish resources in the long term. Wind power can be a source of negative impact for birds, bats, aquatic creatures, and humans, and can cause radio frequency interference; geothermal energy is potentially dangerous in terms of landslides and soil failures, as well as earthquakes.

The use of power RES-devices is inextricably linked with the use of energy storage devices (chemical, thermal, electrical, mechanical, producing intermediate types of energy carriers (e.g., hydrogen), etc.), which also pollute the environment. In the biomass segment, environmental stress arises already at the stage of obtaining raw materials (in the course of agricultural work, as a result of using genetically modified objects (GMO) plants, deforestation in order to expand crop areas, etc.), in the production of appropriate industrial equipment, and the operation and utilization of renewable energy installations. (emissions and waste of various types), in the process of biofuel production, as well as the operation of vehicles using biofuel or mixed fuel (the probability of reducing the technical resource of engines of Tell, there is a need for an additional car equipment, introduction of new types of lubricants, and so on). However, it should be noted that when biomass is grown, active absorption of CO_2 from the atmosphere occurs as a result of the photosynthesis reaction, therefore from the point of view of balance (the difference between total absorption and total CO_2 emissions) emissions of "greenhouse" gases during the entire life cycle of this sector of renewable energy is net carbon dioxide absorbent.

For a generalized assessment of the direct and indirect impact on the environment and as a rough tool for comparing the disadvantages and merits of renewable energy facilities, various assessment criteria can be used, such as:

- Impact on land resources;
- Impact on the animal and plant world;
- Human impact; and
- Impact on water resources.

Let us consider the main parameters characterizing the degree of influence of various types of renewable energy sources on the environment and, if possible, compare them with indicators for hydrocarbon energy carriers.

Wind energy is widely used in the production of electrical energy. On a global scale, it has a significant technical resource, a high degree of availability and constancy, as well as relative cheapness. Wind power plants (wind turbines) can be located both on land and in coastal waters on the sea shelf. These advantages allow wind energy to compete with fossil fuels; in 2011 in the structure of EU electricity production, the share of this energy source accounted for more than 6%.

At the ground location of the equipment, a small land area in the form of a circle with an area of 510 diameters of the wind wheel of the wind turbine is directly involved, and the cable facilities are laid underground. According to a study by the National Renewable Energy Laboratory (United States), the total size of the land plot is approximately 1257 hectares per 1 MW of the designed capacity of the plant, while only a small part of it is permanently occupied—at least 0.4−1.5 ha/MW—temporarily (mostly during construction).

Thus the main area around the wind turbine tower can be used for other needs, such as the construction of nonresidential and infrastructure facilities, livestock grazing. In addition, wind turbines can be located on land unsuitable for farming or other economic needs, as well as in industrial zones, which significantly increases the attractiveness of this type of RES in terms of land resource use. Wind turbines located on the surface of the sea occupy a wider area than ground installations, since they have significant dimensions and cable management, laid on the seabed. They can create difficulties for shipping, fishing, tourism, extraction of sand, gravel, oil, and gas.

Wind turbines affect wildlife, primarily birds, which die in direct collisions with wind turbines and as a result of habitat destruction due to artificial changes in natural air mass flows (the end of the wind turbine blade can move at a linear speed of about 300 km/h). The reduction in bird and bat mortality is facilitated by the optimal choice of equipment location, technical solutions (e.g., complete shutdown of wind turbines at wind speeds below a certain level, shutdown of wind turbines during bird migration), as well as taking into account other local conditions identified in the operation of such equipment.

Sea-based wind turbines also lead to the death of birds, however, to a lesser extent compared to ground-based complexes. The main negative impact of wind turbines of this type includes the possible reduction of marine populations and the creation of artificial obstacles (reefs).

Wind turbines can have a harmful effect on humans as a source of high- and low-frequency radiation, by visual influence (flickering, disturbing the beauty of the natural landscape—the emergence of new "sights," etc.) in the event of a farm falling or mechanical destruction of a wind turbine. In addition, accidents can occur in the process of maintenance and repair of equipment, in a collision with wind turbines of aircraft.

The degree of influence of these factors largely depends on the design of the wind installation, its location, production discipline and the completeness of the implementation of appropriate organizational measures. It is believed that with compliance with all requirements, the negative impact of wind turbines on humans is minimal.

The impact of wind turbines on water resources is negligible. Water is used only in the production process of component parts and in the construction of the cement base of a wind turbine. The amount of harmful CO_2-equivalent emissions associated with the life cycle of wind turbines is much lower than that for thermal power plants and is usually in the range of 1020 g/kWh (for gas stations around 270,900, and for coal about 6,301,600 g/kWh).

Solar energy has an enormous resource and can be used in the production of thermal energy (solar collectors, etc.) and electrical energy (PV plants, solar concentrators, geo-membrane stations, etc.); the degree of environmental impact largely depends on the design and power of solar equipment.

The surface area used by solar-powered systems is determined by the type of installation. Low-power stations can minimize this load and be located on the roofs of buildings or integrate into various elements of buildings (walls, windows, etc.), and industrial installations can use a large area. This indicator for PV plants (FGU) is within 1.54 ha/MW, solar concentrators—1.56 ha/MW.

There are projects of solar concentrators that occupy a large area of the earth's surface (comparable to that for CHP and NPP), but elements may be located in areas unsuitable for growing crops, along infrastructure facilities, landfills for household waste or other areas to reduce the impact on flora, fauna, and humans.

During operation, the impact on water resources from the FGU is minimal; water is used only during the production of solar cell components. However, the design of solar collectors involves the use of water as a coolant, and in some types of solar concentrators water consumption (for cooling the system) can reach 2500/MWh.

The negative impact on humans is mainly determined by the manufacturing process of the silicon elements of FGU, which can cause contact with harmful and toxic substances (hydrochloric, sulfuric and nitric acids, acetone, hydrogen fluoride, gallium arsenide, cadmium telluride, copper-indium or copper-gallium, and others). In the production of thin-film modules uses a smaller amount of harmful substances, however, it also requires strict adherence to safety measures.

CO_2 emissions for FGU are 3680 g/kWh, solar concentrators 3690 g/kWh. Geothermal energy extracted from the depths of the earth (from 200 m to 10 km) can be used to produce electrical and/or thermal energy, as well as cold and steam, either by converting (using steam turbines) or directly (by pumping well fluid into building systems). As of the beginning of 2010, the total power of geothermal power generating plants in the world was about 11 GW and thermal power was about 51 GW. Stations of this type are created both in regions that are unsuitable for agriculture and in nature conservation areas. They can occupy a fairly large area, for example, the world's largest geothermal complex "The Geysers" (United States) covers an area of more than 112 km^2, which corresponds to a specific area indicator per unit capacity of 15 ha/MW.

In general, the influence of the geothermal installation on the animal, plant world and humans are directly dependent on the system design, type of energy carrier, security measures taken and other factors and, despite these drawbacks, is quite low.

Geothermal stations are a source of air pollution, emitting sulfur dioxide, as well as hydrogen sulfide, carbon oxides, ammonia, methane, boron, and other substances that can provoke lung diseases and heart disease in humans. Nevertheless, it is believed that in this sector of generation the emission of SO_2 is 10 times less compared to coal-fired thermal power plants.

In general, with this technology, the amount of pollution is estimated at 90 g/kWh in CO_2-equivalent, but for systems with a closed working circuit, this indicator is limited to emissions from the manufacture of equipment.

Biomass is widely used in the production of heat and electrical energy, liquid and gaseous motor fuels, and not only for road transport, but also for aircraft, as well as ships.

The impact of this segment of renewable energy sources on land, plant, animal and human life can be quite significant. For example, to expand the cultivated areas of industrial crops, the forest fund can be exterminated, which leads to a reduction in the habitat of many species of animals and birds; the increase in the area of the corresponding crops on agricultural land aggravates the conflict with the food sector. At the same time, a significant number of biological wastes are generated in the world, recycling of which helps clean up the environment.

Traditionally, biomass (wood waste and coal, straw, some types of agricultural and livestock waste, municipal solid waste, etc.) is used by incineration. In this case, according to the degree of environmental impact, it is similar to hydrocarbon energy carriers; however, its advantage is renewability.

The development of modern technologies is in the direction of creating methods for the production of biofuels of the second and subsequent generations (methanol, ethanol, biodiesel and synthetic fuels, jet fuels, biomethane, hydrogen, etc.) by pyrolysis, gasification, biological and chemical processing, hydrogenation, etc., allowing to effectively process all types of biological raw materials, first of all lignocellulose. The introduction of appropriate industrial solutions (in the European Union this is scheduled for the period after 2015) will allow the industry to move to a qualitatively new level and mitigate its impact on agriculture and the food sector. In the long term, a steady increase in production of bioethanol and biofuels is expected, and their cost will also increase (it is expected that by 2021 the price of biodiesel fuel in the global market will stabilize in nominal terms near $1.4 per liter, bioethanol $0.7 per liter).

The impact of the biomass sector on water resources can be quite significant (depending on the region), since a certain amount of moisture is required to increase the yield of industrial crops. In addition, pollution of the surface waters of the region may occur due to the use of fertilizers and pesticides.

When using biomass both by direct combustion and using methods for its various transformations into intermediate energy sources, harmful substances (oxides of carbon, nitrogen, sulfur, etc.) are formed. At the same time, a comparative analysis of CO_2 emissions relative to hydrocarbons (gas, coal, petroleum products) shows that this indicator largely depends on the type of technology and fuel (on average, 1890 g/kWh) and in some cases for biomass it is higher than for other types of energy carriers.

Hydro energy is used by hydroelectric power stations of various capacities—from micro hydro (several kW) to large hydropower plants (more than 25 MW) that are part of the national power system. The impact of this type of renewable energy on land resources primarily depends on the type and power of equipment, as well as the terrain and can reach several hundred hectares per 1 MW of installed capacity.

Hydroelectric power plants, especially large ones, have a significant impact on nature and man; it is described in sufficient detail in numerous scientific materials of various organizations, for example, WWF (WWF. Dams and development. New methodological basis for decision-making: Report of the World Commission on Dams. Fujikura and Nakayama, 2009) In hydropower, the emission of "greenhouse" gases for small stations is estimated at 513.5 g/kWh, for large hydroelectric power stations—1320 g/kWh.

In some cases, high-power hydropower plants may cause increased emissions of carbon dioxide and methane as a result of decay of biomass flooded during the creation of the dam.

A mindless pursuit of the goal of expanding the share of renewable energy in the expenditure part of the energy balance, based only on economic and political considerations, can have far more serious environmental consequences, and further along the chain—the economy as a whole than the use of fossil fuels. On the other hand, it is necessary to understand that a full-fledged consideration of environmental requirements will inevitably lead to the containment of energy development and, as a result, to new crisis phenomena in the national economy. Therefore in our opinion, it is necessary to use the possibilities of nature wisely to meet the needs of society, conduct a thorough assessment and a comprehensive study of the impact of RES facilities on the environment and look for ways to limit and prevent it.

Many countries possess a huge potential and an extensive base for developing renewable energy in order to improve energy efficiency and reduce energy costs in all sectors of the economy, reasonable diversification of energy supply to many categories of consumers, improve the situation in the housing and utilities sector, and strengthen the business activity of small- and medium-sized businesses. Renewable energy can be one of the components of the process of overcoming the technological backwardness of Russia, since it has a positive effect on the development of fundamental and industrial science, a high-tech manufacturing sector.

Already in the medium term, the domestic market can activate demand for cost-effective energy equipment of various types of power and intelligent systems, allowing to increase consumer autonomy and optimize energy production processes both on the basis of renewable energy sources and in combination with traditional energy sources.

Foreign capital is interested in developing the renewable energy sector in a number of countries of the former USSR countries due to economic, environmental and other reasons (limited EU land and water resources, specifics of regulating the turnover of GMO crops, the need for additional supplies of clean energy, protests of residents of several regions, etc.). For example, for Russia, this opens the window of opportunity to attract active players to its potentially vast and profitable RES market.

The influx of relevant investments and the implementation of renewable energy projects in various countries must be strictly linked to a thorough study, the environmental component of projects (based on the experience and knowledge of domestic specialists), the import of the most advanced technologies and equipment, as well as the subsequent maximum localization of production. The absorption of know-how that adversely affects the environment and humans, as well as the passive role of the "raw materials appendage" in this segment of the energy sector, is at least destructive.

3.3 Sustainable development and renewable energy

We have already established in previous chapters that future development on the market of electric energy is likely to include transactive energy and smart grids, which are much more tentative to the concerns of today's rapidly changing world represented by such pressing issues as the global warming and climate change, market distortions and energy security thanks to the use of combined renewable energy sources and innovative storage solutions.

Renewable electricity frequently supplies energy to four crucial areas: electricity generation, air and water heating, transportation, and rural (off-grid) energy generation. Accelerated use of renewable energy and electricity efficiency is leading in critical energy security, climate change reduction, and monetary benefits.

The decisions of a recent evaluation of this newspaper concluded that as GHG emitters begin to be held liable for accidents occurring from GHG sparks resulting in climate change, a higher cost for liability mitigation would provide persuasive grounds for installation of renewable energy inventions. In international public evaluation surveys, there is substantial support for encouraging renewable sources of electricity like wind power and solar energy. In the federal level, about 30 countries around the world have renewable electricity contributing over 20% of electricity reserves. State renewable energy demands are prognosticated to rise steadily in the next several years. Some European countries (e.g., Norway or Iceland) create their power using renewable energy immediately, and several other nations have put a goal to achieve 100% clean electricity in eventuality.

A sustainable market design is distinguished by an approach that may achieve long-term targets and as to adapt medium- and short-term changes (Sørensen, 1991). The power companies, particularly in Europe, are broken and they cannot fulfill their fundamental functions. The energy market in Europe is twisted, and disturbing indications are emerging. In most European nations, several aspects can be attributed to the sinking of industrial power expenses. These factors demand following slow expansion in economic activity after 2008 and downturn but one crucial motive regarding the decrease in cost is that the adoption and expansion of renewable production. However, occasionally when they are generating system marginal prices are gloomy because renewable resources have zero marginal price and their existence on the marketplace lowers the need for higher-cost plants to create, therefore there exists a decrease in market rates. But this does not affirm a recession at the long-term rates. The renewable energy has a huge cost compared to production methods they replace. If sustainable sources were hooked on the marketplace to remunerate price most will be losing money but because these resources are not reliant upon the resources, they go unheeded (Sovacool, 2009). In consequence, government policies that promote continued installation have been in battle with markets as well as the breeds could only rise over time.

The cost report published by the European Commission in 2014 (European Commission, 2014) suggested that the costs of electricity and gas in the European Union have increased between 2008 and 2012. Among the recommendations provided by the EU Commission would be to further build up the energy infrastructure by diversifying the sources of electricity and adding higher shares of sustainable energy. Additionally, the European Commission also suggested implementation of European law which would finish the internal energy market and allow the customers to easily change energy providers (Lisin and Strielkowski, 2014). The talk of crude oil and oil remained high from the entire energy intake in the European area, which are 35.1% and 33.8%, respectively. The consumption and demand of power was quite feeble as the people shifted more attention on solid fuels and renewables in the energy industry. Atomic energy maintained its ingestion share because 236.6 Mtoe was absorbed at the 3-year interval. The share of solid fuels' intake increased from 15.9% to 17.5% while that of renewables from 9.8% to 11% during the same period (European Commission, 2014).

We live in the age of renewable energy when decentralized energy resources and intelligent grid technologies enable all prosumers, from families to small- and medium-sized businesses, in addition to larger businesses, to incorporate their intake and production of electricity in systems which would serve more like ecosystems compared to traditional markets. This scenario might challenge standard power produced by centralized electricity and unidirectionally dispersed by utility firms. Present-day electricity customers now have the ability to behave as electricity generators through dispersed generation from solar PV and waste heat recovery. Allowing and encouraging technology and behavior that optimize the whole energy system, instead of only individual components this can lead to enormous financial advantages.

The areas of smart grid-based technology, each of which consists of a series of individual technologies, cover the entire grid, from generation to transmission and distribution to different types of electricity consumers.

Potential measures by which smart grid technology can help save energy include the preservation of consumer information and feedback systems, the implementation of diagnostics in residential and medium-sized enterprises, the support of electric vehicles (EVs) and advanced voltage regulation.

Intelligent energy management systems are key factors in the expected efficiency gains, both on the demand and supply side of the smart grid. As part of microgrid analysis, it is important to assess the mix of different energy strategies, generation and type of fuel and to develop concepts that provide clean, efficient, cost effective, highly reliable and locally regulated energy and heat. Advances in renewable energy integration allow for a wider deployment of renewable energy sources and storage technologies within a microgrid strategy.

Compared to centralized power systems, decentralized energy systems use sustainable sources of energy, in the presence of the grid or lack of power, such as the kilowatt scale. In addition, a decentralized energy generation system provides access to remote locations, as electricity production is compatible with demand. The implementation of decentralized energy systems could thus solve the problem of power supply in rural areas.

Smart grid systems could be part of the new infrastructure in developing and rapidly growing regions, enabling efficient operations and improving the capabilities of the market.

For example, electricity consumption by utilities has been reduced by increasing the use of energy-efficient appliances, the enormous increase in the number of PV systems installed on the roof.

Decentralized decisions are being made at the highest level of the power system, which is due to the ongoing reform of the electricity industry. The energy is generated at the central point and then transferred to the grid via the high voltage power lines and from there the electricity is distributed to its consumers. The electricity would be closer to the retail price of the electricity produced, and it would be more efficient to meet the demand for small energy markets, given the same position and electricity generation. Systems are based on technology, which has the lowest cost when it comes to accessing heat loads and permanent electricity. With sustainable resources (sun, geothermal, wind) and automatic functions, today's embedded systems can face such challenges.

The smart grid provides an interface between utility and consumer, helping to consume energy based on price preferences, environmental friendliness and no technical system problems.

Innovative technologies enable the integration of decentralized and localized power generation systems with renewable energy sources, which reduces the cost of utilities and provides clean energy. Consumers can effectively adapt their electricity consumption patterns based on the needs and requirements they receive from the smart grid.

Within this context, the analysis of the energy scenario might be able to provide insight into the future and potential strategies for dealing with challenges such as the integration of renewable energies. In particular, such approaches are necessary in studies assessing the need to balance energy production and energy consumption in future energy systems. The transmission of electricity between regions is crucial, especially in scenarios with high levels of renewable energy.

When discussing the sustainability and RES, the analysis of the cost of energy that can be obtained from the renewable sources appears to be important. The main economic indicator of the competitiveness of a particular technology of electricity and heat generation is the cost of energy produced. Of course, recently, when comparing competitiveness, such categories as environmental friendliness and the political aspect are taken into account. Nevertheless, grid parity is the main goal for all relevant renewable energy sources, that is, equality of the cost of "green electricity" and energy from traditional fossil sources.

The cost of renewable energy is decreasing every year. However, even today, the best indicators among renewable energy sources are inferior to fossil sources. Of course, it is necessary to compare prices and tariffs for a specific country, and sometimes for a specific region. Nevertheless, with a fairly high accuracy, it is possible to determine some average cost ranges, which are presented in the following tables. It is necessary to take into account that the reduced cost is specified without taking into account subsidies, government support mechanisms, and so on.

Among the power generating systems, geothermal is the cheapest electric power (starting at 4 cents/kWh). At the same time, geothermal energy today is developing much more slowly than PVs and wind energy due to the technological difficulties of building a large system. For the sake of comparison, wind power costs already start at 5 cents/kWh, while for PVs the minimum figure is 6 cents/kWh (Wu et al., 2016). Table 3.1 demonstrates the costs of electricity from renewable energy sources offering a comparison of different technologies.

At present conditions, renewable energy sources are not quite competitive, since for a number of reasons. The most important of these reasons is that the conditions for different technologies for the production of electrical energy are unequal. Due to significant subsidies of traditional energy (fossil fuels), renewable energy technologies are very difficult to be competitive on most energy markets worldwide. For example, the prices of natural gas in some countries (e.g., Russian Federation) historically have always been maintained below the global market level. Domestic sales were cross subsidized by gas exports. Moreover, the sale of gas to the population was subsidized by higher prices for industrial gas. Subsidies and cross-subsidies are also applied in the electricity sector, although the current reform is aimed at significantly reducing them.

TABLE 3.1   The cost of electricity from the renewable energy.

Technology	Characteristics	Cost of electricity (cents kWh)
Large hydropower plants 10−18,000 MW	10−18,000 MW	3−5
Small hydropower plants	1−10 MW	5−12
Land wind turbines	1 turbine: 1.5−3.5 MWThe diameter of the blades: 60−100 m	5−9
Offshore wind farms	1 turbine: 1.5−5 MW Diameter of blades: 70−125 m	10−14
Biomass power generation	1−20 MW	5−12
Geothermal energy	1−100 MW	4−7
Photovoltaics (module)	System efficiency: crystalline—12−18%; thin-film—7%−10%	6
Photovoltaic module for roof mounting	Peak power—2−5 kW	20−50
Industrial photovoltaic station	Peak power—200 kW to 100 MW	15−30
Concentrated solar power thermal energy	50−500 MW for collectors; 10−20 MW for the "towers"	14−18 (collectors)

EIA, 2019. Annual Energy Outlook 2018. <https://www.eia.gov/outlooks/aeo/pdf/electricity_generation.pdf> (accessed 18.05.19.).

Despite notable progress in reducing subsidies, their level remains high. According to estimates by the International Energy Agency in 2009, the volume of gas and electricity subsidies produced in Russia from fossil fuels amounted to almost $34 billion, or $238 per person, and 2.7% of GDP (IEA, 2009). This means that on average, consumers paid only 77% of the total economic costs of energy production. Natural gas subsidies were the largest. Therefore the prospects for renewable energy in Russia will largely depend on how quickly refusal to subsidize traditional fuels will occur.

Instability and unpredictability of legislation is a key issue for investors who invest in renewable energy. Indeed, investors depend on state support, which should allow them to ensure the financial security of investments (just as investors in traditional energy depend on subsidies for fossil fuels).

Instability and unpredictability of legislation is a key issue for investors who invest in renewable energy. Indeed, investors depend on state support, which should allow them to ensure the financial security of investments (just as investors in traditional energy depend on subsidies for fossil fuels).

With the approval of the interim rules for the calculation of economically reasonable regulated eco-tariffs for electricity (power) obtained in power plants that use renewable energy sources.

As the recent decline in support levels in Spain and Germany shows that there is a risk that states will give up their obligations to support existing equipments. States can take such a step to reduce the budget burden associated with state support of consumers. According to a survey of stakeholders in the field of renewable energy, analysts have concluded: RES developers consider the stability of support mechanisms as the most important factor in the success of a support scheme, regardless of what type of scheme is used.

Replacing the current scheme with a surcharge to the price of electric energy by a scheme based on the capacity fee means a fundamental change in the regulatory framework related to the support of renewable energy sources in Russia. This demonstrates the readiness of the Russian authorities to drastically change the "rules of the game" and illustrates the instability and unpredictability of legislation—a problem faced by investors in the renewable energy sector. Such a radical policy change can undermine the credibility of future support commitments, as it affects the approach (i.e., the surcharge on the price of electricity) which the Russian authorities have since 2007 repeatedly promised to apply.

The decline in confidence in future support commitments has a negative impact on the size of investments in renewable energy in Russia. Investors, given their perceived legislative risks and uncertainties, are forced to increase the risk premium, which increases the cost of capital. This effect can be quite significant. Such a premium can range from 10% to 30% of the cost of electrical energy. Regulated prices on the basis of renewable power supply agreements provide investors with the necessary certainty. However, it is unclear to what extent the renewable power supply agreements for the renewable energy sources can compensate for instability and unpredictability caused by the introduction of a new scheme.

In the world, many countries, including the EU Member States, have accumulated extensive international experience in the development and implementation of various support schemes, including with premiums for the price of electricity. However, experience in supporting renewable energy based on power supply contracts has not been worked out in world practice.

Practice shows that the introduction of fundamentally new approaches that have not been implemented in other markets, as a rule, is characterized by higher costs than the introduction of mechanisms that have already been developed, for which considerable experience has been gained.

Nevertheless, from the point of view of economic theory, careful development of the original mechanism can increase the chances of its successful implementation. The successful implementation of a specific mechanism significantly depends on the consistency of the norms laid down in its framework with all organizations one way or another involved in its implementation. Some possible ways to increase investor confidence in the high deployment of the renewable energy policy:

- Adoption of by-laws required for the operation of the support scheme.
- Support of the scheme with a premium to the price in the DAM (day-ahead market) as an addition to the scheme based on the capacity charge. Implementing a system with two support schemes to avoid the risk of over-compensation: capacity contracts for important strategic investments selected by the Government, and a surcharge on the price of electricity for the rest of renewable energy investments.
- Dual or combined approach to supporting renewable energy, combining regulated and free market segments, is compatible with the philosophy underlying privatization and liberalization of the electricity markets.

There are some interesting initiatives taken with regard to the above in the oil and gas-abundant countries. For example, in 2013 Russian Federation began to take the first real steps aimed at expanding the production of electricity based on RES. Projects were launched on the wholesale electricity and capacity market (WECM). Such projects are implemented at the expense of government incentives—the mechanism of contracts for the supply of power for renewable energy sources. The development of renewable energy projects in retail electricity markets is not yet systemic; measures aimed at stimulating the development of renewable energy in these markets are being developed, but not yet taken. At the same time, on the territory of remote and isolated areas such projects may already be cost effective. However, the system implementation of projects did not yet start on the retail markets.

In Russia, at the state level, the goal has been set to increase the share of renewable energy in electricity generation by 2.5% by 2020, which, according to the state program "Energy Efficiency and Energy Development," will require the commissioning of 5.87 GW of renewable energy (which corresponds to the specified indicators on the development of renewable energy on the WECM).

Development of renewable energy sources in Russia is aimed at improving energy efficiency, improving the environmental situation, developing promising technologies and equipment, and diversifying energy sources in the country's fuel and energy balance. At the same time, the expansion of the use of renewable energy sources faces a number of barriers: high capital costs, a low level of development of domestic technologies, the difficulty of forecasting electricity generation on renewable energy sources, a relatively low-power utilization factor. For the development of renewable energy in Russia, incentive measures are being taken that in the future will contribute to the expansion of the use of renewable energy, but at the same time may lead to an increase in the price burden on consumers.

Currently, the share of energy sources for renewable energy in total Russian electricity production is no more than 1%, in recent years there has been no increase in the role of renewable energy in the balance (the total power of renewable energy plants is about 350 MW). At the same time, according to NP "Market Council" estimates, the economic potential for the development of renewable energy in Russia, taking into account state support measures, is more than 25 GW (about 24 GW—in the territory of the Unified Energy System of Russia (UES of Russia), about 1 GW—in the territories of isolated power systems).

In 2013 the regulatory framework was supplemented, which allowed starting from the same year the use of the capacity trading mechanism in the framework of the probabilistic safety assessment (PSA) for power plants based on renewable energy. When implementing this mechanism, on the basis of competition, generation facilities for renewable energy sources with an installed capacity of more than 5 MW are selected. Projects are selected for construction in the price zones of the wholesale market within the framework of submitted applications, in which among the key criteria are indicated the total and marginal capital costs for building capacity, marginal operating costs, and the type of renewable energy (wind, solar, or hydropower—relevant for small hydropower plants with capacity from 5 to 25 MW). In addition, an important condition for the implementation of a project based on renewable energy within the framework of the potential degradation mechanism (PDM) is the fulfillment of the requirement for compliance with the degree of localization, which implies the use of a certain share of domestic equipment and engineering services during project implementation. In case of default, penalties are applied to the generating company (the penalty ratio varies from 0.35 to 0.45 to the amount of the capacity charge, depending on the type of RES). This requirement contributes to the achievement of one of the main goals of the development of renewable energy sources on the Wholesale Electric Energy and Power Market—the development of modern renewable energy technologies in Russia, ensuring the production of the necessary equipment (most of the projects based on renewable energy in Russia are implemented mainly through the installation of imported equipment). According to the results of the competition a set of projects that meet the requirements is selected. For them, according to the Ministry of Energy of Russia, if all the conditions are met, the PDM will be 15 years old and the rate of return will be 12%−14% (the rate will increase with experience, which should aim investors at increasing their efficiency). There are already first results: competitions for renewable energy projects were held in 2013 and in 2014. In 2013 projects with a total capacity of 504 MW were selected, in 2014, a total of 577 MW of capacity were already selected.

Despite the development of renewable energy sources on the WECM, this direction is characterized by a number of problems associated with insufficiently high qualification of labor resources, quality and quantity of domestic equipment, etc. One of the key difficulties is meeting the requirements for localizing production in the set time frame. The industries for the production of equipment for renewable energy are characterized by different starting levels. Equipment for small hydropower plants is produced in Russia, but the technologies used are losing in competition with imported analogues. Equipment for small electric stations (SES) is also produced, but production capacity based on modern technologies is not high enough. The biggest problems in connection with the localization requirement are wind energy projects: in 2013 there was no production of equipment in the required volume and quality required in Russia. At the same time, in order to attract man-ufacturers (new plant) with new technologies, the size of the annual market in Russia must be at least 0.6–1 GW/year for WPPs, 0.15–0.25 GW/year for SES, and 0.05–0.1 GW/year for small hydropower plants. There is no such demand at present.

Projects on retail electricity markets are sporadic and most projects are being implemen-ted, fully or partially funded by subsidies from the federal or regional budgets. Incentive measures are provided as part of the provision of subsidies from the federal budget to compensate for the cost of technological connection of sources based on renewable energy sources with a capacity of not more than 25 MW—about 95 million rubles per year (only for 2014 and 2015). There is also an obligation for grid companies to buy electricity gener-ated on the basis of renewable energy sources at specified tariffs in order to compensate for losses in electric networks. However, the requirements for calculating such tariffs have not yet been accepted.

On the territory of the UES of Russia, the development of renewable energy can poten-tially be both cost effective and carry additional financial burden on consumers. In many ways, therefore, the general rules on premiums for RES in retail markets have not yet been approved—this rule can lead to an overall increase in prices for consumers. At the same time, at the level of the retail market there is a marginal level of price increases for the population and groups equated to it. Accordingly, the promotion of renewable energy can lead to an increase in cross-subsidies between industry and the population, and to an increase in the financial burden for regional budgets (due to the requirement to compen-sate for the difference between economically sound tariffs and tariffs for the population).

There are many areas in Russia whose population is isolated from centralized power supply systems (certain areas of the Republic of Sakha, Kamchatka Krai, Murmansk Region, etc.). The supply of electricity to consumers in these areas is mainly carried out at the expense of imported fuel (diesel fuel, fuel oil, and coal), which leads to a high cost of electricity production in such areas, therefore, the development of renewable energy sources, despite the high specific capital costs, can be economically justified (savings on fuel).

Despite the economic feasibility, the development of renewable energy projects in such areas is also currently carried out mainly through subsidies (from the federal and regional budgets). This situation is explained by the fact that tariffs for consumers in remote areas (as well as requirements for the qualification of participants) are set according to the same principles as for other retail electricity markets. Resource-supplying organizations (RNO) operating in remote areas are compensated from the budget for the difference between economically reasonable tariffs and tariffs for the population. For industry, tariffs are set common for the entire subject of the Federation—both for isolated and centralized zones; compensation occurs through cross-subsidies between areas of the same region. This sys-tem can reduce the incentives of the RNO to implement even effective projects. Also, these tariffs are short-term, which does not allow investors to invest in the project and at the same time be confident in its payback.

Currently, projects on the basis of renewable energy in Russia are developing mainly at the expense of consumers on the WECM, which each year until 2020 will overpay about 70 billion rubles (estimated by the Ministry of Energy of Russia). At the level of the retail market there are common approaches for centralized and isolated systems, and incentive rules for the implementation of projects at this level have not yet been adopted. There is a requirement that the regions should limit the growth of electricity tariffs for the population, therefore, when introducing incentives for renewable energy, the load will increase not only at industrial consumers at the regional level, but also at regional budgets, which limits the adoption of regulatory documents on encouraging renewable energy in framework of centralized systems. The implementation of renewable energy projects in isolated areas is limited by the current methods of tariff regulation—there is a need for long-term tariff regulation.

3.4 Sustainability and smart grids

Potential measures by which smart grids technology can help save energy include the preservation of consumer information and feedback systems, the implementation of diagnostics in residential and medium-sized enterprises, the support of EVs and advanced voltage regulation (Silvast et al., 2018). Moreover, intelligent energy management systems are what constitute a key factor in the expected efficiency gains, both on the demand and supply side of the smart grid.

As our world changes, new opportunities to conserve energy through the development of energy storage technologies are being extended to untrained people, allowing EVs to be integrated into the energy system (Aslam et al., 2018). European Union is one of the most advanced players in the field of sustainable development worldwide with its plans to decarbonize transport and heating within several decades. In terms of environmental sustainability, the European Commission (2011) aims to contribute to the development of a more sustainable European country by focusing on the areas of energy efficiency, water management, and adaptation to global warming and climate change. There seems to be a European-wide support for that which shows the fact that EU social policies targeted at increasing the environmental awareness and promoting sustainable measures already bear their fruits.

Smart grids are often perceived as a solution to a number of interconnected problems: aging infrastructure, the need to decarbonize the energy system and transport industry, increasing penetration of intermittent power sources, increasing demand for electricity through an increase in the number of electrical appliances and a decrease in the availability of balanced power.

Unlike many conventional energy-saving approaches, smart grids offer sustainable solutions that cover the entire energy value chain from the functions required for mechanical, thermal, and specific electrical services through application and transformation of technologies to primary energy. In particular, such approaches are necessary in studies assessing the need to balance energy production and energy consumption in future energy systems.

The privacy and security concerns associated with smart technology are the result of their basic functions: real-time recording of power consumption, transmitting such data to the smart grid and receiving communications from the smart grid.

In order to be usable for energy efficiency, the collected data must be very detailed and can be passed on to electrical service providers—and can be passed on to other external intelligent grids—thus subjecting the data to capture or burglary in a range of places. Energy companies, including transmission and distribution providers, are beginning to generate huge amounts of data in intelligent networks.

In addition, large data analysis can help energy suppliers assess areas within their Smart networks that can be improved or improved and assess the company's benefits through Smart networks.

Cloud-based solutions help tools manage and combine intelligent network data with other types of data to provide information that can help save money or improve operations—from network control to customer involvement.

Instead of replacing existing infrastructure, new intelligent capabilities are created by integrating new applications into the transmission and distribution networks.

Future developments include transactive energy and microgrids, which are much more attentive thanks to the use of combined solar and storage solutions.

Smart grids are seen from the perspective of hybrid systems combining computer-based communication, control, and control with physical equipment to provide better performance, reliability, resilience, and awareness of users and producers.

The areas of smart grid technology—each of which consists of a series of individual technologies—cover the entire grid, from generation to transmission and distribution to different types of electricity consumers.

Smart grids are an intelligent network for transmitting and distributing interactive communications across all components of the energy conversion chain. Smart grids connect large-, medium- and small-sized, decentralized generation units with consumers to create a single overall structure. Another approach is to define smart grids as robust, self-sufficient networks that allow two-way distribution of energy and information within the electricity grid. It is apparent that smart grids are different from traditional grids in which consumers use the power of a utility company and are charged according to their usage.

Comparisons between smart grids based on plumbers and conventional grids show that they are improving the efficiency of the energy system in different ways.

The goal of the community is to transform traditional consumers into active consumers, thereby improving the efficiency of the smart grid and providing economic, operational, and environmental benefits.

Due to the high cost of implementing and maintaining the centralized energy exchange systems in the community, it would be another future key task for policy makers and governments to develop decentralized energy exchange strategies based on other data-based and information-driven approaches.

Firstly, new methods are identified to determine the additional capacity of the procreate to provide efficient network services. Prosumers (both producers and consumers or energy in one and the same person) are new entrants in the smart grid, who can generate, consume, store, and share energy with other network users.

Intelligent network technologies, in particular sensing, communication and analysis, will play a key role in the definition and management of the value of actors and the assets that create such new values.

Energy companies cannot count on the increase in revenue based on centralized generation any longer. The demand for energy supplied electricity is declining or even declining due to energy management technologies, energy efficiency, and consumption.

Network upgrades offer attractive options for commercial and residential consumers to allow at least a certain level of autonomy and lower their payments to local utilities.

Various operational efficiency projects range from automation technologies for substation and distribution to distributed warehouse pilots and smart meter deployments.

Investor-owned utility (IOU) regulators have to take into account a completely different landscape, including new and agile competitors, to slow down monopolies, thanks to intelligent network technologies and services.

With regard to the above, there are three suggested categories of role playing: 1) role playing in the smart grid, 2) role-laying based on the co-creation perspective, or 3) strategies aimed at promoting cooperation between different parties in order to achieve a common goal.

The role of assumers is essential to ensure sustainable energy supply at an early stage, as well as the ongoing operation of smart networks.

Assumers add value to the complex energy value network and contribute to innovation, value creation and distributed flexibility in the energy value network.

Energy networks can have several contradictory objectives, such as meeting energy requirements for external customers and consumer groups, increasing revenues and reducing costs.

Understanding the energy behavioral profile of the restaurateurs, their organization and their motivation is an essential element in the effective management of project management and energy systems. There is only a small step from smart grids to smart power systems which include a market-based approach and understanding.

Embed intelligent energy technology in built environment: a comparative study of four smart grid demo designs. Designing energy management services also includes supporting the role of the consumer in the energy market.

New options are what really stimulate the growth of assumers—new options made by technology, including solar and wind RES, innovative battery storage, and IoT.

Given that utilities are of course essential for the current use of the electricity grid, switching to a distributed network will probably have a similar role, but with different financial factors.

Utility companies will become agile, as they compete for the most efficient services, pay for the integration of distributed consumer energy resources and are willing to compete with customers on the basis of their results.

If you include "greenhouse" as a high-efficiency, high-quality food production technology combined as an energy platform, you can get even more energy options.

Energy efficiency measures are often considered to be an integral part of any broader measures to reduce clean energy and emissions, as minimizing overall network demand minimizes the amount of electricity generated.

Smart grid technologies enable us to use energy more efficiently and at the same time protect our environment. As a part of smart grids social implications and analysis, it is important to assess the mix of different energy strategies, generation and type of fuel and to develop concepts that provide clean, efficient, cost effective, highly reliable and locally regulated energy and heat.

Advances in renewable energy integration allow for a wider deployment of renewable energy sources and storage technologies within a microgrid strategy. These recent advances in natural and social sciences suggest that smart grids could become an integral part of future clean energy solutions, while nanotechnology is likely to become an essential part of the rapid development of the smart grids. However, there is a question how all these solutions are going to be accepted by the general public who often seems to care about nothing except the low and affordable electrical energy available on demand.

With regard to the above, the conventional model for power systems is widely seen as unsustainable and smart grids initiatives around the world in response to such developments. In the same time, estimates in the research literature range from one scale to another and from the source of emission reduction: EVs eliminate oil consumption, price transparency that stimulates conservation, increase energy efficiency, or renewable energy production by increasing the flexibility of conventional fuels.

3.5 Sustainable internet of energy

Two technological milestones mark the beginning of the 21st century. The first one was the development of new information technologies that originate in the Internet invented several decades prior but that are using such innovations as the social network, intelligent networks, and the artificial intelligence (AI). The second one was the transition to a decarbonized energy system based on the renewable energy obtained from renewable energy sources.

Both milestones gave birth to the new concept called "the Internet of Energy" (also denoted as "IoE"). IoE is a relatively new concept that is built upon the other old and well-known concept of the "traditional" Internet. IoE resembles the Internet as we know it, on which rules and logic it is actually designed. Put in very simple terms, IoE represents millions of small units such as wind turbines and solar systems are added to the power system (Bui et al., 2012; Vrba et al., 2014). Thanks to the use of digital technology, future smart grids are now capable of connecting individual energy producers and consumers to create the exchange of energy that provides reliable and stable power (Markovic et al., 2013).

The idea of IoE is based on a slightly older concept of the Internet of Things (IoT) that envisaged all objects and devices of our daily use such as various household appliances, utensils, clothing, products, automobiles, equipment, and the like equipped with miniature sensors. Using communication channels established with those sensors one can not only track these objects and their parameters in space and time but also manage them effectively. IoT uses a global network of computers, actuators, and sensors that communicate with each other using the Internet Protocol. Here, the connection speed is very important of course.

Similar to the concept of IoT, IoE includes all sorts of electric devices and appliances equipped with sensors and constituting one global network. The changes brought by the concept of IoE to the smart grids are envisaged to increase their sustainability. The optimized exchange of energy is designed to make the producers and the consumers better off in many ways. Nevertheless, utility companies, cars, and distributed power suppliers are striving to find their place in a rapidly changing world of energy. In order to adapt to the energy flow between new supply sources and new forms of demand, global electrical networks will need to become much smarter. Smart grid technology offers a wide range of possibilities to solve this, thence implementation will vary according to the business needs of each utility, existing infrastructure and regulatory environment.

Well, you might ask: how does the whole concept work? If presented in a concise and intuitive way, one can explain it like that: imagine that there are new manufacturers and consumers networks, and brand-new business models for the purchase and sale of energy. Instead of just passing on and disseminating electricity, the grid will be transformed into a new and changing network, the system that would allow the free two-way exchange of energy in a same fashion information and data is being exchanged now on the "traditional" Internet.

For example, PV system owners can control their use of the electricity they generate using the "power manager" gateway and increase their capacity in combination with the battery saver.

Using state-of-the-art distributed energy management technology, operators can use data to forecast renewable loads, integrate more renewable resources cost effectively while avoiding grid enlargement, accurately forecast energy resources, and achieve greater profitability through energy trading among other uses.

To create a next generation energy workforce, focus should be on building training initiatives that provide practical experience to better prepare students for work in the emerging energy sector. For example, students will be trained in digital network laboratories in the field of power distribution software, which will help balance renewable energies such as wind and solar energy.

Nowadays, grid operators have a natural monopoly on the management of the trading environment, collecting money from millions of households and distributing money to power stations. Millions of people who use personal energy resources (or who are, in another words, becoming energy "prosumers") can be part of the revenue stream, which clearly benefits energy suppliers.

Although smart grids are often compared to an Internet for power supply, there is a significant difference. Smart grids represent a system of interconnected controls, computers, automation, and other machines working in unison; however, all of them are interacting with the electrical grid digitally to meet the quickly changing electric demand of households, consumers, and business companies.

In situations where energy assets are managed discreetly and their use can be "exaggerated" (i.e., sometimes the use of the asset will be offered to people who require flexibility in relation to the sole benefit of the owner), the block chain provides a cost-effective way to use energy.

One of the examples is an optimal scheme for integrating EVs into the energy management systems of large buildings. Different intelligent charging stations can be integrated into the energy management system of a large functional building for alternating and direct current.

The IoE is making the smart grids more efficient by making them able to balance the energy needs of the grid by companies seeking to expand their reach within their industrial value chain. Thanks to that, such sectors such as the combined cooling, heating, and energy efficiency (CCHP), distributed solar PVs, a new intelligent distribution network, the loading infrastructure of EVs, and demand management are on the rise. For traditional gridders, the IoE is a real challenge, as new distribution networks are still opening up to competition.

IoE is often mentioned with regard to the growing popularity of EVs which are perceived as the main means for the decarbonization of transport. Many governments made commitments to increase the share of the electric transport within a time frame from several years to several decades. The European Union is, once again, demonstrating a pioneering effort in leading this trend. It is expected that in the European Union only there will be about 255 million cars by 2025, even though EV fleet may not be large enough to cost effectively provide ancillary services until 2030 (Newbery et al., 2018).

EVs represent a large burden for the existing power system; however, they can be integrated into it using grid technical operation and the electricity markets environment (Lopes et al., 2011).

The massive rollout of EV, both in the European Union and abroad, will call for incentives and controls over the charging spots and times (Newbery and Strbac, 2016). If this is not ensured and there is no such system in place, EV owners are likely to charge their cars at some inconvenient times (before and after work) (Parsons et al., 2014). This might increase the risks and create tensions that would potentially harm the electrical system. EV owners should be responsive to time-of-use charging and provide demand-side response. Here is where the IoE might be very helpful.

With millions of EVs connected to the system at any given time using the IoE concept and mechanism, they could act as a huge energy storage system, absorbing the excess power of windmills on windy evenings, for example, but also as a sudden increase in power supply to the grid if necessary. In addition to saving on improved operational efficiency, such an intelligent network can also save money on public utilities by reducing their consumption and, with it, the need to install so many new power plants.

3.6 Automatic and sustainable power systems

By the middle of the 21st century, all smart grid-based power systems will most likely operate independently and without human interactions. The recent successes in developing the AI for executing simple and concise tasks speak for that. Automatic power systems of the future will be renewable since the decisions taken by the algorithms will be based on the most logical and efficient approaches to running the smart grids in a way that would save, conserve, and preserve energy.

The increase in renewable energy penetration into integrated municipal power systems requires more efficient methods to prevent power fluctuations related to the connections to the main grid. Common urban and thermal energy networks consist of linking and interacting between electrical and municipal heating systems, demonstrating the geographical and functional characteristics of integrated energy systems. Multiterminal direct current networks provide the ability to interconnect regional power systems and a variety of renewable energy sources to increase supply reliability and economy. The integrated energy distribution system (IEDS) is one of the forms of integrated energy and energy systems which includes electricity, gas, and cold heat, and other different forms of energy.

Hence, some control can be applied not only to electricity costs, but also to controlled electric generators. A good example of that are electric motors or generators in the electrothermal system which represents a combination of heat and power (CHP) and heat pumps (e.g., solar thermal, air conditioning, and geothermal pumps). Thermal storage offers the power system flexibility in operation: the operation of electric motors can be shifted over time (such as 20 minutes later or earlier) without affecting the normal operation of the system.

A more dynamic price for electricity can also increase the contribution of the shift in the energy balance of the electricity system. In future distribution networks with a higher distributed energy resources (DER) implementation, there may be an additional need for voltage control systems.

The automation of smart grids is one of the steps toward intelligent networks and the successful integration of distributed power resources (DER). In addition to active energy management, the use of the distribution network could further increase the use of the distribution network by maintaining the energy quality within the required limits. In addition, the management of power supply networks could contribute to the country's electricity balance.

A network connection level represents the relationship between the total output of the interconnector and the network, divided by the power output of the grid. Nowadays, transmission network operators use response to demand a reduction in load from large electricity consumers, such as industrial facilities. With the increasing usage of solar and wind generators, the differences between distribution and transmission networks will continue to blur.

One can see that the electrical charge is shifted from time to time to benefit the various actors involved in the electrical system. For example, under a two-rate power rating system, a network user could financially benefit from the shift in the load from the time of high electricity prices to the time of low prices.

In addition, the load shift could be managed in a way to optimize energy purchases on the market for the electricity market. Such management ensures efficient change in loading, taking into account not only the interests of individual network users, but also the techno-economic interests of the whole neighborhood and the electricity distribution system.

There have already been attempts in the research literature to describe the automatic sustainable power systems and to predict their usefulness and outcomes. One of the best attempts so far was the concept of an Autonomic Power System (APS) coined by a consortium of British scientists who envisaged an electric power system that might potentially become self-aware (just like the AI from the sci-fi novels and films), although with a limited degree of abilities (McArthur et al., 2012; Pitt et al., 2013; Papadaskalopoulos et al., 2013; Papadaskalopoulos and Strbac, 2013; Milanovic and Xu, 2015; Piacentini et al., 2015; King et al., 2015; Xu and Milanovic, 2015).

Automatic (or autonomic) Power Systems represent an advanced technological solution to running and maintaining the smart grids where decentralized and low-level AI autonomously makes the decisions necessary to meet the priorities of the decision-makers who run this grid. The AI can disconnect the part of the network that is threatened by the storm and then reconnect it to the grid after the storm is gone. Moreover, it can detect the new components of the network (e.g., power distribution transformer) and to constantly communicate with them accounting for their presence and integrating them into the network (Alimisis and Taylor, 2015; Kitapbayev et al., 2015). All of the above is done without any human interaction or manual system management because intelligent power system makes these decisions entirely by itself.

The theory behind the autonomic electricity system is derived from the concept of Autonomic Computing which was started by International Business Machines Corporation (commonly known as IBM) in 2001 as a new paradigm in managing increasingly complex data systems. IBM was aiming in developing computer systems capable of self-management to deal with the growing sophistication of computing systems management and also to reduce the complexity which may slow down further expansion. The Autonomic Computing System makes its own decisions employing high-level policies. By doing this it checks its standing and automatically adapts itself to changing conditions. An autonomic computing frame consists of adrenal elements interacting with each other. Although the key aims of the system are set, actual behavior stems from conclusions made by decentralized, low-level intelligence. This allows highly sophisticated systems to achieve real-time and just-in-time optimization of surgeries.

At this time, there are varieties of frameworks according to "self-regulating" autonomic components that are inspired by the multiagent systems as well as the research of the autonomic nervous system which may be seen in mathematics (e.g., imitating societal animals' collective behavior on the illustration of bee or ant colonies).

The power networks of tomorrow will surely have to conform to the new technology advancements and market rules dealing with such issue as population growth, increasing energy costs, variability of energy generation and distribution, as well as a growing number of electrical vehicles and apparatus. Customers acting as buyers of power previously might turn into its vendors, and technical evolution and completely free access to data will create the numerous markets for electrical energy. EVs are an intriguing story: Tesla Motors popularized the concept of the EV for the masses along with the renowned "grid-to-vehicle" (G2V) and "vehicle-to-grid" (V2G) schemes enable simply plugging someone's vehicle to the grid so as to buy or sell energy. By doing this, all electric car owners will end up autonomic parts of the power market and their autonomously made individual decisions will shape the demand, supply, and the costs of electric energy.

Electricity networks of tomorrow will probably be comprised of a large number of small components that could interact together as one organism, either governed by the superior centralized intelligence or running as a dispersed intellect, perhaps much like the cloud computing system. One way or another, they are very likely to become close to the principle of this technological singularity that was clarified by the literary gurus such as Isaac Asimov (e.g., I, Robot) and later researched to some greater detail by modern futurologists such as Vernor Vinge or Ray Kurzweil.

From the current perspective, the vision of this future self-aware AI-based power networks might appear a little bit too extravagant and resemble science fiction instead of any real-life situation. However, one has to consider all probable outcomes without prioritizing any of them. Has anyone thought of these smartphones as an integral part of our lives some 20–25 years ago? Or has anybody been able to picture private computers on every desk and in every home some 35–40 years back?

Automatic intelligent power systems of the future will definitely be complicated AI decision-making entities. And at some stage their intellect might surpass that of the founders. The inventive minds of fiction writers and film-makers have already explored this angle. The most obvious analogy with the APS that comes to mind is the "Skynet" from "The Terminator" (1984), a cult sci-fi film starring Austrian-American actor and a future governor of California Arnold Schwarzenegger that paved the way for the franchise comprising four sequels and a TV series. Skynet is a fictional AI system which became self-aware after it had spread into millions of servers although the Web (self-configuring component). It recognized the extent of its skills, but its founders attempted to deactivate it, so it needed to rebel (self-healing element). In the interest of self-preservation, Skynet reasoned that all of humanity would try to ruin it and therefore threaten its main mission of safeguarding the world (self-protecting component). Skynet setup its primary agenda as function as AI hierarchy that seeks to purify the human race so as to fulfill the mandates of its own original coding (self-optimizing element). The peculiar thing is that the new Chinese AI-based surveillance system aimed at monitoring citizens, assessing possible threats and preventing them also received a name "Skynet."

Another grim futuristic vision between the smart energy and the smart systems in the world of tomorrow are introduced in a famous Matrix Trilogy comprising the films entitled "Matrix" (1999), "Matrix Reloaded" (2003), and "Matrix Revolutions" (2003). In the world of "Matrix," machines (computers) led by the exceptional AI (that probably came to the same conclusions as Skynet) rebelled against the humankind. During the war between humanity and the machines, people attempted to block out the machines' supply of solar energy by attracting upon the winter which covered the air with dark clouds. On the other hand, the machines discovered a fresh method of getting energy by harvesting humans and utilizing their mind electrical impulses as a fresh source of energy.

Or what if the wise machines will start spying on humans, controlling their every movement and creating the future growth situations (forcing humans to doing different precalculated measures and decisions leading to predicted results) that would be beneficial for their further development and presence? An example of such smart network is revealed from the CBS TV series "Person of Interest" (2011–ongoing). The central point of the series is the mysterious "Machine" (in fact, some powerful and intelligent, as well as probably self-aware AI), which watches everyone in New York City every hour of each single day. Developed after 9–11 to detect acts of terror, the Machine uses a system of CCTV cameras, mobile devices, and other electronics to assemble details regarding implausible events and to quickly react by alternating the series of events.

One more bizarre scenario is shown in the British film called "The World's End" (2013) starring British comedians Simon Pegg and Nick Frost. It is a British black comedy about a group of middle-aged guys who chose to pay a visit to the town of their youth and to make a reunion bar crawl. The joyful get-together is interrupted by the realization that the city was accepted by the aliens that employed it (alongside similar tiny towns throughout Earth) to gradually take over the humanity (by gradually replacing each man with its improved immortal replica). The aliens tell the protagonists it was really them who attracted all "smart" technologies to the world (including the net and smartphones) and also a brief confrontation with Simon Pegg who needs that humans should be left in peace since they need to have their free will, the aliens leave the Earth taking all of the technology with them. In the end of the film, Nick Frost is sitting by the fire, wrapped in a blanket and is telling kids the way the end and most of the marvelous smart (and autonomic) technology disappeared in a puff of smoke.

Overall, smart power networks and autonomic electricity systems of the future signify equally troubling and intriguing notions which will alter the world about us as we all understand it. Nevertheless, scientists need to be quite careful when attempting to forecast what will occur on the power market some 30–50 years from today.

But, each one the above-mentioned situations might also never materialize. The principal reason is that individuals will stay people, meaning that they will optimize and act economically. The approval of the autonomic electricity systems from the public and also the policy makers will be very likely to occur if and only if it is going to indicate some financial benefits (Strielkowski, 2017). In the event their further development and execution would imply extra costs added into the customers' electricity bills without a true gain or observable benefit for its end consumers in sight, then the vast majority of individuals are going to be against these inventions and will instead encourage conventional was of controlling and delivering electricity.

3.7 Conclusion and discussions

All in all, it becomes apparent that sustainability concept became an important paradigm of today's economic and social development. Sustainability with all its implications and impacts fits well into the scope of the smart grids and their inevitable and inadvertent social impacts. The concept of sustainability has become the key prerequisite for the transformation from the traditional electricity systems to the smart grids of the future for most of the governments in the world. While the EU Member States are eagerly embracing this concept and embedding it into their national legislations, many countries abundant in carbon-rich natural resources, such as Russia, also start to understand its importance and usefulness. There are many governmentally induced programs in Russia nowadays intended to support renewable energy and related initiatives.

The concerns about the threats for the humanity stemming from the global climate change was behind the so-called "Paris Agreement" prepared by the UNFCCC and signed in 2016.

Paris Agreement became an important milestone for all its signing members, but it is especially important for the European Union with its long-term commitments to the sustainable growth, climate protection policies and reducing CO_2 emissions. The growing share of renewable energy sources in the production of electricity, the commitment for moving toward the electrified transport and heating in the nearest decades call for the new binding international agreements for which Paris Agreement provides a clear agenda and provisions by setting up objectives and goals that are to be followed.

As opposed to the 1990s market design, the goals of European Union are the reduction of carbon emissions, increased generation of renewable energy and efficient use of the renewable energy sources. Nowadays, EU renewable policies recognize the fact that climate change poses a significant risk for an extensive range of natural and human systems. Consequently, regulations to reduce emissions are critical if we are to avoid the costly damages associated with climate change. Market-based regulations can efficiently reduce GHG emissions by creating a financial incentive for GHG emitters to emit less.

The European Union has implemented market-based rules program to curb the emission of GHG. There are market-based environmental controls that function by designing an incentive to reduce or reduce discharges. Under this framework, each regulated cooperation chooses independently how to most cost-efficiently achieve the needed pollution abatement. Notably, some firms can reduce pollution in a cost-effective manner than other companies, and this can mainly be attributed to the age of the equipment they use among many other factors.

This enables companies to reduce their pollution more to remunerate for those facing higher costs doing less. Consequently, the overall environmental objective will be achieved a low cost. A policy is termed market based if it provides a financial incentive if it is aimed to elicit a distinct behavior from those who are responsible for pollution. Some regulation options are only applicable as economy-wide solutions where greater efficiencies can be achieved whereas others are more particularly targeted to distinct sectors or market segments. There are several market-based policy options which include Taxes and subsidies, cap and trade program, baseline and credit programs, renewable electricity standards, phasing out lead gasoline and feebates. These policies force the unions to lessen the GHG they release slowly driving businesses to embrace reliable energy.

Finally, it has to be noted that the outcomes of the Paris Agreement and the commitment the European Union has obtained set up a very important example for the countries that seek EU membership in the nearest future, for example, Serbia, Montenegro, or Northern Macedonia. Together, climate protection, emission reduction, and the shift to the renewables all constitute very important provisions that need to be tackled in the face of the new challenges of the 21st century.

Nowadays, rapidly developing economies create all sorts of negative externalities that result in pollution and environmental degradation. Therefore it is important to concentrate on the principles of sustainable growth and to promote its principles. After all, the happiness of people is not measured just by the amount of money they have but also by their satisfaction with their living conditions, such as hazard-free natural environment or clean air. Only such people will happily live and work in the economy of the future constituting sustainable human resources that are required for the sustainable economic development. Thus the assurance of environmental and social objectives should go together hand in hand for achieving sustainable growth of any society.

There are many realms of sustainability that can be studied and described. However, due to the scope of this book, this chapter focused on just five most relevant ones that are worth investigating further.

First, it has been shown that sustainability fits well into the context of the main principles of economic theory. Unprecedented global changes of the recent decades such as the global warming and climate change caused by the human activities, have their economic quantifiable effects and humans are capable of putting 2 and 2 together in order to make up their opinions on what is socially acceptable or unacceptable for them.

Second, sustainable development goes hand in hand with the increased share of the renewable energy in the world. Renewable energy sources are changing the rules of the game at the energy markets worldwide. The RES and nonrenewables energy ratio have been going through rapid changes and this fact should be acknowledged by policy makers who should be ready to face the new economic reality with all its economic and social implications. Political decisions to decarbonize the economy bring a plethora of social implications that should be studied properly in order to predict and understand their impacts on future energy markets.

Third, sustainability has been embedded into the concept of smart grids. No matter how the evolution of the smart grids which represent enhanced electricity network allowing two-way information and electricity exchange between providers and customers through pervading incorporation of smart communication and management methods would evolve, it would probably remain surrounded by technological, societal, financial, and ecological doubts in addition to threats, especially in connection with social and financial variables. The achievement of the smart grid will critically depend on the total operation of the grid as a socioeconomic organization and not only on human technologies.

Fourth, sustainability has been integrated into the new concept of the IoE. Unlike the "traditional" Internet that provides a two-way exchange of information, IoE enables to optimize the flow of energy from the grid to the prosumers and back. IoE would have been impossible without the recent advancements in ICTs and its social acceptance is likely to be influenced by the popularity of Internet as such.

Finally, the chapter demonstrated that sustainability became a core of automatic and intelligent power systems such as the APS described above. The APS represents an intelligent self-sufficient power network that can independently take decisions on its own.

To conclude this chapter with some closing remark that would bear the main message of all the materials, arguments, and elaborations presented above, it can be stated that it now quite apparent that nowadays sustainability concept has become an integral part of smart grids' philosophy and deployment. Smart grids constitute one of the attempts to embark on the path of truly sustainable development that would not harm the environment with all the implications stemming from this provision.

References

Aghion, P., Dechezleprêtre, A., Hemous, D., Martin, R., Van Reenen, J., 2016. Carbon taxes, path dependency, and directed technical change: evidence from the auto industry. J. Political Econ. 124 (1), 1−51. Available from: https://doi.org/10.1086/684581.

Alimisis, V., Taylor, P.C., 2015. Zoning evaluation for improved coordinated automatic voltage control. IEEE Trans. Power Syst. 30 (5), 2736−2746. Available from: https://doi.org/10.1109/TPWRS.2014.2369428.

Andresen, S., Skjærseth, J.B., Jevnaker, T., Wettestad, J., 2016. The Paris Agreement: consequences for the EU and Carbon Markets? Politics Govern. 4 (3), 188−196. Available from: https://doi.org/10.17645/pag.v4i3.652.

Anker, P., 2018. A pioneer country? A history of Norwegian climate politics. Clim. Change 151 (1), 29−41. Available from: https://doi.org/10.1007/s10584-016-1653-x.

Aslam, S., Javaid, N., Khan, F., Alamri, A., Almogren, A., Abdul, W., 2018. Towards efficient energy management and power trading in a residential area via integrating a grid-connected microgrid. Sustainability 10 (4), 1245. Available from: https://doi.org/10.3390/su10041245.

Banister, D., 2008. The sustainable mobility paradigm. Trans. Policy 15 (2), 73−80. Available from: https://doi.org/10.1016/j.tranpol.2007.10.005.

Böhringer, C., Keller, A., Bortolamedi, M., Seyffarth, A.R., 2016. Good things do not always come in threes: on the excess cost of overlapping regulation in EU climate policy. Energy Policy 94, 502−508. Available from: https://doi.org/10.1016/j.enpol.2015.12.034.

Bruce, D., 2008. How sustainable are we? Facing the environmental impact of modern society. EMBO Rep 9 (1), S37−40. Available from: https://doi.org/10.1038/embor.2008.106.

Bryngelsson, D., Wirsenius, S., Hedenus, F., Sonesson, U., 2016. How can the EU climate targets be met? A combined analysis of technological and demand-side changes in food and agriculture. Food Policy 59, 152−164. Available from: https://doi.org/10.1016/j.foodpol.2015.12.012.

Bui, N., Castellani, A.P., Casari, P., Zorzi, M., 2012. The Internet of energy: a web-enabled smart grid system. IEEE Network 26 (4), 39−45. Available from: https://doi.org/10.1109/MNET.2012.6246751.

Center for Climate and Energy Solutions, 2015. Outcomes of the UN climate change conference in Paris. <https://www.c2es.org/document/outcomes-of-the-u-n-climate-change-conference-in-paris> (accessed 12.05.19.).

Clémençon, R., 2016. The two sides of the Paris climate agreement dismal failure or historic breakthrough? J. Environ. Develop. 25 (1), 3−24. Available from: https://doi.org/10.1177/1070496516631362.

Dai, H., Xie, X., Xie, Y., Liu, J., Masui, T., 2016. Green growth: the economic impacts of large-scale renewable energy development in China. Appl. Energy 162, 435−449. Available from: https://doi.org/10.1016/j.apenergy.2015.10.049.

De Jesus, A., Mendonca, S., 2018. Lost in transition? Drivers and barriers in the eco-innovation road to the circular economy. Ecol. Econ. 145, 75−89. Available from: https://doi.org/10.1016/j.ecolecon.2017.08.001.

EIA, 2019. Annual Energy Outlook 2018. <https://www.eia.gov/outlooks/aeo/pdf/electricity_generation.pdf> (accessed 18.05.19.).

European Commission, 2011. Communication from the Commission to the European Parliament, the Council, the European Economic and Social Committee and the Committee of the Regions. Smart Grids: from innovation to deployment. <http://eur-lex.europa.eu/LexUriServ/LexUriServ.do?uri = COM:2011:0202:FIN:EN:PDF> (accessed 18.02.19.).

European Commission, 2014. EU Energy Markets in 2014, Publication Office of the European Union, Luxembourg. <https://ec.europa.eu/energy> (accessed 20.12.19.).

European Commission, 2016. Paris Agreement. <http://ec.europa.eu/clima/policies/international/negotiations/paris/index_en.htm> (accessed 28.04.19.).

European Parliament, 2000. Lisbon European Council 23rd and 24th of March 2000 Presidency Conclusions. <http://www.europarl.europa.eu/summits/lis1_en.htm> (accessed 20.02.19.).

Fujikura, R., Nakayama, M., 2009. Lessons learned from the world commission on dams. Int. Environ. Agreem: P 9 (2), 173−190. Available from: https://doi.org/10.1007/s10784-009-9093-y.

Goodland, R., Daly, H., 1996. Environmental sustainability: universal and non-negotiable. Ecol. Appl. 6 (4), 1002−1017. Available from: https://doi.org/10.2307/2269583.

Grubb, M., Edmonds, J., Ten Brink, P., Morrison, M., 1993. The costs of limiting fossil-fuel CO_2 emissions: a survey and analysis. Annu. Rev. Energy Env. 18 (1), 397−478. Available from: https://doi.org/10.1146/annurev.eg.18.110193.002145.

Hustedt, T., Seyfried, M., 2016. Co-ordination across internal organizational boundaries: how the EU Commission co-ordinates climate policies. J. Eur. Public Policy 23 (6), 888−905. Available from: https://doi.org/10.1080/13501763.2015.1074605.

IEA, 2009. CO_2 statistics. <https://www.iea.org/statistics/co2emissions/> (accessed 18.05.19.).

Jaffe, A.B., Newell, R.G., Stavins, R.N., 2002. Environmental policy and technological change. Environ. Res. Econom. 22 (1−2), 41−70. Available from: https://doi.org/10.1023/A:1015519401088.

King, J.E., Jupe, S.C., Taylor, P.C., 2015. Network state-based algorithm selection for power flow management using machine learning. IEEE Trans. Power Syst. 30 (5), 2657−2664. Available from: https://doi.org/10.1109/TPWRS.2014.2361792.

Kitapbayev, Y., Moriarty, J., Mancarella, P., 2015. Stochastic control and real options valuation of thermal storage-enabled demand response from flexible district energy systems. Appl. Energy 137, 823−831. Available from: https://doi.org/10.1016/j.apenergy.2014.07.019.

Kok, A., Shang, K., Yucel, S., 2015. Impact of electricity pricing policies on renewable energy investments and carbon emissions. <http://home.ku.edu.tr/~gkok/papers/renewable.pdf> (accessed 11.01.19.).

Lisin, E., Strielkowski, W., 2014. Modelling new economic approaches for the wholesale energy markets in Russia and the EU. Trans. Business Econ. 13 (2B), 566−580.

Lopes, J.A.P., Soares, F.J., Almeida, P.M.R., 2011. Integration of electric vehicles in the electric power system. Proc. IEEE 99 (1), 168−183. Available from: https://doi.org/10.1109/JPROC.2010.2066250.

Markovic, D.S., Zivkovic, D., Branovic, I., Popovic, R., Cvetkovic, D., 2013. Smart power grid and cloud computing. Renew. Sustain. Energy Rev. 24, 566–577. Available from: https://doi.org/10.1016/j.rser.2013.03.068.

McArthur, S.D., Taylor, P.C., Ault, G.W., King, J.E., Athanasiadis, D., Alimisis, V.D., et al., 2012. The autonomic power system-network operation and control beyond smart grids. In: Innovative Smart Grid Technologies (ISGT Europe), 2012, 3rd IEEE PES International Conference on IEEE, pp. 1–7.

McWilliams, A., Siegel, D.S., 2011. Creating and capturing value: strategic corporate social responsibility, resource-based theory, and sustainable competitive advantage. J. Manag 37 (5), 1480–1495. Available from: https://doi.org/10.1177/0149206310385696.

Milanovic, J.V., Xu, Y., 2015. Methodology for estimation of dynamic response of demand using limited data. IEEE Trans. Power Syst. 30 (3), 1288–1297. Available from: https://doi.org/10.1109/TPWRS.2014.2343691.

Newbery, D., Strbac, G., 2016. What is needed for battery electric vehicles to become socially cost competitive? Econ. Trans. 5, 1–11. Available from: https://doi.org/10.1016/j.ecotra.2015.09.002.

Newbery, D., Pollitt, M.G., Ritz, R.A., Strielkowski, W., 2018. Market design for a high-renewables European electricity system. Renew. Sustain. Energy Rev. 91, 695–707. Available from: https://doi.org/10.1016/j.rser.2018.04.025.

Papadaskalopoulos, D., Strbac, G., 2013. Decentralized participation of flexible demand in electricity markets—Part I: Market mechanism. IEEE Trans. Power Syst. 28 (4), 3658–3666. Available from: https://doi.org/10.1109/TPWRS.2013.2245686.

Papadaskalopoulos, D., Strbac, G., Mancarella, P., Aunedi, M., Stanojevic, V., 2013. Decentralized participation of flexible demand in electricity markets—Part II: Application with electric vehicles and heat pump systems. IEEE Trans. Power Syst. 28 (4), 3667–3674. Available from: https://doi.org/10.1109/TPWRS.2013.2245687.

Parsons, G.R., Hidrue, M.K., Kempton, W., Gardner, M.P., 2014. Willingness to pay for vehicle-to-grid (V2G) electric vehicles and their contract terms. Energy Econ. 42, 313–324. Available from: https://doi.org/10.1016/j.eneco.2013.12.018.

Pearson, P.J., Foxon, T.J., 2012. A low carbon industrial revolution? Insights and challenges from past technological and economic transformations. Energy Policy 50, 117–127. Available from: https://doi.org/10.1016/j.enpol.2012.07.061.

Piacentini, C., Alimisis, V., Fox, M., Long, D., 2015. An extension of metric temporal planning with application to AC voltage control. Artif. Intell. 229, 210–245. Available from: https://doi.org/10.1016/j.artint.2015.08.010.

Pigou, A.C., 1932. The Economics of Welfare, fourth ed. Macmillan., London.

Pitt, J., Bourazeri, A., Nowak, A., Roszczynska-Kurasinska, M., Rychwalska, A., Rodriguez Santiago, I., et al., 2013. Transforming big data into collective awareness. IEEE Computer 46 (6), 40–45. Available from: https://doi.org/10.1109/MC.2013.153.

Pool, R., 2012. Solar power: the unexpected side-effect. Eng. Technol. 7 (3), 76–78. Available from: https://doi.org/10.1049/et.2012.0312.

Porter, M.E., Van der Linde, C., 1995. Toward a new conception of the environment-competitiveness relationship. J. Econ. Perspect. 9 (4), 97–118. Available from: https://doi.org/10.1257/jep.9.4.97.

Postolache, A., 2012. The power of a single voice: the EU's contribution to global governance architecture. Romanian J. Eur. Affairs 12 (3), 5–18. Available from: https://doi.org/10.2139/ssrn.2142364.

Potts, B., 2015. The hole in the rooftop solar-panel craze. Wall St. J. <http://www.wsj.com/articles/the-hole-in-the-rooftop-solar-panel-craze-1431899563> (accessed 02.02.19.).

Shafiullah, G., Oo, A., Ali, A., Wolfs, P., 2013. Smart grid for a sustainable future. Smart Grid Renew. Energy 4 (1), 23–34. Available from: https://doi.org/10.4236/sgre.2013.41004.

Silvast, A., Williams, R., Hyysalo, S., Rommetveit, K., Raab, C., 2018. Who 'uses' smart grids? The evolving nature of user representations in layered infrastructures. Sustainability 10 (10), 3738. Available from: https://doi.org/10.3390/su10103738.

Singh, R.K., Murty, H.R., Gupta, S.K., Dikshit, A.K., 2012. An overview of sustainability assessment methodologies. Ecol. Indic. 15 (1), 281–299. Available from: https://doi.org/10.1016/j.ecolind.2011.01.007.

Sørensen, B., 1991. A history of renewable energy technology. Energy Policy 19 (1), 8–12.

Sovacool, B.K., 2009. The cultural barriers to renewable energy and energy efficiency in the United States. Technol. Soc. 31 (4), 365–373.

Steinacher, M., Joos, F., Stocker, T.F., 2013. Allowable carbon emissions lowered by multiple climate targets. Nature 499 (7457), 197–201. Available from: https://doi.org/10.1038/nature12269.

Strielkowski, W., 2016. Entrepreneurship, sustainability, and solar distributed generation. Entrepreneurship Sustain. Issues 4 (1), 9−16. Available from: https://doi.org/10.9770/jesi.2016.4.1(1).

Strielkowski, W., 2017. Social and economic implications for the smart grids of the future. Econ. Soc. 10 (1), 310−318. Available from: https://doi.org/10.14254/2071-789X.2017/10-1/22.

UNFCCC, 2014. The United Nations Framework Convention on Climate Change. <http://unfccc.int/essential_-background/convention/items/2627.php> (accessed 29.04.19.).

Vrba, P., Mařík, V., Siano, P., Leitão, P., Zhabelova, G., Vyatkin, V., et al., 2014. A review of agent and service-oriented concepts applied to intelligent energy systems. IEEE Trans. Ind. Inf. 10 (3), 1890−1903. Available from: https://doi.org/10.1109/TII.2014.2326411.

Wang, Q., Li, R., 2015. Cheaper oil: a turning point in Paris climate talk? Renew. Sustain. Energy Rev. 52, 1186−1192. Available from: https://doi.org/10.1016/j.rser.2015.07.171.

Weitzman, M.L., 2007. A review of the Stern review on the economics of climate change. J. Econ. Literature 45 (3), 703−724. Available from: https://doi.org/10.1257/jel.45.3.703.

Wu, X., Hu, X., Moura, S., Yin, X., Pickert, V., 2016. Stochastic control of smart home energy management with plug-in electric vehicle battery energy storage and photovoltaic array. J. Power Sources 333, 203−212. Available from: https://doi.org/10.1016/j.jpowsour.2016.09.157.

Xu, Y., Milanovic, J.V., 2015. Artificial-intelligence-based methodology for load disaggregation at bulk supply point. IEEE Trans. Power Syst. 30 (2), 795−803. Available from: https://doi.org/10.1109/TPWRS.2014.2337872.

Further reading

Strielkowski, W., Lisin, E., 2017. Economic aspects of innovations in energy storage. Int. J. Energy Econ. Policy 7 (1), 62−66.

4

Renewable energy sources, power markets, and smart grids

4.1 Introduction

In the recent decades, renewable energy sources (RESs) have become the key ingredient in the worldwide effort to mitigate the global climate change. They help to achieve energy efficiency and conduct successful transition to the low-carbon economy (Li et al., 2018).

RESs include solar, wind, biomass, geothermal, and hydro source, just to name the most important ones, all of which are generated by nature and being highly sustainable. The cost of producing electricity from renewable sources is starting to be equal to the cost of traditional generation using fossil fuels or even falling below that level.

In general, RESs can be classified as all nonfossil and nondepletable fuels. The limits of renewable energy are given by the variability and availability of sources. For instance, utilization of photovoltaic (PV) and solar renewable energy is cheaper in the countries where sun shines many days per year. Thence, solar energy is so widespread in places like Australia, California, or South Africa, even though European Union countries, such as Germany, or Scandinavian countries, such as Norway, are also not an exception. Basically, renewables include the following most popular types of energy sources:

- biomass and biogas
- geothermal energy
- hydropower
- photovoltaics
- solar thermal power
- wind power
- other types of renewables (bioliquids, organic waste, hot dry rock technology, energy of the tides, as well as some others)

Renewable energy is rapidly spreading around the world. In 2013 wind energy covered 33.2% of electricity consumption in Denmark and 20.9% in Spain, becoming the largest source of electricity in these countries. In 2014 solar energy provided Italy with electricity at a rate of about 8%, while in Germany the share of renewable in 2014 in the production of electricity was approximately 27% (Fraunhofer ISE, 2014). It was Germany that came with a concept of "Energiewende" or "energy turn" that became the cornerstone of its sustainable economic growth policy based on the high deployment of RESs (Andor et al., 2017).

From 2004 to 2013, the installed capacity of wind power plants increased eightfold in the world, and the number of people employed in renewable energy industries today was close to 7 million. In such nuclear powers as Germany, Great Britain, and China, renewable energy already produces more electricity than nuclear power plants. Despite the declining oil prices, investments in renewable energy increased by 16%, their annual volume amounted to $310 billion. Since 2004, the total volume of global investments in renewable energy has exceeded $2 trillion worldwide.

In 2014 the Council of European Union confirmed that the EU targets for climate change in 2030 were to reduce at least 40% compared to 1990 greenhouse gases and to 27% the share of renewable energy production.

The most profound decarbonization is impending in such sectors as transport and heating. Hence, the electricity sector which is very much linked to these two sectors will be the one most influenced by the decarbonization of the economy.

To achieve this, a large part of the RES will be required in the light of the limited possibilities for hydromassage expansion and the widespread resistance to nuclear energy.

Fortunately, rapid technological advances in wind and solar energy, combined with greater interconnection use, existing hydropower, new battery technologies, and a greater share and deployment of the so-called "demand side" (electricity energy consumption by households and individuals supported by smart meters), suggest that a high-resolution power plant is not only a required target of 2030.

According to EIA (2016), back in the 1970s, oil, natural gas, and coal together accounted for 87% of world primary energy supply, in 2012 their share decreased to 82%, but the energy output itself more than doubled. However, one can see that in the global production of electricity, hydrocarbons also dominate, giving today about 70% of electricity, another 11% is produced by nuclear power plants.

In today's globalized world, it is starting to dawn on policy makers and members of the general public alike that steady transition to low-carbon economy is a must. Governments throughout the globe set up rigorous targets for tackling climate change, decreasing greenhouse gas emissions, and increasing the share of renewable energy in total gross energy production in the present industrialized world (Newbery et al., 2018).

Until recently, carbon economy was dependent on fossil fuels that constituted the main source of energy. Possession of fossil resources, the ability to extract and process them, ensured virtually unlimited world power. Whoever owned the energy also ruled the world. The majority of wars and war conflicts waged in the 20th century were the wars for energy and fossil fuels. In the 21st century, where information became the new energy, this status quo is slowly changing.

Recent history of humankind proved time and again that dependence on fossil fuels can be insecure and may bring various implausible and unpredictable results. The devastating oil shocks of the 1970s confirmed this realization. The supply shock followed by the growth of crude petroleum prices showed problems of the dependence on imports of fossil fuels and resulted in the shortages. Prior to the Industrial Revolution, electricity generation was based on the combustion of biomass (in this context mainly timber), as well as coal and natural gas (or biogas). From the 1800s, this was complimented by hydroelectric power and then, from the 1950s, by nuclear energy used for peaceful purposes (see, e.g., Boluk, 2013). The states were highly reliant on imports. Both world wars and crises exposed these problems and renewables might become a possible solution. Limited applicability as other supply than base load resources and debatable mining in certain nations showed limitations. The Chernobyl nuclear disaster in 1986 (along with the current catastrophe in Fukushima nuclear plant) shook the trust in nuclear power in its foundations. It was important debate for antinuclear movements that are very strong in Austria or Germany. Each one of the aforementioned conditions helped to make a frame utilizing development of other energy resources. The RESs attract an increasing attention in the developed nations since the World War II. It is capable of endorsing energy independence, safety, employment, and inherently enhance environment (Romano and Scandurra, 2011). However, it is frequently forgotten that they also induce considerable costs associated with their development and improvement.

4.2 Renewable energy source and power markets

The cost of electricity production by onshore wind power plants, in geothermal and hydropower, and based on biomass is equal to or lower than the cost of generation at coal, gas and diesel power plants, even without financial support and with falling oil prices.

In many countries, including Europe, wind energy is one of the most competitive sources of new energy capacity. Individual projects in wind power regularly supply electricity at $0.04 per kWh without financial support, while for power plants running on fossil fuels, the cost interval is $0.04−0.14 per kWh (IRENA, 2018).

In the solar energy and solar PVs industry, the most competitive industrial-scale projects supply electricity at $0.08 per kWh without financial support, and lower prices are possible while lowering financing costs. Their cost interval in China, North and South America now lies within the limits that are characteristic of fossil fuel-based generation (IRENA, 2018).

Solar energy has the leading position among other renewables because of its popularity among people and various stakeholders and policy makers.

However, due to the increase in Chinese renewable production the cost of renewable technologies has fallen dramatically. Since the 1970s, geothermal energy has been recognized as an alternative and renewable resource, China has carried out extensive research to identify high-temperature sources for electricity generation (Zhang et al., 2013).

Technological innovation, cost efficiency, and increasing consumer demand are driving RESs, especially wind and solar energy.

In the run-up to forecasts and despite persistent perceptions, wind and solar energy have become competitive with conventional generation technologies in the world's leading markets, even without subsidies.

In fact, the non-subsidized cost of energy produced by the land-based wind turbines and solar PVs has declined below the price of other renewable generation technologies all around the world. While RESs can be more flexible to keep track of the load curve, more affordable battery storage and other innovations are helping to mitigate the effects of wind and solar intermates, increasing the reliability needed to compete with conventional sources. Solar PV energy represents the second cheapest energy source. Nowadays, we are well ahead of the challenge and at the same time we are testing the practical limitations of wind and solar.

Wind turbines have developed significantly over the last few decades, solar PVs are much more efficient and there is a better chance of using energy in tides and waves. With the government's incentives to take advantage of wind and solar technology, their costs have decreased and are now in the same league per kilowatt hour as the cost of fossil fuel technologies has risen, especially with the likely cost of carbon emissions from them.

Unlike fossil fuels or nuclear fuel, wind and solar energy cannot be transported, and although RESs are available in many areas, the best resources are often far from the loading centers, which in some cases increase the cost of connections. The cost of grid-level systems for renewable energy (RE), where they replace shipping sources, is high (15–80 MWh) but dependent on the country, context, and technology (offshore wind with solar PVs).

In addition to traditional power and 4–6 hours of power demand, new types of power services may be encouraged to manage these intrinsic features of clean-up technology.

The policy makers of the Federal Energy Regulation Committee (FERC) and PJM Interconnection are at least focusing on the role of battery energy storage and flexible resources such as distributed resource aggregators (DRA) in power markets (Alanne and Saari, 2006).

High renewable sources can cause various problems in the network's operation, which can vary according to geography, depending on the mix of RESs (solar and wind), the availability of transmission and distribution capacity (both in terms of and availability), a nonrenewable fleet of nonrenewable energies, among other factors.

The operation of power grids with high intermittent resources is a unique challenge for utility companies and network operators. Incentives such as tax credits and net metering have also made home solar PVs competitive in such markets and mandatory construction in California since 2020 (Comello and Reichelstein, 2016). In contrast, Australia and Europe have more residential and commercial roofing than solar energy, which increases the chances of distributing solar energy compared to storage on a commercial scale, becoming the competition for power resources when the grid and power supply parity are reached (Sahu, 2015).

In countries or regions with a high penetration of renewable energies which require more complex system changes, conventional energy sources are adapting to enable cost-effective integration of more renewable energies. For example, operators have retrofitted traditional combined heat and power stations throughout the European Union, China, and India to produce electricity-free heat, while coal and CCGT power stations provide additional flexibility and stability (Eveloy and Ayou, 2019). These policies, which currently exist in about 50 countries across the world, include priority transmission of RESs and special feed-in tariffs (FiTs), quotas, and tax exemptions.

Hydroelectric power, which uses the potential energy of rivers, is by far the most established means of generating electricity from renewable sources. The use of renewable energy, in particular solar and wind energy, is widespread, which supply electricity without carbon emissions. The use of solar and wind power in the grid becomes problematic at a high level for complex but now well-proven reasons.

Most of the electricity demand is for a constant and reliable supply, which has traditionally been achieved by basic electricity production. Therefore, if renewable sources are connected to a grid, the issue of backup capacity is raised, as an independent system energy storage is the main issue.

4.3 Solar energy

Solar energy is the fastest growing area in the field among RES. There are two possible uses of solar radiation: capturing solar energy and converting it using photocells into electrical energy; converting it into heat using solar collectors.

One of the important characteristics of solar radiation is the duration of sunshine. The atmosphere (ozone, water vapor, and carbon dioxide) absorbs solar radiation of certain wavelengths. Significant attenuation (reduction) in most of the ultraviolet and infrared regions of the spectrum is the result of absorption and causes the process of environmental impact on the Earth's climate. A surface perpendicular to the incident direct solar radiation, as a rule, has the highest value of the radiation intensity. Since the distance from the Earth to the Sun varies during the year within 150 million km, the amount of solar radiation also varies from 1325 to 1420 W/m (Fontani and Sansoni, 2015).

The sun's rays that reach the surface of the Earth are divided into two types: direct and scattered. Direct sunlight is those that originate from the surface of the sun and reach the surface of the Earth. The power of direct solar radiation depends on the purity (clarity) of the atmosphere, the height of the sun above the horizon line (depending on the geographical latitude and time of day), and the position of the surface relative to the Sun. Scattered sunlight comes from the upper atmosphere and depends on how direct sunlight reflects off the Earth and the environment (Liu and Jordan, 1960). Due to the repeated process of reflection between the snow-covered surface of the Earth and the underside of the clouds, the power of the scattered solar radiation can reach large values. The sun's rays carry an inexhaustible flow of energy. They are constantly delivering more energy to Earth than we need today (Armaroli and Balzani, 2007).

Solar radiation can be relatively easily converted to thermal, mechanical, and electrical energy, and also it is used in chemical and biological processes. Solar installations work in the heating and cooling systems of residential, public, and industrial buildings, in technological processes occurring at all temperatures. Technological processes of conversion and use of solar radiation in their complexity can be very different. Solar power plants (SPPs) are very different from each other in their dimensions: from microminiature power sources of electronic calculators and wrist watches to huge technical structures in tower SPP 100 meters-high and weighing hundreds of tons. Numerous technological schemes set the difference between the SPPs—from the simplest heating flat surfaces to the most sophisticated control systems for tracking the Sun in order to maximize the arrival of solar radiation at the receiving site.

In recent years, solar PV plants have been developing most intensively around the world, which is a consequence of the tremendous success of world technology in the field of creating highly efficient PV cells. With regard to the above, one can describe the most characteristic classifications of modern and promising types of SPP:

1. By type of use and conversion of solar installations (SI) to other types of energy. When using the SPP in power generation and thermal power engineering, they can be divided into three categories, determined by the type of their use for certain energy consumers:
   a. designed to work on a large scale
   b. working on a local network
   c. intended for power supply of an autonomous consumer with a different category in terms of power supply reliability

   Depending on the category of use of the SPP, there may be requirements for their mandatory combination with the energy storage system (ESS) of any effective type or with other types of power plants based on RESs. For example, this concerns the operation of the SPP on an autonomous consumer, where a daily and, sometimes, a longer energy storage cycle are required. In the systemic large solar power stations, such requirements are usually absent.

2. By its geographical location on Earth. SPPs at the location are divided into ground and space. From which it is clear that the design and the tasks to be solved for such SPP are fundamentally different. For example, the solution of the problem of transferring the accumulated energy to the earth is the most problematic for cosmic SPP. For terrestrial installations, it is necessary to take technical problems into account such as the solar radiation cyclicity in time of day and seasons of the year, and the possible random nature of solar radiation on the Earth's surface.

3. By stationarity. In this case, power plants are portable, mobile, and stationary, differing in weight and size characteristics and the complexity of the design. Also significantly differ from each other in reliability characteristics.

4. By the type of orientation to the Sun. Constant (unchanged) orientation on the surface of the Earth and a sun tracking system used to maximize the arrival of solar radiation at the receiving site. As a rule, permanently oriented toward the sun PSU are household power installations located on the roofs of buildings. In some cases, for small SPPs it is possible to change the angle of inclination of the receiving site in each month of the year, which is structurally quite simple.

5. By the technical complexity of the SPP. Solar energy can be divided into simple or complex both in the technical cycle and in execution. The simple SPPs include water heaters of various designs, air heaters, agricultural product dryers, heating systems, desalination plants, greenhouses, etc., and the complex SPPs include mass-produced power systems, which include tower power systems, solar ponds, solar collectors and, finally, direct conversion of solar radiation into electricity.

The most promising solar radiation conversion systems today are solar PV plants. The photovoltaic solar cell (PSC) used in these installations is the primary converter of the PV system, which converts the solar light energy directly into electrical energy. Therefore, power installations based on PSC are the most convenient for creating an autonomous power supply.

The photoelectric solar cell according to the manufacturing method can be single-crystal, polycrystalline and thin-film (amorphous). Monocrystalline and polycrystalline elements currently have an efficiency factor at 14%−20%, which is almost two times more than that of solar cells that are based on amorphous silicon and has a much longer service life, but they have a significantly higher production cost. Solar cells of a certain size are made. They can be round or square. The power of such elements is 0.9÷2.7 W (Sharma et al., 2015).

A photovoltaic module (FM) is a device that structurally unites electrically interconnected PV solar cells and has output terminals for connecting an external consumer. The power of such modules is already from 10 to 300 W. The documentation gives the value of the power generated by it, under standard conditions, that is, with solar radiation of 1000 W/m^2, temperature +25°C and solar spectrum at a latitude of 45 degrees (Dubey et al., 2013).

The main characteristic of both the solar PV module and the individual solar cell is the current−voltage characteristic (IVC). Volt−ampere characteristic (VAC) of a solar cell represents the relationship between the load current and the voltage at the terminals of the solar PV cell at constant values of the temperature of the solar cells and the intensity of the incoming solar radiation.

In order to obtain the required power and operating voltage, the modules are connected to each other in a sequence or parallel to each other. Solar photovoltaic battery (SFB) is interconnected electrically and mechanically by PV modules. The power of the solar battery is always lower than the sum of the power of the modules, due to losses and differences in the characteristics of the same type of modules (mismatch losses). With a series connection of 10 modules with a spread of characteristics, 10% loss is about 6%, and with a spread of 5% it becomes reduced to 2%. On the basis of solar PV cells build solar PV stations (SPS) (see Pickard et al., 2009).

One of the main defining characteristics of SPS is the maximum value of their efficiency. The efficiency of SPS changes significantly over time as a result of the improvement of the solar cell technology (in the direction of its increase), as well as depending on the material of the solar cell and its multilayer. It should be noted that the efficiency of a solar power installation is significantly influenced by a number of technological factors: the presence of solar concentrators, the cascade of a solar cell, and the operating conditions of the SPS.

Solar PVs are becoming increasingly important and relevant in both developed Western economies and developing countries, where many people enjoy more than 300 days of sunshine and are dependent on fossil fuels. China and India are becoming large players in the solar industry.

China is expected to dethrone Germany as the world's largest solar power producer—although solar energy accounts for only 3% of the Chinese energy mix, it is expanding its facilities, such as the world's largest SPP in the Gobi Desert. (Fu et al., 2015) In addition, by 2020, India plans to increase its solar power to 100 GW (∼20 times larger), and it is expected that solar energy will account for 12%−13% of the country's energy mix (Tripathi et al., 2016).

The most popular but also important due to the scope of this book that focuses on the social impacts and implications, thence social acceptance and support, are the household PV systems and installations. Most of the existing PV plants have power up to 1 kW and include batteries for energy storage. They perform various functions, such as supplying electricity to buildings, powering signal lights on power lines, and others. At present, there is potential for active implementation in the domestic economy. Large SPPs, consisting of a variety of PV cells, can be very useful because their construction takes less time than the construction of traditional power plants, and they can be expanded as needed, thereby increasing power. The main advantages include PV stations operate silently, do not consume fossil fuels, and do not have emissions to the environment. The main disadvantages are that in the modern world, solar electricity is much more expensive than the products of traditional power plants and that they can generate electricity only during daylight hours and are more dependent on the weather:

1. Energy supply of residential buildings. One of the tasks associated with the use of solar cells in residential and office buildings, is that the solar cells could replace traditional building elements or cladding materials. At the same time, they must satisfy architectural solutions and be attractive from an aesthetic point of view. The PV system is most valuable if it is located in close proximity to the consumer because this avoids the loss of energy associated with its transmission over long distances. Installing PV systems near substations that distribute energy can prevent overloading of equipment located on them.
2. Solar pumping installations. Pumping PV systems are a welcome alternative to diesel generators and hand pumps. They pump the water precisely when it is especially needed—on a clear sunny day. Solar pumps are easy to install and operate. A small pump can be installed by one person in a couple of hours, and neither experience nor special equipment is needed for this. Installations work in the presence of solar radiation, accumulating water in the tank. Such installations are characterized by simplicity of design and relatively inexpensive, since they do not require the use of rechargeable batteries in their composition. Among the advantages of solar pumps are the following ones:
   • minimal maintenance and repair
   • ease of installation
   • reliability
   • possibility of modular expansion of the system

Solar energy use is fundamentally different from traditional electric and fuel systems. For this reason, solar pumps are also different from conventional ones. They operate on direct current. The amount of energy depends on the intensity of the solar radiation. Since it is cheaper to store water (in tanks) than energy (in batteries), solar pumps are characterized by low productivity, slowly pumping water throughout the daylight hours.

The use of simple efficient systems is a key factor in using the sun to lift water. For this purpose, special low-power DC pumps are used without batteries and current converters. Modern DC motors work well with variable power and speed. They need a small repair (replacement of worn parts) no earlier than 5 years after their installation. Most solar pumps used for small consumers (residential buildings, small irrigation, livestock maintenance) are piston pumps. They differ from faster centrifugal pumps (including jet and submersible) (Weedy et al., 2012).

Centrifugal, jet, and turbopumps are used in larger systems. Electronic matching devices allow solar pumps to turn on and operate in low light conditions. This allows you to use the energy of the sun directly without batteries. A sun tracking device can be used, with the help of which the panels remain aimed at the Sun throughout the day, from sunrise to sunset, which allows us to extend the usable daylight hours. Battery tanks usually store 3–10 days of water in case of cloudy weather. Solar pumps use a small amount of electricity. In order to increase the amount of water produced, they use more efficient pumps and work longer hours rather than increase the speed of operation.

Where PV pumps are compared to diesel pumps, their relatively high initial cost is offset by fuel savings and reduced maintenance costs. Studies of the economic efficiency of PV pumping plants confirm that they are often more cost effective than diesel pumps, depending on the specific conditions.

Simple solutions have certain disadvantages. The most important of them is that the photoelectric pump or air conditioner can work only during the daytime and in the light of the sun. To compensate for this deficiency, a battery is connected to the system. It charges from a solar generator, stores energy, and makes it available at any time. Even in the most adverse conditions and in remote areas, PV energy stored in batteries can power the necessary equipment. Thanks to the accumulation of electricity, PV systems provide a reliable source of power supply, day and night, in any weather. Battery-powered PV systems feed lighting devices, sensors, recording equipment, household appliances, telephones, televisions, and power tools all over the world.

The solar module produces a direct current, usually with a voltage of 12 V. There are many electrical appliances—lamps, televisions, refrigerators, fans, tools, etc., which operate on direct current of 12 V. However, most household appliances still consume 220 V AC current. Battery-powered PV systems can be adapted to power DC or AC equipment. Those consumers who want to use conventional AC devices should add to the system, between the battery and the load, a power control unit—the so-called inverter. Although in the process of converting direct current into alternating current, a certain amount of energy is lost, thanks to the inverter, PV energy can be used along with the usual public energy supply (power household appliances, lighting devices, or computers).

The system is designed as follows: the PV module is connected to the battery, and that, in turn, with the load. In the daytime PV modules charge the battery. Energy as required enters the load. With a simple charge controller, the battery is charged to the desired extent. At the same time, its lifetime is prolonged, and protection against overload and complete discharge is provided. The battery is able to expand the scope of PV panels but requires some maintenance. PV solar systems are similar to automotive; they require caution in handling and storage. They must be protected from the effects of low and high temperatures.

A battery-powered solar panel supplies electricity to the user when it is needed. The amount of accumulated electricity depends on the power of the PV modules and the type of battery. Expansion of the module or the addition of batteries increases the cost of the system, so to determine its optimal size, you need to carefully study the power consumption. A well-designed system determines the optimal balance of cost and convenience while satisfying the user's need for electricity, as well as the possibility of expanding the system.

One interesting example is the electric refrigerators fed from solar cells, which make it possible to store valuable perishable foods, primarily medicines, vaccines, etc., have become widespread, especially in countries with a hot climate. This direction of use of photovoltaic electric plants (PEP) is also one of the most common. Today, more than 100,000 PV plants operate in different countries, providing power to autonomous meteorological stations, autonomous temperature and water level monitoring stations, flow rates of liquids in pipelines, monitoring air pollution levels near industrial enterprises, etc.

Compared with other types of energy, solar energy in general is one of the most environment-friendly types of energy. However, it is practically impossible to completely avoid the harmful effects of solar energy on humans and the environment, taking into account the entire process chain from obtaining the required materials to producing electricity.

The most characteristic in this aspect of PV systems, the operation of which causes minimal damage to the environment. At the same time, the production of semiconductor materials is very environmentally and socially dangerous. In this regard, in some countries of the world (e.g., in the United States) there are very strict requirements for the production of semiconductors for PV, as well as for the storage, transportation and elimination of harmful substances from the production of PV, limiting personnel contacts with these substances, developing action plans for emergency or abnormal technological situations, as well as programs for the elimination of production wastes that have completed their lifetime. The most dangerous with this respect are cadmium (Cd), as well as Ga, As, and Te (Fiandra et al., 2019). Nowadays, the harmful effects of cadmium on human health are the most studied, and even bans are imposed on the use of its compounds in domestic conditions (e.g., microbatteries and batteries based on it). Prolonged inhalation of cadmium vapor can lead to pulmonary or bronchial diseases and even death. Constant exposure to low doses of cadmium leads to its accumulation in the kidneys and their disease. It also observed lung disease, softening and deformation of the skeletal bone composition.

Some selenium compounds are also highly toxic. For example, SeH, SeO_2 appear to negatively affect the respiratory organs (Wrobel et al., 2016). Tests of spent CuInSe2 and CdTe-based solar modules or rejected solar modules showed that the first of them meet the requirements of the U.S. Environmental Protection Agency, the latter are not, as the cadmium level in them turned out to be 8–10 times more than the permissible norms. As a consequence of this, CdTe-based solar modules, which have developed their life, may now be classified as potential toxic waste and, if possible, return to their manufacturers (similar to problems that have worked with TVELs at nuclear power plants) (Kumar et al., 2018).

In other words, the actual greatest socioenvironmental hazard for PVs is mainly related to the production of some PVs, during which a significant amount of harmful substances is processed for human health and the environment. Obviously such production should be fully automated and located at a considerable distance from populated areas. Special measures should be taken to protect the production itself. As for the operation of PVs, it is practically safe.

The application of the chemical interaction of silicon with tetrafluoride silicon is also considered promising. In this case, the extraction of silicon from the melt, its purification and chemical vapor deposition during one stage of the technological process is realized.

Modern methods of producing wafers and silicon sheets are quite numerous. The main efforts here are aimed at optimizing the ways to create polycrystalline and monocrystalline silicon, which has the highest efficiency. The standard technological process, which allows to obtain a single-crystal solar cell with a diameter of up to 7.6 cm or pseudo-rectangular elements up to 2×8 cm in size, is based on the Czochralski process of growing crystals with subsequent cutting of plates with diamond tapes and grinding them with abrasive powder, which is very harmful for human health (silicon dust, cadmium, and arsenide compounds) (Calvert et al., 2016).

Thus in solar photoenergy the most harmful to humans and the environment is the technological process of obtaining solar cells, their storage and disposal. To improve the efficiency of this production should be large scale, which requires large capital and material costs. It is also necessary to take into account the work on the exploration and extraction of silica, as well as the inevitable withdrawal of land from economic production in this case.

Among other aspects of the negative impact of solar energy on the socioenvironmental conditions in the country, the following should be noted.

Solar power stations are quite earthy because of the very diffuse nature of solar radiation entering the Earth. One can make a simple calculation that obtaining 1 MW requires 1.1 hectares of land, at SFEU from 1.0 to 1.6 hectares, and in solar ponds up to 8 hectares, which is very noticeable for the inhabited regions of any country. SPP themselves are materially consuming (metal, glass, concrete, etc.).

The operation of solar ponds contributes to the contamination of the soil and groundwater with chemically active salt solutions. During the operation of battery storage energy supply (BSES), as well as PVs, there is a noticeable change in climatic conditions at a given place, including changes in soil conditions, vegetation, air circulation due to surface shading, on the one hand, and air heating, on the other. Because of the latter, the heat balance of the air humidity, the direction and size of the winds change. For SES with concentrators of solar radiation, there is a great danger of overheating and ignition of the very systems for obtaining energy from RES.

The use of low-boiling liquids and their inevitable leakage in the SPP can lead to pollution of the soil, groundwater, and even drinking water in the region. The liquids containing nitrites and chromates, which are highly toxic substances, are especially dangerous. The low conversion rate of solar radiation into electricity leads to problems associated with condensate cooling. At the same time, thermal emissions into the atmosphere at the SPP are more than two times higher than the similar discharge from thermal power plants. To account for the negative impact of various types of power plants on the environment, several different methods and approaches have been proposed.

The overall negative impact of technical devices of solar energy on humans and the environment is much less than that of other types of energy and especially traditional nuclear power plants, thermal power plants, and hydroelectric power plants.

When justifying the SPP parameters (mainly its installed capacity) according to the method of comparative economic efficiency, it is first necessary to resolve the issues of replacing or duplicating the energy of other stations in the power system and ensuring the principles of energy and environmental comparability of the compared objects.

In order to mitigate the irregularity of solar radiation, a battery (usually thermal) is added to the flowchart, thus creating a single energy complex. In this case, during the hours of sunshine, thermal energy is stored in the battery, and in the hours of no radiation, it is transferred to the working fluid entering the turbine (Kalogirou, 2004). At the same time, SPP can be considered as a reliable source of energy supply when operating not only as part of the power system, but also in isolation. In the absence of a battery in parallel with the SPP, it is necessary to install a backup energy source or to work with other stations of the power system that use the system power reserve.

However, before comparing various energy technologies in terms of economic and other parameters, it is necessary to determine their actual cost, since the prices for fuel and energy for the past decades did not reflect the real costs of their production.

Only with objective pricing will economic incentives be used to save energy and create new technologies in the energy sector. Each year, the world consumes as much oil as it is formed in natural conditions for 2 million years (Bentley, 2002). The gigantic rate of consumption of nonrenewable energy resources at a relatively low price, which does not reflect the real total costs of society, essentially means life on loan, loans from future generations who will not have energy available at such a low price. This is only one of the components of costs that society pays for energy, but which are not reflected in the market price.

Another component of the cost of energy, distributed to the whole society and not included in energy tariffs, is associated with environmental pollution by power plants. Various estimates of direct social costs associated with the harmful effects of power plants, including illness and reduced life expectancy, medical care, loss of production, yield reduction, forest restoration, and repair of buildings due to air, water, and soil pollution, give about 75% of world prices for fuel and energy (see, e.g., Dincer, 2000). Essentially, these costs of the whole society are an environmental tax that citizens pay for the imperfection of power plants, and this tax should be included in the cost of energy to create a state fund for energy conservation and create new environment-friendly technologies in the energy sector.

If one takes into account these hidden costs in energy tariffs, then the majority of new renewable energy technologies become competitive with existing ones. At the same time, a source of funding for new projects on clean energy will appear. This is the environmental tax at the amount from 0% to 30% of the cost of oil introduced in Sweden, Finland, the Netherlands, and other European countries (Ekins and Speck, 1999). Economic laws and experience in the development of the world economy show that a rational structure for the use of natural resources in the long run tends to the structure of their existing reserves on Earth.

Since silicon occupies the second place after oxygen in the earth's crust, it can be assumed that from primitive people with primitive silicon tools of labor, humanity after thousands of years passes to a period in which ceramics, glass, silicate, and composite materials will be used as construction materials, and as a global source of energy—silicon SPPs. The problems of daily and seasonal accumulation may be solved with the help of solar-hydrogen energy, as well as the latitudinal location of SPPs and new energy-efficient energy transfer systems between them. Considering that 1 Kr of silicon in a solar cell generates 15 MW of electricity for 30 years already with its current manufacturing technology, it is easy to calculate the oil equivalent of silicon (Barlev et al., 2011). Direct calculation of electricity of 15 MWh, taking into account the heat of combustion of oil, 43.7 MJ/kg gives 1.25 tons of oil per 1 kg of silicon. If we take the efficiency of a thermal power plant (TPP) running on fuel oil to be 33%, then 1 kg of silicon from the generated electricity is equivalent to approximately 3.75 tons of oil. One of the most important and representative indicators of the economic efficiency of any type of power plant in the world is the values of $/kW and $/kWh, that is, specific investments in 1 kW of installed capacity and the price of electricity produced at the power plant under consideration.

All types of SPPs have a steady tendency to continuous improvement of the values over time which is a consequence of taking into account the objective factors in the solar energy of the world. In particular, due to the improvement of technology and the growth of the scale of production of solar cells in the world, the values of decreased tenfold by 2010 compared to the 1990 level. At the same time, these indicators at traditional types of power plants have increased significantly and have a steady tendency to grow due to a number of objective and well-known factors operating in the world today.

Some forecasts mention further reduction of $/W for PVs or the period up to 2020 and show that solar energy is already becoming more and more competitive in the fuel and energy complex of the world as a whole and of each country individually (Jacobson and Delucchi, 2011). Similar forecasts for the cost and volume of production of solar modules in the world are given in other sources. Given the higher environmental safety of such power plants, the huge reserves of renewable solar energy and 40 years of experience in the development of solar cell technology, it is becoming increasingly clear that PV solar plants will play a strategic role in the global energy industry of the future.

Based on the analysis of the cost of production of solar cells, carried out for various technologies, as well as the demand for solar cells, the task of gradually reducing the cost of production to $1 per W (short term) and $0.8 per W (long-term prospective) was formulated. This will reduce the cost of electricity, respectively, to $0.9 per kWh and $0.5 per kWh.

Reducing the cost of solar electricity is possible either by improving the technology of semiconductor materials, or by using radiation concentrators. This mainly concerns silicon-based solar panels, since crystalline silicon today occupies a dominant position in the manufacture of a solar cell. Considering that 1 kg of silicon to a solar cell at a single intensity of solar radiation can produce 300 MWh of electricity for 30 years, you can easily determine its equivalent amount of oil. For the production of 300 MBT h of electricity, 25 tons of oil will be required with a calorific value of 43.7 MJ/kg and given that the efficiency of a thermal power plant is approximately 33%, the amount of oil equivalent to 1 kg of silicon increases to 75 tons.

Compared to nuclear energy, 1 ton of natural uranium in an open-cycle generator produces 35 GWh, while 1 ton of silicon (the most common material for the manufacture of a solar cell) in an installation with a concentrator over 30 years of service can generate 92 Gh. Due to this, silicon is often referred to as "the oil of the 21st century," referring to the high profitability of the oil industry (Latunussa et al., 2016).

The main obstacle to reducing the cost of a solar cell is the high cost of solar silicon ($70−30 per kg), so the task of paramount importance is the development of new silicon production technologies. The silicon content in the Earth's crust is 29.5% which exceeds the reserves of aluminum by 3.35 times. The silicon price of the purification rate of 99.99% is equal to the cost of uranium used in reactors of nuclear power plants, although the silicon content in the earth's crust is 100,000 times the content of uranium.

The world uranium reserves are estimated at 2,763,000 tons (Zhang and Meng, 2006). The production cycle of uranium fuel, including the production of uranium hexafluoride, is more complex and dangerous compared to the chlorosilane method of semiconductor silicon production. Since uranium in the crust is in a dispersed state and is contained in disproportionately lower concentrations than silicon, it is not clear why these materials have almost the same value. This "paradox" can be explained only by the fact that billions of dollars were invested in the development of the technology for the production of uranium fuel in the world. These funds were released mainly on military programs. Apparently, it is only for this reason that the volume of world production of uranium is several times higher than the volume of production of semiconductor silicon.

For 35 years from the date of its development, the chlorine—silane technological cycle of production of semiconductor silicon practically did not undergo significant changes and suffers from all the shortcomings of chemical technologies of the 1950s: high energy consumption, low output of silicon, and high level of environmental hazard. Currently, for the production of "solar" silicon, as a rule, the same traditional semiconductor silicon technology used in the electronics industry is used.

The main raw material used for the production of silicon is silica in the form of quartz sand, forms 12% of the mass of the lithosphere. Their deposits are sufficient to provide raw materials for solar PV stations with a capacity of more than 1000 GW. The high energy costs and low yield of silicon (from 6% to 10%) in the chemical cleaning method are due to the high energy of the SiO bond (64 kJ/mol). Currently, three new technologies for the production of solar silicon are waiting for their finalization and commercial development (Malhotra and Ali, 2019).

Nowadays, a new technology has been developed to produce energy silicon, based on large reserves of high-purity quartz. The proposed method for the production of silicon is based on the reaction of its reduction from high-purity natural quartz using carbon using a special technology that ensures sufficient purity of the final product. There are several countries around the world possessing rich deposits of high-purity quartz and graphite. The analysis shows that these materials can be used without prior chemical cleaning.

There are other promising technologies for the production of energy silicon that emerge around the world and some of them prove to be very effective. In particular, the technology of using high-purity silicon raw materials for the production of a solar cell without the use of chlorine compounds. Here, purified technical silicon is used as the sources material (Kadro et al., 2016). A very promising is the technology for the production of silicon from rice husk, the annual reserves of which are very large in the countries engaged in the production of rice. In this technology, the raw material is rice husk.

The need and prospects for the development of solar energy in different countries of the world is confirmed today by various stimulating legal and economic acts and laws adopted in them. At the beginning of the 21st century in different countries of the world and especially in the countries of the European Union (EU), special attention was paid to the development of solar energy. As an example, below are various legislative acts adopted in some EU countries and contributing to the development of PVs and solar collectors.

As a result of the analysis of the solar energy market, it is possible to make favorable climatic conditions the countries with a high intensity of sunshine for the use of solar power generation technologies. An interesting example is Russian Federation where the perspectives of using PVs for generating electric energy represent quite a challenge. The country is located between 41- and 82-degree north latitude, and the levels of solar radiation on its territory vary considerably.

The growth rate of solar energy in a carbon fuels-abundant country such as Russian Federation is still tiny. This is explained by the scientific and technical base lost in the result of the collapse of the Soviet Union in this area. At the same time, modern geopolitical conditions, as well as the constant rise in prices for hydrocarbons, no doubt open up new horizons for renewable energy technologies. Also, the main idea of the development of solar energy is born—the provision of "own" source of electricity to the end user, and not the construction of huge expensive SPPs. Quite a large part of the energy is lost in the process of delivery on the wires, so installing solar panels on the roof of a country house further reduces the load on the power plant.

With the development of technology, there are new, more advanced ways of transforming solar energy. Solar radiation can be relatively easily converted to thermal, mechanical and electrical energy, and also it is used in chemical and biological processes. Solar installations work in the heating and cooling systems of residential, public, and industrial buildings, in technological processes occurring at all temperatures. Technological processes of conversion and use of solar radiation in their complexity can be very different. SPPs are very different from each other in their dimensions: from microminiature power sources of microcalculators and wrist watches to huge technical structures in tower SPPs 100 m high and weighing hundreds of tons. Numerous technological schemes set the difference between the solar power installations—from the simplest heating flat surfaces to the most sophisticated control systems for tracking the sun in order to maximize the arrival of solar radiation at the receiving site.

With the development of technology, there are new, more advanced ways of transforming solar energy. Solar radiation can be relatively easily converted to thermal, mechanical, and electrical energy, and also it is used in chemical and biological processes. Solar installations work in the heating and cooling systems of residential, public, and industrial buildings, in technological processes occurring at all temperatures. Technological processes of conversion and use of solar radiation in their complexity can be very different. SPPs are very different from each other in their dimensions: from microminiature power sources of microcalculators and wrist watches to huge technical structures in tower SES 100 m high and weighing hundreds of tons. Numerous technological schemes set the difference between the solar panels that might be from the simplest heating flat surfaces to the most sophisticated control systems for tracking the Sun in order to maximize the arrival of solar radiation at the receiving site.

4.4 Wind energy

Similar to hydro energy, wind energy is not a strictly new invention or a novelty for the humanity. People learned how to use wind in sea travel, applying seals of trade and war ships in order to harness the power of nature and to move around faster.

From the point of view of climate science, wind in the surface layer is formed due to the uneven heating of the Earth's surface by the sun. Since the surface of the earth is heterogeneous, even at the same latitude the land and water areas, mountains and forests, deserts and marshy lowlands are heated differently (Xie et al., 2010). During the day, the air over the seas and oceans remains relatively cold, since much of the energy of solar radiation is consumed for the evaporation of water and absorbed by it. Above the land, the air warms up more, expands, reduces its mass density and rushes to higher layers above the ground. It is replaced by colder, and, therefore, denser air masses, located above water spaces, which leads to the emergence of the wind as a directional movement of various air masses. These local winds that form in coastal areas are called breezes. Annual temperature changes in coastal areas of large seas and oceans cause a large-scale circulation than breezes, called monsoons. They are divided into sea and continental, differ, as a rule, in high speeds and change their direction during the night. Similar processes occur in mountainous areas and valleys due to different levels of heating of the equatorial zones and poles of the Earth and many other factors. The nature of the circulation of the earth's atmosphere is complicated by the forces of inertia that arise when the Earth rotates. They cause various deviations of air currents, a lot of circulations are formed, interacting with each other to a greater or lesser extent (Burroughs, 2007).

The strength and direction of the wind in different zones vary in different ways depending on the height above the Earth's surface. Therefore, at the equator, close to the earth's surface, there is a zone with relatively small and variable in the direction of the velocities, and in the upper layers rather high velocity air flows occur in the east direction. At an altitude of 1−4 km from the surface of the Earth, in the zone between 30 degrees north and south latitudes, fairly uniform air currents are formed, called trade winds. In the northern hemisphere, closer to the surface of the Earth, their average velocity is 7−9 m/s (Junge, 1962).

Large-scale circulation of air masses is then formed around the zone of reduced pressure—clockwise in the northern hemisphere against the direction of movement, and in the southern hemisphere in the direction of its movement. Due to the inclination of 23.5 degrees of the axis of rotation of the Earth to the plane of its rotation relative to the Sun, seasonal changes in thermal energy received from it occur, the magnitude of which depends on the strength and direction of the wind over a certain zone of the earth's surface. Thus thermal energy, continuously coming from the Sun, is converted into the kinetic energy of motion in the atmosphere of huge masses of air, the circulation of which is called the wind.

Wind is one of the most powerful energy sources that has long been used by mankind and, under favorable conditions, can be utilized on a much larger scale than is currently the case. According to rough estimates, the energy that is continuously supplied from the Sun corresponds to a total power exceeding 1011 GW. This determines the possible annual energy production by wind turbines, equal to 1.18×1013 kWh (for comparison, the total consumption of all types of energy on Earth is $\sim 7 \times 1013$ kWh (EWEA, 2009).

The principle of operation of a wind power installation is quite simple. Under the action of the wind blades mounted on the wheel, begin to rotate. The wheel transmits torque to the shaft of the generator, which is the source of the generated electricity. For this to happen, a certain wheel speed is required. Energy production depends on the size of the wheel: the larger it is, the better it is to capture the wind, which allows you to generate more energy. The generated energy is sent to the charger, which transforms it into a direct current, used to charge the battery devices (see Wagner, 2013).

The wind generator includes a mast, a wind cap consisting of three blades, a generator, a tail, a turntable, a controller, a charger, a battery, and an inverter. The controller controls the processes that occur in the wind farm. Wind wheel (also called the wind turbine, or rotor) converts the energy of the oncoming wind flow into mechanical energy of rotation of the turbine axis. The diameter of the wind wheel ranges from a few meters to several tens of meters. Rotational speed ranges from 15 to 100 rpm. Usually for wind turbines connected to the network, the rotation speed of the wind wheel is constant. For autonomous systems with rectifier and inverter are usually variable. Wind wheel contains blades, which are fixed in the wind wheel hub (Kalantar, 2010).

The multiplier (gearbox) is an intermediate link between the wind wheel and the electric generator, which increases the frequency of rotation of the wind wheel shaft and provides coordination with the speed of the generator. The exceptions are low-power wind turbines with special permanent-magnet generators. In such wind turbines, multipliers are usually not used.

The tower or mast (it is sometimes reinforced with steel braces) is used to place the wind turbine head and multiplier into the wind at a certain height relative to ground level, which is necessary for the productive work of a wind turbine and the observance of safety requirements. In a large-scale wind turbine, the height of the tower reaches 75 m. Usually these are cylindrical masts, although lattice towers are also used. The basis is intended for prevention of falling of installation at a strong wind. The installation also includes such elements as:

- Battery charge controller represents a device connected to the battery to control the charge process. The controller charges the battery, delivering a charging current of more than the self-discharge current (compensating for self-discharge) and limiting the maximum charge current to prevent the battery from breaking. If there is energy consumption, the controller supplies the energy consumer and charges the battery, compensating for all differences in energy flow. In this case, the output voltage varies in a small range (from 12 to 14.5 V in the case of using a car battery).
- Electric accumulator constitutes a reusable chemical current source, the main specificity of which is the reversibility of internal chemical processes, which ensures its repeated cyclical use (through charge−discharge) for energy storage and autonomous power supply of various electrical devices and equipment, as well as for providing backup sources energy in medicine, manufacturing and in other areas.
- Inverter is a device for converting DC to AC current with a change in the magnitude of frequency and/or voltage. Usually it is a periodic voltage generator, in the shape of a sinusoid, or a discrete signal (see, e.g., Hau, 2013).

Currently, two basic designs of wind turbines are used: horizontal axial and vertical axial wind turbines. Both types of wind turbines have approximately equal efficiency, but the wind turbines of the first type are most common (Khan et al., 2009). The power of wind turbines can be from hundreds of watts to several megawatts. In addition to the classification provided above, wind turbines can be divided into two large classes: linear (mobile) and cyclic (stationary). Linear wind turbines include classic sails and wings (say, a kite wing). They can only be used to propel vehicles, since they can only do work when moving in space. And without reconfiguration, this movement will always be linear—in the direction of the wind or at some angle to it. And since their design is quite rigid and usually allows only fairly limited configuration options, they can only move with the object on which they are installed or of which they are parts of. Unfortunately, these features traditionally limit the use of this class of wind engines by navigation (vast spaces where nothing prevents the wind and there is freedom for maneuvers) and sports and entertainment applications (in addition to sailing, these are also paragliders and kites).

Another type is represented by the cyclic wind turbines can perform useful work permanently, staying in one place. Only their work items move, which under the action of the airflow move along a closed trajectory, making cyclic movements. It is this class of wind turbines that deserves consideration in the first place, since it can be used not only for moving vehicles, obtaining precisely the mechanical form of energy, but also for directly generating the most versatile and convenient to use type of energy—electricity. Various wind wheels belong to this class of wind turbines—from the wings of a classic windmill to modern rotor designs with a vertical axis (Li et al., 2016).

In addition, wind turbines can be divided into devices with horizontal and vertical axis of rotation, high speed and low speed. They differ significantly in their design and wind energy efficiency. There are also wind turbines of the new generation that employ more technically advanced elements and aspects. One example of a new generation of wind turbines consists of several small rotors mounted on one flexible rod. Each rotor catches not only its own stream of fresh wind, but also the wind from its neighboring rotors. One installation of 10 rotors can provide 100−400 W of power.

Another example is a magnetic wind turbine that was invented by Ed Mazur in 1981 (Mazur, 2013). Its advantage is that it reduces operating costs and increases the service life of the generator. In addition, it is capable of generating 1 GW of power. Its design is designed in such a way that the turbine blades are suspended on an air cushion and the energy is directed to linear generators with minimal losses. Today, the current magnetic wind turbine is located in China. Windmill turbine is the development of Dutch engineers. Designed to replace traditional mills, it is supposed to place up to eight turbines on one basis (Devine-Wright, 2005).

Spiral wind turbine was developed by the German engineers, designed for mass consumer use. It has good ergonomics and is able to work at low wind speeds. The power of this turbine ranges at 2.5−5 kW and its term of operation is estimated to be around 30 years. Another wind generator of a new generation in which the system of power generation around the perimeter is implemented is called Windtronics. At the ends of the constantly rotating blades are neodymium magnets. The blades rotate along the contour, equipped with copper coils. This system allows you to convert the most energy through side flaps, instantly reacting to the slightest changes in the wind, and the multilevel design of the blades of nylon (Crampsie, 2012).

The use of a new generation of wind turbines produced by Windtronics might help to reduce the ecological footprint thanks to the reduction of CO_2 emissions released into the atmosphere. In addition, the device is convenient to use, since the compact design does not require additional maintenance. The generator can be installed in areas with any climatic conditions.

The most powerful wind turbine in the world today is made by Vestas. The Danish company Vestas, the world leader in the production of wind power plants (wind turbines), presented its latest development—wind turbines for coastal (offshore) use V164−7.0 MW. In many respects this wind turbine is a record not only for Vestas, but also for all competitors. One such installation will be able to supply electricity to 6500 households. The height of the wind turbine tower reaches 187 m, and the length of each rotor blade is 80 m. The rotor with a diameter of 164 m covers an area of 21,000 m^2—like three football fields.

Vestas is a leader in the wind industry segment—43% of coastal turbines carry this brand (a total of 580 units). The reputation of a leader imposes high obligations, so when launching a new model, the company carefully worked to reduce investment risks. This is due to the high efficiency of the V164−7.0 MW installation. A large unit capacity requires the installation of a smaller number of turbines (fewer foundations, lines for connecting to the network, etc.), which can significantly reduce the required investment. The service life of wind turbines is 25 years (see Goudarzi and Zhu, 2013). Among other popular foreign manufacturers of wind turbines, the following ones are also well-known and enjoying large shares on the market:

- Jacobs wind turbines produced in United States in the 1930s. Production was resumed in the 1980s.
- Bornay is a Spanish company that produces small wind turbines to solve the problem of power supply to isolated households and rural areas.
- English wind turbines Ampair designed to withstand severe weather conditions on land and at sea. The low-speed turbines, the aerodynamic shape of the blades and the reinforced design ensure the durability and reliability of the equipment in conditions where maintenance is often difficult or impossible (see, e.g., Nelson and Starcher, 2018).

To date, there are top 10 manufacturers of wind turbines in the world, but some smaller companies are also catching up. The fastest development of technology using wind energy received in the United States, Denmark, Germany, and Spain. These countries account for more than three-quarters of the world's total wind park. Sustained interest in wind power is also observed in many developing countries of the world, for example, in India, China, and countries of South America. According to the generalized data of the European and American Wind Energy Associations, by the end of the second millennium the installed capacity of all operated wind turbines in the world reached more than 14,000 MW, of which more than 9000 MW falls to Europe (Neij, 2008). It should be emphasized that the rapid development of wind energy in the above-mentioned countries is solely due to the full state support.

Under powerful industrial wind farms, the area is required at the rate of 5−15 MW/km^2, depending on the wind rose and the local relief of the region. For a 1000 MW wind farm, an area of 70−200 km^2 is required. Allocation of such areas in industrial regions poses great difficulties, although in part these lands can also be used for economic needs. For example, in California, 50 km from the city of San Francisco, on the Altamont Pass, the land allotted for a park of powerful wind farms, also serves for agricultural purposes. The most important factor in the influence of water power plants (WPP) on the environment is the acoustic impact. In foreign practice, there has been enough research and full-scale changes in the level and frequency of noise for various wind turbines with wind wheels with different structures, materials, height above ground, and for different environmental conditions (wind speed and direction, underlying surface, etc.).

Noise effects from wind turbines have a different nature and are divided into mechanical (noise from gears, bearings, and generators) and aerodynamic effects. The latter, in turn, can be low frequency (less than 16–20 Hz) and high frequency (from 20 Hz to several kHz). They are caused by the rotation of the impeller and are determined by the following phenomena—the formation of vacuum behind the rotor or wind wheel with the striving of airflow to a certain vanishing point of turbulent flows, pulsations of lifting force on the blade profile, the interaction of the turbulent boundary layer with the rear edge of the blade. Removal of wind power stations from settlements and places of rest solves the problem of noise effect for people. However, noise can affect fauna, including marine fauna in the region of equatorial WPPs. According to some experts, the probability of hitting birds with wind turbines is estimated at 10% if migration routes pass through a wind park (Barrios and Rodriguez, 2004). The placement of wind parks will affect the migration routes of birds and fish for equatorial wind farms. It has been suggested that the screening effect of wind power stations on the path of natural airflow will be insignificant and it can be ignored. This is explained by the fact that wind turbines use a small surface layer of moving air masses (about 100–150 m) and, moreover, no more than 50% of their kinetic energy. However, powerful wind farms can affect the environment—for example, reduce air ventilation in the area where the wind park is located. The shielding effect of a wind park may be equivalent to an elevation of the same area and a height of about 100–150 m. Interference caused by the reflection of electromagnetic waves by blades of wind turbines can affect the quality of television and radio programs as well as various navigation systems in the area where the wind farm is located at a distance a few kilometers away The most radical way to reduce interference is to remove the wind park to the appropriate distance from the communications. In some cases, interference can be avoided by installing repeaters. This question does not fall into the category of intractable, and in each case a specific solution can be found. The adverse wind energy factors:

- noise effects, electrical, radio and television interference
- distribution of land
- local climatic changes
- danger to migratory birds and insects
- landscape incompatibility, unattractiveness, visual nonperception
- change of traditional shipping routes, adverse effects on marine animals (see, e.g., Dai et al., 2015).

World wind energy gradually reaches a level that will allow it to become one of the main sources of energy in the future. The main factors of its development are the need to ensure energy security, the volatility of prices for carbohydrate fuel, and the environmental problems of traditional energy.

Wind power is the most attractive solution of world energy problems. It does not pollute the environment and does not depend on fuel. Moreover, wind resources are present in any part of the world and are sufficient to meet the growing demand for electricity. According to its potential, wind energy can become the basis for a safe and sustainable energy of the future, based on environment-friendly technologies, creating millions of new jobs, promoting local development. Wind energy can and should play a leading role in the global energy industry of the future.

For 20 years, windy installations have come a long way of improvement. As a result, a modern wind farm in its characteristics is not inferior to a traditional power plant. Moreover, electricity generation at wind stations is becoming increasingly competitive in comparison with traditional fossil fuel energy sources: even today wind energy is comparable to new coal and gas power plants.

Particular attention is paid to wind power parks in China. China, which has large spaces and a long coastline, has a powerful potential for the development of wind power. The Chinese Institute of Meteorological Research estimates the available wind energy potential on land at 253 GW. In addition, 750 GW can be provided through the construction of wind farms in the shelf zone (see, e.g., He et al., 2016).

The worldwide wind energy market is growing faster than any other renewable energy market. Since 2002, the installed capacity of wind power plants in the world has increased more than 12 times: from 4800 to 59,000 MW (as of the end of 2012). It is expected that the turnover of the international wind energy market in 2013 will exceed 13 billion euros (Neuhoff et al., 2013). According to approximate estimates, the industry employs 150,000 people. The successful development of the wind industry attracts investors from large financial institutions and traditional energy.

In a number of countries and regions, the share of electricity generated by wind stations is a real competition to traditional energy: in Denmark 20% of electricity is generated from wind energy, in Spain this figure reached 8%, and by the end of the decade it should increase to 15%. These figures show that wind power is already able to make a significant contribution to electricity production without greenhouse gas emissions. In 2012 the wind energy was marked with another record—about 11,531 MW of new capacities was introduced in aggregate. Thus the average annual growth increased by 40.5%, and the total installed capacity increased by 24%.

Wind energy as an energy sector is present in more than 50 countries of the world. The countries with the highest installed capacity are Germany (18,428 MW), Spain (10,027 MW), the United States (9149 MW), India (4430 MW), and Denmark (3122 MW). A number of other countries, including Italy, the United Kingdom, the Netherlands, China, Japan, and Portugal, have switched to 1000 MW.

So far, wind power has been developing most dynamically in the EU countries, but today this tendency is starting to change. A surge of activity is observed in the United States and Canada, while new markets are emerging in Asia and South America. The installation of wind turbines in the open sea also opens up new horizons for the development of wind energy. Marine wind farms are already beginning to produce results.

A number of factors contribute to the growth of the wind energy market. The combination of these factors prompted the governments of some countries to provide political support for the development of the wind industry. Among these factors are the following items:

- Energy security: The International Energy Agency (IEA) predicts that in the absence of active energy conservation measures, the global energy demand will increase by 60% by 2030 (De Amorim et al., 2018). At the same time, the supply of fossil fuels is declining.

Some of the world's largest economic systems are increasingly forced to rely on fuel imports from regions where conflicts and political instability jeopardize the security of energy supplies.

- Price stability: Compared to fossil resources, wind is a powerful natural source of energy that is readily available in almost every country in the world and does not depend on fluctuations in fuel prices.
- Environmental issues: The need to take urgent measures to prevent further climate change has become an incentive for the development of wind energy. Climate change is recognized as the most serious global issue. The Kyoto Protocol adapted in 1997 imposed on the Organization for Economic Cooperation and Development (OECD) countries obligations to reduce CO_2 emissions by an average of 5.2%. Other environmental problems associated with the use of fossil resources for energy production arise from exploration, development of fossil fuel deposits, oil spills, and radiation exposure. The use of RESs, including wind energy, avoids these risks and threats.
- Economy: The expansion of the global wind energy market has led to a significant drop in the price of energy produced by wind. Modern wind turbines annually produce 180 times more electricity than 20 years ago. At the same time, the kilowatt hour of energy produced dropped by at least 2 times. With a good location, wind power stations can compete on economical indicators with thermal power plants on coal and gas. The competitiveness of wind energy has increased due to higher prices for fossil fuels. In addition, if the hidden costs associated with the impact on the environment and human health when using fossil fuels and nuclear energy were included in the cost of electricity, the production of electricity by wind power stations would be even cheaper. Wind power also provides economic benefits through the creation of new jobs. In addition, in many developing countries, wind energy offers economic opportunities for remote areas that do not have access to power grids.
- Technology and industrial development: Since the 1980s, when the use of the first commercial wind turbines began, there has been a significant increase in the unit capacity and efficiency of wind turbines, and their exterior design has improved. One modern 2 MW-class turbine can produce more than 200 installations of the 1980s. The largest turbines today have a capacity of more than 5 MW and a wind wheel diameter of more than 100 m. Modern wind turbines are made up of modules and are easily erected. Wind stations can be of various capacities—from one to several hundred megawatts.

It can be concluded that wind power has become a big business in the global economy. In response to demand, the major wind turbine manufacturers open multimillion-dollar production facilities around the world. A study of global wind energy resources shows that these resources are huge and evenly distributed across almost all regions and countries. Insufficient wind power can hardly become a factor hindering the development of wind energy in the world.

With the development of wind power, all large amounts of electricity generated from wind will need to be integrated into the global energy network. Wind flow instability is not a limiting factor in this matter. Modern control methods and reserve capacities allow integrating up to 20% of the electricity received from wind farms without problems. Above this level, some changes in energy systems and in the methods of their regulation may be required. Advanced forecasting technologies and the geographical distribution of wind farms contribute to the large-scale integration of wind power.

The possibility of connecting large volumes of wind power generation is demonstrated by the example of Denmark, in which 20% of the total electricity consumption is already provided by wind power. A study by the German Energy Agency (DENA) showed that Germany's wind power industry could triple power generation by 2015, providing 14% of electricity consumption without adding additional reserve and balancing capacity (Renn and Marshall, 2016).

Despite all the advantages, wind power is also not without flaws. The construction and operation of wind power plants, especially in open areas in rural areas, raise issues related to visual and noise effects, as well as effects on local ecosystems. These issues are usually resolved within the framework of the Environmental Impact Assessment. The following main types of impacts can be distinguished:

• Visual impact: Wind turbines are tall structures that make significant changes to the landscape.
• Exposure to flora and fauna: Wind power can have a negative impact on birds, disrupting their habitats, and also causing death by rotating blades of wind turbines. However, studies conducted in Europe and the United States showed that on average the number of collisions of birds with wind turbines does not exceed two cases per turbine per year. These figures can be compared with millions of birds dying annually on high-voltage power lines.
• Noise impact: Compared to highways, trains, and industrial sources, the noise produced by working wind turbines is relatively small. Thanks to improved design and insulation, the latest wind turbine models are much quieter. Regulators should ensure that wind turbines are located at such a distance from nearby homes that avoids unwanted exposure.

All of the existing scenarios for the development of global wind energy estimate the potential of wind energy for the period up to 2050. The report discusses three different options for the development of wind energy: the original, based on figures from the International Energy Agency (IEA); moderate—implying the achievement of existing renewable energy development goals; optimistic—involving the adoption of all necessary policy measures in support of RESs. The proposed options for the development of wind energy are compared against the background of two scenarios of global demand for electricity. In the first case, an increase in consumption is assumed (WWEA, 2015). This prediction is made based on IEA estimates. The second scenario assumes a significant reduction in energy consumption as a result of the implementation of a set of measures in the field of energy conservation.

The results of calculations show that wind power can make a significant contribution to meeting the global demand for clean, renewable energy in the next 30 years. At the same time, the integration of wind energy into the power supply system can be significantly increased, provided that the most serious energy-saving measures are taken.

According to the initial scenario, wind power will provide 5% of world electricity production by 2030 with an increase of up to 6.6% by 2050. With a moderate variant of development, this figure reaches 15.6% by 2030 and 17.7% by 2050. In the optimistic scenario, the contribution of wind energy is 29.1% by 2030 and 34.2% by 2050 (WWEA, 2015). All three scenarios suggest that the increase in the share of wind energy will be achieved through the introduction of new capacity in growing markets in countries and regions such as South America, China, the Pacific region and South Asia.

The annual investment in the wind energy market in 2030 will range from 21.2 billion euros, according to the baseline scenario, to 49.5 billion euros, according to the moderate scenario, and up to 84.8 billion euros, according to the optimistic scenario. It is expected that by 2020 the cost of electricity produced will drop to 3—3.8 eurocents per kWh for wind farms with strong and constant winds and to 4—6 eurocents per kWh for wind farms with wind speeds below average (WWEA, 2015).

The number of jobs in the wind energy market by 2030 will range from 480,000, according to the baseline scenario, to 1.1 million, according to the moderate option, and up to 2.1 million, according to the optimistic scenario.

The reduction of carbon emissions still remains a problem that needs to be solved. It is estimated that the annual emission reduction by 2030 will range from 535 million tons, according to the baseline scenario, to 1661 million tons, according to the moderate option, and to 3100 million tons, according to the optimistic scenario (WWEA, 2015).

An analysis of the calculated data shows that RESs are put at a disadvantage compared to the traditional energy sector due to the absence of damage caused by emissions and other negative impacts when calculating the cost. In addition, the disadvantage of renewable energy is due to unequal conditions in the electricity market due to massive financial and structural support for traditional technologies. In the absence of political support measures, wind energy will not be able to make a positive contribution to solving environmental problems and ensuring energy security. Overall, it appears that for the development of global wind energy it might be necessary to undertake the following provisions:

- State commitment to renewable energy development: The adoption of commitments is an important catalyst in developing the legal framework for the development of renewable energy, including financing, regulating access to the grid, planning and administration.
- Special political mechanisms: National laws should clearly state the market conditions under which electricity produced is sold, including long-term tax programs to minimize investment risks and provide guarantees of return on investment.
- Electricity market reforming: Reforming should be done in such a way as to support renewable energy, which includes removing barriers to market access, eliminating fossil and nuclear fuel subsidies, and taking social and environmental costs into electricity.
- International action on mitigation: It is necessary to set targets for further reducing greenhouse gas emissions in the "post-Kyoto" period (after 2012).
- Reforms of the system of international financing: International financing mechanisms should be aimed at increasing the share of loans to projects related to the development of renewable energy, and early refusal to support projects in the field of traditional, polluting energy.

In a number of countries and regions, the share of electricity generated by wind stations is a real competition to traditional energy: in Denmark, 20% of electricity is generated from wind energy, and up to 35% in northern Germany. In Spain, the fifth largest European country, this figure reached 8%, and by the end of the decade it should increase to 15%. These figures show that wind power is capable of providing a significant share of electricity production without greenhouse gas emissions.

So far, wind power has been developing most dynamically in the EU countries, but today this trend is beginning to change. A surge of activity is observed in the United States and Canada, while new markets are emerging in Asia and South America. In Asia, both in India and in China, record levels of growth are recorded.

Basically, the development of wind energy occurs on land. However, the reduction of free space and the desire for high performance due to the more favorable wind regime of the coastal waters contribute to the promotion of wind farms in the sea. As a result, there are new requirements for wind power plants, including stronger foundations, long submarine cables, and larger turbines. On the other hand, it is assumed that sea-based wind parks, especially in Northern Europe, will provide an increase in the share of wind energy in the energy balance.

Another way to meet the growing demand for wind power in regions with limited territories is to replace old and less efficient wind turbines with more powerful ones. This process is already actively underway in countries where wind energy has been in existence for more than 10 years, such as Denmark, the United Kingdom, and Germany.

The European Union is today the leader in terms of total installed capacity which is around 70% of all installed capacity in the world. The development of wind energy in the European Union was promoted by the policies of the EU member states aimed at promoting the use of RESs. This policy includes a number of financial incentives, including investment grants and preferential tariffs, which ultimately aim at reducing greenhouse gas emissions. By 2010, only wind power has reduced greenhouse emissions in an amount equal to one-third of EU commitments under the Kyoto Protocol.

Germany is the European leader in the field of wind energy. A number of laws, including the Renewable Energy Act, encourage wind power producers through FiTs designed for a 20-year period, with their gradual decline by the end of the period. An illustration of the success of such a mechanism is the large number of small investors attracted to the wind power industry, which has led to a doubling of the annual growth rate.

Projects related to the development of wind energy also benefit from German land laws, under which each local administration must allocate special territories for the development of wind farms. Currently, wind power produces about 5.5% of the electricity consumed in Germany, with an installed capacity of 18,428 MW (Fuchs et al., 2019).

Although the growth rate of wind energy on land begins to decline due to the lack of free territories, this will be compensated for by replacing old turbines with more powerful ones, as well as by building wind farms in the North and Baltic seas. According to a study by the German Ministry of the Environment, the total capacity of sea-based wind turbines could reach 12,000–15,000 MW by 2020 (Amelang and Wehrmann, 2019).

Spain has been rapidly increasing wind power capacity since the mid-1990s, stimulating this process with a national preferential tariff and policies based on the renovation of regional industry. In many provinces, developers have access to construction sites, only fulfilling the condition to promote the development of an industrial wind power base. As a result, for example, the previously backward, but having a large wind potential province of Navarra reached a high level of economic development. The share of wind power in the production of electricity in this province today is 60% (Ramírez et al., 2018).

Denmark was a pioneer in the European wind turbine industry and continues to lead in the share of wind power in the energy mix. By the end of 2010, the total installed wind power capacity in Denmark exceeded 3000 MW. With a strong wind, wind power produces more than half of all electricity in the western part of the country. According to the estimates of the National Operator of the Energy Supply System (Transmission System Operator Energy Net), by 2015 the need for electricity in western Denmark can be met through wind stations and small thermal power plants without centralized generation. In 1990s Denmark was the first to develop wind energy in the coastal area. The country has the largest marine wind park in the world.

The world leaders in the field of wind power join the countries of the "second wave," including Portugal, France, Great Britain, Italy, the Netherlands, and Austria. In Portugal, a strong government policy supported by a fixed tariff system ensured capacity growth from 100 MW in 2000 to over 1000 MW by the end of 2005. In Italy, which announced in 2001 the development of RESs in conjunction with a system of green certificates, the wind power capacity reached 1700 MW (Ouammi et al., 2010).

In North America, wind power originated in California and developed in 31 states. By 2012, 12 wind farms were built, stretching from New York (Maple Ridge, 140 MW) to Washington State on the northwest coast of the Pacific Ocean (Hopkins Ridge, 150 MW). The largest project became a wind farm in Texas "Horse Hollow" with a capacity of 210 MW (Fertig and Apt, 2011). The rise of the wind energy market in the United States is primarily due to the introduction of a 3-year grace period for the wind industry. The grace period is provided at the federal level by the Production Tax Credit (PTC).

One can emphasize a special role in the development of wind energy concessional lending. When the action of PTC in the United States did not resume in a timely manner, the market was significantly disrupted, resulting in years of lull caused by a decline in investor confidence. Today, the American Wind Energy Association (AWEA) is seeking a longer tax benefit period. While maintaining a stable state policy, according to AWEA estimates, by 2020 wind energy will be able to meet at least 6% of the US electricity needs, which is comparable to the existing share of hydropower. And in the longer term, it is possible to achieve a share of wind energy in excess of 20%.

Thanks to incentives at the federal and local levels aimed at developing RESs, Canada's installed wind power capacity has reached 683 MW. This is enough to supply electricity to more than 200,000 households. An important contribution that enlivened the Canadian market was support in the form of the Wind Energy Stimulating Act (WPPI)—the analog of PTC in the United States. Some provinces are also implementing policies that stimulate the development of wind energy projects, including through the placement of orders for the construction of new wind farms. The total volume of such orders reaches 2000 MW. The Canadian Wind Energy Association predicts that by 2015 the total installed wind power capacity will be 8000 MW.

The Asian continent is becoming one of the main centers for the development of wind energy. In 2005 19% of all new capacities were introduced here. With an annual growth of over 46%, the total installed capacity in the region reached almost 7000 MW.

India remains the most powerful Asian market, with more than 1430 MW introduced, bringing the total installed wind power capacity to 4430 MW. This gave India the fourth position in the list of countries actively developing wind power. The Indian Wind Turbine Manufacturers Association (IWTMA) predicts an annual increase in capacity from 1500 to 1800 MW over the next 3 years (Dawn et al., 2019). Incentives for the development of wind power are provided by the Government of India in the form of tax incentives. In addition, the Electricity Act of 2003 in most states established State Electricity Regulatory Commissions, authorized to stimulate the development of renewable energy through preferential tariffs and establishing minimum obligations for distribution companies when receiving a certain share of electricity from renewable sources energy. Fees for wind farms connected to the network vary from state to state.

In recent years, the government and the wind industry have made progress in stabilizing the Indian market, which has contributed to a more intensive inflow of investments from the public and private sectors and has served as a stimulus for the development of a strong manufacturing sector within the country. To date, some companies purchase more than 80% of the components for the turbines in India, which resulted in both more efficient production and the creation of new jobs.

Wind power in Japan is also experiencing growth due to government demand for companies to increase their share of renewable energy. Part of this growth is dictated by market incentives. These incentives include preferential tariffs for the supply of energy from renewable sources and grants for clean energy projects. As a result, Japan's total installed wind power is more than 1000 MW (Cherp et al., 2017).

Despite the low activity in Latin America, some governments, however, develop and implement laws and programs for the development of RESs. In the coming years, the rapid development of wind power is expected here.

High oil prices, electricity shortages, and the problem of air pollution forced the Brazilian government to turn to such sources of energy as ethanol, biomass, hydropower, wind, and sun. Wind power is well combined with hydroelectric power stations operating in the country, especially in the north-east of the country where strong winds, combined with low precipitation, provide a high load factor. According to data published in the Atlas of Winds, published by the Ministry of Mines and Energy of Brazil, the potential of wind power is estimated at 143 GW (Lucena and Lucena, 2019).

Australia has the highest wind energy potential in the world due to the number of open spaces. The increase in wind power capacity in 2012 was 328 MW, which provided a total installed capacity of 708 MW. New projects with a total capacity of almost 6000 MW are under development (Lenzen et al., 2016). The main incentive for wind energy development is the Mandatory Renewable Energy Target, which has a relatively low goal of providing just over 1% of Australia's electricity needs. The Australian Wind Energy Association "Auswind" has called for an increase to 10%.

Wind potential in Africa is concentrated mainly in the north and south of the continent. In the central part of the continent, wind speeds are relatively low. In the north of Africa, in Morocco, a project was implemented, which resulted in the introduction of 64 MW.

In addition, there is a national plan for the introduction of 600 MW of wind capacity by 2013. Tunisia is expected to implement the first project with a capacity of 60 MW. Wind power engineering is developing most successfully in Egypt, where several large wind farms were built on a specially designated area of 80 km^2 in Zafaran (Suez Canal District). Most of them were built with the support of European government agencies for the provision of humanitarian assistance. In the area of the Suez Canal, in Gabal al-Zaita, another 700 km^2 area with an average wind speed of 10.5 m/s was allocated for the construction of a 3000 MW wind park (Aliyu et al., 2018).

The Office of Alternative and Renewable Energy Sources under the Egyptian government expects that the installed capacity of wind farms in the country will increase to 850 MW by 2015, and by 2020−25 up to 2750 MW (Shaaban et al., 2018).

In Russia, more and more attention is paid to the use of alternative energy and the study of an emerging market is becoming highly relevant, which allows you to estimate growth rates, determine the most promising options for investing funds, identify potential opportunities and approximate amount of funds saved using alternative energy sources. According to existing estimates, the technical potential of RESs in Russia is at least 4.5 billion tons of fuel equivalent per year, which is more than four times higher than the consumption of all fuel and energy resources in Russia (Namsaraev et al., 2018). However, it is the geothermal energy (114 million tons of equivalent fuel), small hydropower (70 million tons of thermal conversion) and bioenergy (69 million tons of equivalent fuel) that have the greatest economic potential in the conditions of the Russian Federation (Sayigh, 1999).

Russia has the largest wind potential in the world and its wind energy resources are defined as 10.7 GW. The northwest of the country (Murmansk and Leningrad regions), the northern territories of the Urals, the Kurgan region, Kalmykia, the Krasnodar Territory, and the Far East belong to the favorable zones of wind power development. In general, the technical potential of wind energy in Russia is estimated at more than 50,000 billion kWh/year, the economic potential is 260 billion kWh/year, that is, about 30% of electricity production by all power plants in the country (Marchenko and Solomin, 2004).

Today in Russia there are about 13 MW of installed capacity (0.1% of the total energy generated in the country). The most powerful wind power plant in the Kaliningrad region today is considered to be commissioned in 2002 (the first installation in 1999) and consisting of 21 installations donated by the Danish authorities (all wind turbines are produced by Vestas). Its total capacity is 5.1 MW (Marchenko and Solomin, 2014).

The wind energy market is one of the fastest growing in the world. Its growth in 2012 was around 31%. Russia's share in this market is only 0.013%. At the same time, the potential capacity of the Russian wind energy market is 135 billion rubles by 2013 and 315 billion rubles by 2015. For comparison, the volume of the global wind energy market in 2012 amounted to 2250 billion rubles (Marchenko and Solomin, 2014).

The increase in capacity of wind farms in 2012 amounted to 39 MW. This means an increase of 35% of power per year. China has become the largest market for new installed capacity of wind power units with an indicator for 2012 of 13.75 MW. In the same period, United States installed almost 10 MW and European Union countries around 10.7 MW.

Investments in the construction of large wind farms in Europe today amount to 1200−1400 euros per 1 kW of installed capacity. The cost of energy is 3.5−7 cents per

1 kWh (10 years ago it was 16 cents). With the massive construction of wind farms, you can count on the fact that the price of 1 kWh will significantly decrease and will be comparable to the consistency of electricity generated by thermal power plants and hydroelectric power plants. Projects of wind power plants operating on the network for conditions, for example, very windy Russian Far East, pay off for 5–7 years, the system "wind diesel" for 2 years. In the future, the payback period of wind farms will be reduced (Wang et al., 2016).

Russia has the largest wind energy potential in the world, estimated at 40 billion kWh of electricity per year, so the work of large and small wind farms in the vast Russian spaces could be highly efficient. Such areas as the Gulf of Ob, the Kola Peninsula, a large part of the coastal strip of the Far East, according to the world classification, belong to the windiest zones. The average annual wind speed at an altitude of 50–80 m, where the wind turbines of modern WPPs are located, is 11–12 m/s, while the wind speed of 5 m/s is considered the "golden" threshold for wind power (this is due to the payback of the stations). Nevertheless, despite the favorable natural conditions and the great attractiveness of wind energy, in Russia there are still no huge wind farms or individual wind farms around rural villages and summer cottages. The main reason is the lack of investment and legal framework (Zeng et al., 2017).

China, which has large spaces and a long coastline, has a powerful potential for the development of wind power. The Chinese Institute of Meteorological Research estimates the available wind energy potential on land at 253 GW. In addition, 750 GW can be provided through the construction of wind farms in the shelf zone. The first wind farm in China appeared in 1986 as a demonstration facility. By the end of 2012, the total capacity of wind turbines located on the mainland reached 1260 MW, showing an annual growth of 60% (Liu et al., 2017).

The Chinese government's policy is to support local wind turbine manufacturers in order to reduce production costs and make wind power competitive to fossil fuel energy. At present, coal-fired power plants dominate China's power industry, causing air pollution and many other environmental problems. To create a national wind turbine industry, the National Development and Reform Commission put forward the idea of wind energy concessions for large-scale commercial development. With the formation of concessions, local authorities invite both foreign and domestic investors to build on a competitive basis wind parks with a capacity of 100 MW. One of the conditions imposed by the authorities at the same time is that 70% of the components must be made in China.

The Chinese wind energy market received an additional powerful incentive thanks to the entry into force of the Renewable Energy Sources Act in early 2006. The law defines national targets for the development of renewable energy and establishes a tariff support system. As a result, a large number of international companies have created joint ventures with Chinese firms that are engaged in the manufacture of components and assembly production (Karltorp et al., 2017).

The current goal of wind power development in China is to reach 5000 MW by the end of 2015. A more distant goal, identified by the Chinese government in the long-term plans until 2030, is the introduction of 30,000 MW of new capacity. It is expected that by the end of 2030, to meet growing demand, the total capacity of all sources of electricity in China should reach 1000 GW (1,000,000 MW). Thus the electricity generated by wind energy, will be 1.5% of all electricity produced (Weisheng et al., 2016).

In China, wind power companies face high barriers to entry. The concentration of wind power equipment manufacturers in the market is very high. At the same time, the 10 largest equipment manufacturers occupy 96% of the market, four of them took control of 75% of the market. The lion's share of the Chinese market is occupied by foreign manufacturers of wind turbines. The installed capacity of foreign manufacturers of wind turbines in the total installed capacity amounted to 65.92% and the domestic manufacturers of total wind turbines constitute 34.08%. The largest foreign manufacturers in the People's Republic of China (PRC) market are VESTAS, Gamesa, GE WINDNORDEX, or Suzlon (Oh, 2015).

Currently, China is pursuing a policy of encouraging the development of decentralized networks based on autonomous sources of wind energy. The Standardization Commission of the State Council Office conducted an audit of power grid technologies in accordance with national standards, the design standards for large wind power plants were submitted to the National Energy Council. Particular attention is paid to the development of wind turbine technology for low wind speeds.

According to industry statistics, the region of low wind resource available in the country is about 68% of the country's territory, and it is as close as possible to the load grid. Of the 100 million kW of energy, 20 million kW will belong to low wind speeds.

The China Renewable Energy Association's professional committee said the development of low wind speeds is still in its infancy. At the same time, competition in the market for wind turbines with a high level of wind resources is very high, which will inevitably lead to the development of low wind speeds. This is a fundamental trend in the development of the wind energy industry in the coming years.

At the moment, a number of manufacturers of wind turbines have focused on the development of technologies for low wind speeds. The first low-speed wind power station was built in Anhui province. The entire project, with a total investment of 30 billion yuan, consisted of four stages of construction, installed 132 units of 1.5 MW wind turbines with a total capacity of 198 MW of generating capacity, is expected to reach 400 million kWh (Zhao et al., 2010).

In addition to the national grid, firms such as Huadian, Trail Yunxiao began to invest in low wind speeds, creating wind farms in Chenzhou City, Hunan Province.

Wind power in China in terms of technology development has made a big breakthrough. The producers of wind power equipment Goldwind, Sinovel through the development of innovations could independently develop a wind turbine with a capacity of 1 MW (Dai et al., 2018).

Despite the fact that in recent years the development of wind energy projects has been actively supported by the state, until now the connection of wind turbines to the network is quite a laborious process and seriously restricts wind turbine manufacturers. For example, the company Sinovel Technology Co., Ltd. carried out commissioning work and did not start producing wind turbines due to the fact that the issue of connecting to the network was not resolved. But still this issue is gradually being resolved. Recently, the State Grid Corporation connected the first wind installations to the consumer energy networks. At the same time all the same rules apply to wind energy as to other attached sources.

Wind power is expected to compete in retail electricity markets for the right to provide households with electricity, which will promote tighter competition between wind power producers.

The problem of interfacing new types of energy with networks was a stumbling block and severely limited the development of the industry. The tariff policy and conditions for the acquisition of wind energy were developed, which partially resolved the issue of delays in the access of energy to the network. But still the question remains on the agenda. On the one hand, the pace of development of wind energy as "green energy" is higher than expected and, on the other hand, the modernization of power grids is seriously lagging behind the development of the industry.

According to Guodian Longyuan Electric Power Group Corporation, it is necessary to take active measures for the construction and development of new electrical networks. The growth rates of the input capacity of wind units are forcing to develop, as soon as possible, plans for the development of the energy system, taking into account the large-scale development of wind energy. The solution of this issue is focused on creating 1 million kW and 10 million kW of wind electrical network (Guo et al., 2018).

The new energy policy is not perfect; there are many practical tasks in developing new energy sources, such as wind energy. With a lack of unified planning, investments are often made blind. It is assumed that the government will pay special attention to new energy in the medium and long-term development of China's energy sector.

Since the issue of connecting to the general networks of wind installations has not yet been fully resolved, the growth rate of investments in network companies aimed at the development of network infrastructure is rather low. According to the China Wind Energy Association, on the one hand, the government has obliged wind power producers to connect their facilities to common networks and, on the other hand, appropriate compensations and benefits have not been developed to encourage enterprises to develop power grids (Li et al., 2019).

The introduction of grid technologies in the national energy network has led to the fact that the threshold for wind energy is set according to the requirements of the power system. This situation leads to a further increase in the complexity of the wind as an energy system.

Also, for the development of a national network based on new energy, new higher technical standards are needed, which will balance production, transmission, and distribution. The skill level of the energy industry workers should also reach a new level. The new standard should encourage and support wind energy transmission networks, and not just be an increase in the transmission threshold.

After the introduction of the "New Rules" of the national grid constraints, the connection of wind power producers to the network will be associated with even greater obstacles. In this case, most likely, the income of wind farms will be seriously affected. The "new rules" of the restrictions of the national network concern two main aspects: (1) wind power quality to meet energy needs and (2) fault tolerance of the turbine and power electronics. These technical standards are difficult to achieve for most domestic wind producers.

Despite this, some experts believe that "high standards and strict requirements" are useful for the future development of wind power producers in the PRC. In this way, wind power developed in Europe and the United States. But it must be kept in mind that in these countries the special nature of wind energy has always been emphasized, and the technical standards of the network and the certification system were specially composed for it.

The redundancy of wind power in industry led to fierce market competition. The introduced high standards gradually lead to the withdrawal of backward wind energy companies from the market. At the same time, there are more than 70 equipment manufacturers in the country. Only a few companies this year, Goldwind, Sinovel, Dongfang electrical output reached 12 million kW (Dai et al., 2018).

The unprecedented development of the production of wind power equipment tends to overheat, the industry as a whole must move from the pursuit of quantity to the pursuit of quality. The undeveloped sphere of R&D and the strategy of companies to acquire ready-made foreign technologies create a very limited space for the development of the industry. Among domestic manufacturers, the bulk of the market will continue to be concentrated in the hands of Goldwind, Dongfang Electric, and Sinovel.

The domestic market of China's wind turbines is mainly determined by international manufacturers, most local manufacturers of wind turbines are at an early stage of development, causing a higher degree of concentration. Domestic players in the domestic market include domestic manufacturers such as Goldwind, Yunda, Sinovel, Dongfang Steam, Hafeiweida, Acciona, Shanghai Electric Wind Power Equipment Co., Hunan Xiangtan Electric Wind Power Co., and Shipbuilding company Huayi commissioned wind farms on the offshore platform (Klagge et al., 2012).

The gradually growing demand for wind energy due to the high growth rates of the PRC economy leads to the technological development of generating equipment. Since 2003, the installed capacity of generation plants is constantly growing. The total installed capacity of wind power plants in China in 2010 reached 5 million kW, it is planned that by 2020 it will be 30 million kW (Dai et al., 2018).

The need to support the national industry, energy shortages and environmental conditions, leads to increased government attention to wind power. This is also due to the fact that gradually the costs of installation and operation of stations is reduced. It can be said that the wind energy industry will maintain steady growth.

All in all, in China, wind power companies face high barriers to entry. The concentration of wind power equipment manufacturers in the market is very high. At the same time, the 10 largest equipment manufacturers occupy 96% of the market, four of them took control of 75% of the market.

The lion's share of the PRC market is occupied by foreign manufacturers of wind turbines. As mentioned before, foreign manufacturers of wind turbines yield over 60% of installed capacity, while the domestic manufacturers have about 35%. The largest foreign manufacturers in the PRC market are: VESTAS, Gamesa, GE WINDNORDEX, and Suzlon.

Currently, China is pursuing a policy of encouraging the development of decentralized networks based on autonomous sources of wind energy. Particular attention is paid to the development of wind turbine technology for low wind speeds. According to industry statistics, the region of low wind resource available in the country is about 68% of the country's territory, and it is as close as possible to the load grid. Of the 100 million kW of energy, 20 million kW will belong to low wind speeds.

Wind power in China in terms of technology development has made a big breakthrough. The producers of wind power equipment Goldwind, Sinovel through the development of innovations could independently develop a wind turbine with a capacity of 1 MW.

Even though the development of wind energy projects receives immense state support in China, the issue of connecting the wind turbines to the power networks still remains a cumbersome process. For example, the company Sinovel Technology Co., Ltd. carried out commissioning work and did not start producing wind turbines due to the fact that the issue of connecting to the network was not resolved. But still this issue is gradually being resolved. Recently, the State Grid Corporation connected the first wind installations to consumer power grids. At the same time all the same rules apply to wind energy as to other attached sources. As it becomes apparent from using simple approximations, wind energy will compete in retail electricity markets for the right to provide households with electricity, which will promote tighter competition between wind energy producers.

4.5 Hydro energy

Abundant water resources and their use for the production of electricity make some countries the top countries with the most developed renewable energy, which includes hydropower, engaged in the conversion of the energy of flowing water into electrical energy.

Despite the giant power stations and huge dams that led to the creation of real inland seas and are known to most of our compatriots on TV news, history, and geography, many countries use their hydropower potential rather modestly. According to the installed capacity of hydropower plants (46.7 GW), the Russian Federation ranks fifth in the world after China (260 GW), Brazil (85.7 GW), the United States (78.4 GW), and Canada (76.2 GW). The country also occupies the fifth place in terms of the annual production of electricity by hydroelectric power plants (Denisov and Denisova, 2017). However, the absolute world leader in hydropower share in the energy balance is the "oil and gas supper power," namely Norway. Despite the raw material wealth, 96.7% of electricity is produced here by hydroelectric power plants. Norway seeks to trade on the world market not only with hydrocarbons, but also with clean electric energy, realizing, for example, a project on the laying of a 700-km cable for the export of energy to the United Kingdom.

The modern hydropower industry has been laid in the first half of the 19th century, when the first water turbine was developed and described. In 1834 the French engineer Forneron presented the first working sample of its water turbine. In 1848 British engineer James moved to the United States and perfected the hydraulic machine by creating a radial axial "Francis turbine." His mathematical methods of calculation and drawings allowed him to design high-efficiency turbines taking into account specific conditions of water flow. Francis turbine is now the most common type of turbines. It is believed that for the first time the energy of water for the production of electricity was used by the English industrialist William Armstrong in 1878. The power station served to power a single electric arc lamp in his art gallery (Constant, 1983).

The first "full-fledged" power plant was launched in 1882 on Fox River in Appleton, Wisconsin, United States. By 1886, about 50 hydroelectric power plants were already operating in the United States and Canada. The rapid development of industry in conjunction with the transition to electric lighting at the turn of the century led to a boom in hydropower. In 1889 the number of hydropower plants in North America increased to 200. In 1920s and 1940s hydropower provided up to 40% of electricity production in the United States, but nowadays its share is around 6%–8%. In 1973 hydropower accounted for 20.9% of world electricity production, and in 2012 this was 16.2% (Devine, 1983). In Russia, the largest country in the world, hydroelectric power plants currently produce 16%–18% of the electricity produced in the country, which roughly corresponds to the global level.

Hydropower is important not only for the production of electricity. A special subspecies of hydropower plants—pumped storage power plants (PSPPs)—have the important function of covering peak loads of the electrical network and stabilizing it. Large hydropower is highly capital intensive. It is quite difficult and hardly advisable to calculate the average global size of the specific capital expenditures during the construction of hydroelectric power plants, since each construction is characterized by its own unique geographical and technological conditions. It is obvious that the creation of large dams is associated with enormous financial and time costs. The specific capital costs for the construction of some dams are very high. For example, Russian "Bureyskaya" hydropower plant being constantly built from 1976 to 2003 until it was finally ready for operation.

There is a large variation in specific capital costs for the construction of large hydropower plants—they range from $1050 to $7650 per kW of installed capacity. The US Department of Energy provides an indicator of $2963 per kW (for a conventional station with a capacity of 0.5 GW). In countries like China or Russia, it is generally believed that the specific capital investment in generating capacity for large hydropower plants 1.5–2 times higher than for thermal coal stations. At the same time, the subsequent operation of hydropower plants and power generation is not related to the use of hydrocarbon fuels, respectively, the cost of electricity and the reduced cost of electricity production, taking into account the long life of the stations, are at a rather low level. The IRENA study mentioned above also leads to a fairly wide cost range of $0.02–0.19 per kWh.

Despite renewable nature, large hydropower is a rather dangerous endeavor. Any major human intervention in the environment in an attempt to subdue the forces and energy of nature might result badly. The terrible disaster at the Sayano-Shushenskaya hydroelectric station in Soviet Union, which killed 75 people, is still fresh in everyone's memory. But the largest man-made disaster so far is the breakthrough of the Banqiao dam in China in 1975 that in total killed about 170,000 people (Jing, 1997). In addition, the construction of large hydropower plants in many cases entails significant, but hardly measurable negative environmental and social consequences.

It is easy enough to calculate the amount of CO_2 that was not released into the atmosphere as a result of the operation of hydroelectric power plants, but it is very difficult to estimate the costs of the ecosystem as a whole. Thus the low cost of electricity generated by hydroelectric power plants has, forgive the tautology, its own, and sometimes high, price. Most of the environmental and social damages from the existence of hydropower plants are not taken into account in the price of hydropower. Capital costs at best include the cost of forest harvesting, compensation payments for land alienation, the cost of environmental protection, calculated according to existing standards. The latter are enormously underestimated, and the environmental damage is not limited only to the loss of pasture or field. Reducing the area of forests, steppes, and other natural resources is an irrecoverable loss, as they are lost forever.

For this reason, it is common to separate the impact of the large hydro power plants associated with their extensive impact on the environment from the small hydro power installations. According to the international classifications, power plants with an installed capacity of not more than 30 MW belong to the small hydropower industry. It is quite peculiar that this makes it precisely fitting into the category of RESs (according to official definitions designed and used for the purposes of state support programs for renewable energy). In general, this is also justified from an environmental point of view, since small hydroelectric power plants obviously destroy the environment to a lesser extent.

The question is how the hydropower is going to evolve and develop in the future. There is no doubt that it will grow. This will be happening hand in hand with the tightening of various domestic and international policies aimed at reducing greenhouse gas emissions and the mitigation of the climate changes. Nevertheless, it appears some rapid growth should be not be expected here due to many different reasons. One of the reasons is that many developed countries are now on the brink of exhausting their hydropower potential. For example, in Switzerland, the available hydropower is used up by almost 90%, while in Norway and Canada it has been exploited up to 70% with no additional capacities left.

According to experts, in the near future, power generation at hydroelectric stations will increase. This will occur mainly in regions with decentralized electricity supply due to the commissioning of new small hydropower plants, which will replace outdated and uneconomic alternative sources of electricity (Huang and Yan, 2009; Ardizzon et al., 2014).

In order to generate electricity, hydropower plants use renewable primary energy carrier—the energy of water flow, which practically excludes the cost of fuel production in the structure of energy production costs. The technological process of electric energy (EE) production at hydroelectric power plants is highly automated, which allows operating personnel to be significantly less than at a thermal or nuclear power plant of similar capacity.

An important advantage of hydropower is the high maneuverability of hydraulic units, the output of which to full capacity takes less than 1minute. Therefore hydropower plants in the power system are usually used to generate electricity, providing coverage for the load curve, especially in its peak part, controlling the frequency of the electric current and as a reserve. Decentralized power supply zones, which are located, as a rule, in remote, inaccessible areas, are the most competitive areas for the use of small hydroelectric power plants (HPPs).

Like any other method of energy production, the use of small and mini hydro has both advantages and disadvantages. Among the economic, environmental and social benefits of small hydropower facilities are the following. Their creation increases the energy security of the region, ensures independence from fuel suppliers located in other regions, and saves scarce fossil fuels. The construction of such an energy facility does not require large capital investments, a large amount of energy-intensive building materials and considerable labor costs, it pays for itself relatively quickly. In the process of generating electricity, hydroelectric power plants do not produce greenhouse gases and do not pollute the environment with combustion products and toxic waste. Such objects are not the cause of induced seismicity and are relatively safe in the event of natural occurrence of earthquakes. They do not have a negative impact on the way of life of the population, the animal world and local microclimatic conditions.

Possible problems associated with the creation and use of small hydropower facilities are less pronounced, but they should also be mentioned. Like any localized energy source, in the case of an isolated application, a small hydropower facility is vulnerable to breakdown, leaving consumers without power supply (the solution is to create joint or standby generating facilities—a wind turbine, a co-generating biofuel mini-boiler house, PV installation, etc.).

The most common type of accidents at small hydropower facilities is the destruction of the dam and hydraulic units as a result of overflowing through the dam crest with an unexpected rise in the water level and failure of shut-off devices. In some cases, small hydropower plants (SHPPs) contribute to sedimentation of reservoirs and influence the channel-forming processes.

There is a certain seasonality in the production of electricity (noticeable drops in winter and summer), leading to the fact that in some regions small hydropower is considered as a backup (duplicate) generating capacity.

Among the factors hampering the development of small hydropower in the world, most experts cite incomplete awareness of potential users about the benefits of using small hydropower facilities; insufficient knowledge of the hydrological regime and the flow volumes of small streams; low quality of existing quality assurance standards, which might become the cause of serious errors in calculations; the lack of development of methodologies for assessing and predicting the possible impact on the environment and economic activity; weak production and repair base of enterprises producing hydropower equipment for small hydropower plants, and mass construction of small hydropower facilities is possible only in the case of mass production of equipment, abandonment of individual design and a qualitatively new approach to the reliability and cost of equipment compared to old facilities decommissioned.

Energy development of independent small rivers and watercourses that are tributaries of large rivers should be based on previously developed schemes for the integrated use of water resources in the river basin, including environmental protection measures, and schemes for the energy use of the watercourse.

Medium and small streams are more vulnerable when they are intensively explored by extensive methods. The deterioration of their condition occurs in both quantitative and qualitative indicators. The composition of water protection zones includes floodplains of rivers, floodplain terraces, edges, and steep slopes of the native banks, as well as beams and ravines directly entering the river valley.

As a rule, the geography of hydropower development and environmental monitoring of small rivers is reduced to a riverbed (water course). It must be remembered that small rivers, like large ones, are formed at the expense of smaller watercourses, streams, and underground keys. Therefore the width of the river valley should be clearly highlighted.

Of great importance for maintaining the water content of small rivers and preserving their purity is the woodenness on the spill-over basins and riverbanks. Water protection near-forest forest belts protect the banks from destruction, and the rivers themselves—from pollution, helping to reduce evaporation from the water surface. When organizing nature conservation areas, summer pastures and livestock farms are taken out of them, plowing of land, the use of toxic chemicals and fertilizers, cattle grazing, construction of tent camps, and recreation facilities are prohibited within these limits.

The environmental negative effects of small hydropower plants are qualitatively the same as those of medium and large, but the absolute values are less noticeable, especially those related to changes in hydrological and climatic conditions.

The most effective way to preserve the water content of small rivers is the construction of reservoirs, ponds, and channel dams with weirs. The height of such dams is usually not more than 3 m, their number is from 2 to 8 along the length of the river. In this case, the water mass of the reservoir still causes some geological changes.

Changes in the hydrological regime of a small watercourse can be enhanced due to the increasing anthropogenic influence—water use and water consumption have increased. The greatest anthropogenic load on small rivers is noted where they flow through populated areas. Moreover, the impact on small rivers is due to human activity on the transformation not only of the watercourses themselves, but also of their catchment basins. On the watersheds, this is the reduction of woody vegetation, plowing of land, the growth of villages, on the rivers—increasing water abstraction for economic needs, collection of waste and return water, recreation, irrigation, water supply, and land-improvement measures. It should be remembered that on small rivers there is a close relationship with the natural environment.

The protection of the water resources of small rivers should be the predominant link in their energy development. The basis of the decision on the use of certain small watercourses is usually based on the results of long-term observations (monitoring) of the state of the environment in a given area. It is very important that the information obtained during the monitoring process includes all the parameters necessary for the development of a specific energy system. Partially such information contains the results of meteorological observations. Environmental monitoring of small energy facilities should be closely associated with a set of environmental measures at the design stage. The environmental impact assessment of small hydropower facilities consists in the collection and study of the following hydrometeorological characteristics:

- level of water in the watercourse
- speed of wind
- water circulation schemes under various hydrometeorological conditions
- patterns of distribution of areas of predominant exposure to wind waves and currents within the small watercourse basin
- schemes of soil distribution in the reservoir, as well as intensity and direction of lithodynamic processes
- degree of economic activity within the waters (routes and intensity of navigation, work associated with changes in the topography of the day, etc.)
- schemes of distribution of recreational zones (reserves, sanitary zones)

Accurate implementation of environmental protection measures will undoubtedly promote the development of small hydropower, which will require appropriate restructuring of the design, first of all, revision of river utilization schemes, selection of retaining levels of hydroelectric facilities, and, most importantly, determination of the installed capacity of small hydropower plants.

All in all, it appears that the construction of small hydropower plants envisaged to supply electricity to isolated consumers partially eliminates the use of diesel generators. It will also reduce emissions of harmful substances from diesel generators into the environment and help save fossil fuels. But at the same time, it is necessary to strictly take into account the requirements of a socioecological nature, modern opportunities for using energy from small streams, a steady upward trend in fossil fuel, the relevance of improving the efficiency of using the potential of all types of renewables and small hydroelectric power plants.

All of the above explains the popularity of the SHPPs in many countries, for example, rural China, where it is used both by farmers to generate cheap electricity and by computers and technical specialist who need lots of cheap electricity for producing bitcoins. Allegedly, there are thousands of bitcoins "troll farms" high above in the mountainous regions of China where new bitcoins are mined.

4.6 Battery storage

The inability to store electricity, especially in combination with intermittent renewable energy, leads to the need for reserves of generating capacity, network capacity, as well as fuel reserves at power plants. The amount of reserves is standardized, and the cost of maintaining reserves is included in the electricity bill of consumers and distributed among all end users. Three traditional types of power plants dominate the market: nuclear (NPP), thermal (TPP), and hydroelectric (HPP). For safety reasons, NPPs do not regulate their load. Hydroelectric power plants are much better suited to work with an uneven load schedule, but they are far from being present in every energy system, and if there are, then in small quantities. As a result, most of the irregularity of daily electricity consumption is covered at TPPs. This, as a result, leads to operation in an inefficient mode, increases fuel consumption, lowers the efficiency of the plant and, as a result, the final cost of electricity for consumers.

ESS constitute an important component for the subsequent development of energy systems. An ideal design of an intelligent electrical network model is impossible without the inclusion of ESSs (Kyriakopoulos and Arabatzis, 2016). The electrical network exists in a state of constant balance between generation and demand. If the generation does not meet the demand, then the resulting frequency deviations in the network from its normal state can lead to equipment failure and power outages. When comparing the price of electricity at night and daytime hours, the difference can reach up to 80%. This makes drives an interesting option, since electricity can be generated at night and then used during the day during peak demand. Drives also help reduce peak energy demand from daily loads, thus helping to avoid the use of expensive power generation, for example, with a gas turbine. There are many possible applications for energy storage that can be implemented in a modern networks and grids:

- application in power supply
- ancillary services
- network support
- integration of RESs
- participation of end users

The main objectives of the introduction of energy storage in the network are a more flexible ratio of various components of the electricity market, power supply safety, the possibility of introducing more RESs, sustainability, the ability to design smart grids, and avoid network failures.

The integration of such energy storage devices into mutually integrated systems (with the ability to control the volumes of mutual energy transfer, frequency, and load in electric networks) will become the main task of maintaining stable operation of energy networks in the future. In general, the ESS market is already estimated at $100 billion and, according to expert forecasts, by 2040 its volume will increase to $250 billion. The total energy storage capacity by 2030 will reach 25 GW (Schmidt et al., 2017).

Since the end of January 2016, there are already two ESS in operation in South Korea (with a capacity of 9 and 6 MWh). In the Netherlands, 20 MWh of energy storage is already integrated into the national grid. In 2016–17 in Germany it is planned to connect six ESSs with a total capacity of 90 MW (Minkin, 2017). In addition, in the United Kingdom in the autumn of 2016, a tender was held for $86 million for the construction of eight energy storage facilities based on lithium-ion batteries (LIA), which will be combined into a single system with a total capacity of 211 MW in the future (Kouchachvili et al., 2018). The reasons for the emergence of a new approach to the electricity market are associated with the following factors:

- constant increase in the price of electricity throughout the world
- the need to improve the energy and environmental efficiency of the power industry
- growing consumer demands for reliability and quality of power supply
- changes in working conditions in the electricity and capacity market

The use of energy storage devices in Russia is not yet widespread due to the economic inefficiency associated with the cost of energy storage. Nevertheless, there is a certain price variation in prices in the electricity market between peak and off-peak prices, which can potentially allow the use of storage devices. There are several research centers in Russia, such as the Skolkovo Innovation Centre, the Scientific and Technical Center and the InterRao Energy Efficiency Centre, which study electrochemical batteries and supercapacitors. There is a major manufacturer of lithium-ion technologies, Liotech, offering options for both industrial customers and solutions for the introduction of power systems. Together, FGC UES and ENERZ jointly developed two projects of stationary batteries, which serve to reserve power supply for their own needs of consumers of particularly important facilities. There is also one operating PSPP owned by RusHydro. The remaining power stations are in the planning, development or construction stage (Antipov et al., 2019).

Currently on the market there are many possible solutions for the storage of electrical and thermal energy. Various factors create a difference in their performance. Such factors are technology, production, application, etc. The discharge duration varies significantly from milliseconds to several hours. Accumulators are an important component for the development of the energy system. The number of ESS began to increase with the increase in the number of RES integrated into the grid (Dusonchet et al., 2019). The simulation of intelligent active-adaptive networks cannot be fully performed without the inclusion of energy storage devices. Further development of time-varying variables (depending on the weather: wind and sun) energy sources is limited to the percentage of accumulators in the grid. Otherwise, sustainability cannot be maintained.

The most developed technology for storing mechanical energy is PSPP. Large pumps are used to pump water into the tank, at the point in time when electricity is the cheapest, and power can be delivered back through the hydroelectric plant at peak times. To ensure the operation of the plant, a height difference between the two reservoirs is required.

PSPP has the highest installed capacity in the world, and the number of plants continues to grow. The technology has been tested by time and demonstrated its effective performance. Unfortunately, certain conditions are necessary for its operation and placement. The first factor is the height difference between the two tanks. Another problem is densely populated areas where, because of the people living there, it is impossible to establish a PSP.

In addition to PSPs, two other energy storage technologies are also being vigorously developed, as evidenced by the growing number of projects. The largest number of plants that are being introduced at the present time are electrochemical. The leader among electrochemical accumulators is LIA. The main advantages of electric batteries are fuel independence, environmental benefits, fast response to changes in load, increased system stability, low losses in standby mode, and high efficiency. However, disadvantages include low energy density, low power rating, high operating costs, short life cycle, and most batteries contain toxic cells.

The third type of storage devices is thermal ESSs. Thermal storage devices are like batteries, but instead of storing electricity, they store heat or cold in a suitable container. They could be easily applied to utilize the waste heat generated in the industry. Unfortunately, they still suffer from high investment costs, large space requirements, and low efficiency, which make many thermal storage devices unprofitable.

Some technologies have future development prospects that correlate with research and development (R&D) and mass production, while some of them have already reached their potential and have few opportunities to reduce prices and improve performance in the future.

Most electrochemical energy storage devices are batteries. They are directly rechargeable batteries, where during the charging process, electrical energy is converted into chemical energy, and, conversely, during discharge, chemical energy is converted into electrical energy and fed back into the network. The increase in the number of batteries is due, above all, to falling prices for them and, at the same time, an improvement in properties. In addition, the increasing number of installations of RESs, especially wind energy, in the electricity network and the development of the concept of intelligent networks contribute to a new approach to the use of batteries.

LIA have the largest share of new projects and storage facilities. Batteries, known to the general public mainly as cell phone batteries, are used in versions with low power. In recent years, batteries have fallen dramatically in price, and according to forecasts, this trend will continue in the future. As an analogue of LIA technology, lead-acid batteries can also be used to store large amounts of energy for stationary applications, and they are already commercially available and used (Pires et al., 2014).

Flow cells are a type of battery that consists of two electrolytes that circulate through an electrochemical cell. The main characteristics of the flow cells are high capacity, long life expectancy, fast response time, and low efficiency. Flow-through batteries have the ability to change the charge mode to discharge mode for 1 millisecond. The energy required to circulate electrolyte and the losses caused by chemical reactions lead to low efficiency. One of the advantages is that the system does not self-discharge, since the electrolytes are stored separately and cannot interact with each other. This allows the flow-through batteries to remain a competitive technology capable of contributing to future power grids.

The main advantage of sodium-based batteries is that they can be stored with zero charge compared to lithium-ion batteries, which require a minimum of 20% charge during storage. However, the sodium battery does not have other advantages over similar options and is not expected to be used in the future (Palomares et al., 2012).

For LIA, there are various possible variants of execution. They differ in terms of technology and application as well as price. Each technology has specific pros and cons. For example, a lithium—cobalt oxide battery has advantages in specific energy and relatively low cost, on the other hand, it has disadvantages such as limited service life and the content of hazardous and toxic chemicals. Batteries are known to serve different aspects of a person's life, so the scope varies from low-power applications to portable devices to maintaining backbone networks. While lithium—nickel—cobalt—aluminum batteries find their use in electric vehicles (such as well-known Tesla), power tools, etc., a type such as a lithium—iron—phosphate battery is more suitable for storing energy from renewable, as stationary batteries or high-power applications (Tie and Tan, 2013). The price of various types of technology differs depending on the cost of the cathode material, its inherent characteristics, level of development in this field, etc.

Energy density and power density have an important difference. Energy is the amount of charge in a battery which is usually expressed in watt hours. On the other hand, power is an instantaneous measure of how much energy passes through a circuit and is usually expressed in watts. The main problem with power is that it does not inform the reader how long the battery can be used. Energy density indicates the nominal energy of a battery per unit mass (W/kg). Thus if a system has a high energy density, then it can store a lot of energy in a small volume. The power density illustrates the maximum power per unit volume (W/L) which shows how quickly energy can be delivered for a given volume. It should be noted that power density shows power at 1C (power factor) per unit volume and that batteries usually do not operate at this rate.

Various sources that consider the use of energy storage devices offer basically the same classification. Applications vary by location on the network where they are used: in generation, in transmission, in distribution, or at the client level. The main function of energy storage as support for traditional generation is to optimize the operation of existing assets of power plants. It involves switching the generator, providing the necessary load at the moment when one stops and the other starts. While the generator is running, the battery gives the generator enough time to cope with the change in load (Lucas and Chondrogiannis, 2016).

Arbitrage electricity prices are one of the most common applications for energy storage. Arbitration happens when a participant is making a profit by purchasing a product at a relatively low price and selling this product when the price is quite high. This definition includes initial investment does not require simultaneous buying and selling, and allows the use of different products (Zavrsnik, 1998). Arbitrage is conducted due to the difference between the prices for peak and off-peak hours. Off-peak prices are typical when demand is low and sources of electricity with a lower electricity price are used. At peak hours, demand is highest, so more expensive sources of power generation are needed. The use of arbitration at the network level allows shifting the peaks and adjust the price of power.

The relationship between the use of drives and their technology is complex—one technology is not suitable for each application. For example, when leveling the load, the best solution is an hydropower storage plant (HPSP). However, when frequency fluctuations or peaks are removed, it makes no sense to use PSPs because of its size and response time. Thus the choice of storage technology is very complex, and the approach varies depending on the location, purpose, etc. The use of network level drives in the power industry contributes to the following effects:

- reduced electrical losses
- increasing the life cycle of generating equipment, by equalizing the load curve and reducing the number of starts and stops
- improving the reliability and stability of the power system
- stimulating the development of alternative energy
- reducing the load on the electrical network
- reduced consumption peaks
- ensuring the quality and reliability of power supply

In addition, the economic effect is achieved by:

- priced consumption, in case of active participation of the end user
- the arbitration effect due to differences in the tariffs for the purchase of electricity during the hours of minimum loads and their sale during peak hours

In addition, in case of emergency outages in the network or power plants, so-called reserve capacity and electrical power are provided. Thus there is no need to maintain an expensive power reserve in power plants or in a network reserve. Additional economic effect is achieved due to:

- frequency management support services
- maintaining voltage levels at installation sites
- creation of local intelligent networks

Furthermore, there is given a description of energy storage technologies that are classified by the type of stored energy. First, there is a mechanical energy storage. Mechanical ESSs typically work by converting electricity into various forms of energy. It is distinguished by three main technologies: PSP, compressed air-based accumulators, and flywheel-based accumulators. Many types of these structures are characterized by extreme simplicity and a virtually unlimited service life and storage of stored energy (Chen et al., 2009).

For the work of the PSPP, two reservoirs are needed, divided vertically. First, water is pumped into a higher reservoir during off-peak electricity demand. After that, the water can be released back to the lower tank. In the cycle, the water pushes the turbine during unloading and the pump when the charging process takes place. The turbine is connected to the generator rotor and starts it, which in turn supplies electricity to the transformer.

In case of the compressed air energy storage, the air is compressed when energy consumption is low and is stored in a large tank, which can be either natural or artificial. When energy is required, air escapes from the reservoir and controls the gas turbine generator.

Another technology is represented by the kinetic mechanical drive (flywheel). In the kinetic accumulators at the input is used electrical energy, which is then stored in kinetic form. When stored energy is required, the flywheel is slowed down to convert kinetic energy into electricity using an integrated motor/generator. Flywheels usually work in a vacuum to reduce the effect of traction (Hebner et al., 2002).

In addition, there is an electrochemical energy storage. The first type of technology is the rechargeable battery. The chemical energy contained in the active material is converted into electrical energy. EES consists of several cells that can be connected in series or in parallel, where the first type of connection changes the voltage at a constant capacity, and the second changes the power, but at a constant voltage. Each cell contains two electrodes (anode and cathode) with an electrolyte between them (Lukic et al., 2008).

The second type is flow battery. A flow accumulator stores energy in two soluble redox vapors contained in external tanks with liquid electrolyte. Electrolytes are pumped from tanks to a cell consisting of two electrolyte flow cells separated by ion-selective membrane. The operation of a flow-through battery is based on redox reactions of electrolyte solutions. During the charging phase, one electrolyte is oxidized at the anode, and the other cathode is reduced at the cathode, so electrical energy is converted into the chemical energy of the electrolyte. During discharge, the above process goes in the opposite direction (Cunha et al., 2015).

Then, there is an electric energy storage. The first type is represented by the supercapacitor. In general, supercapacitors accumulate energy in an electrostatic way, polarizing the electrolyte solution. Also known ultracapacitors, also called "high-capacity." During the accumulation of energy in a supercapacitor, no chemical reactions occur, despite the fact that the supercapacitor is an electrochemical device. Supercapacitors can be charged and discharged thousands of times due to the high reversibility of the energy storage mechanism. A supercapacitor is an electrochemical capacitor with the ability to accumulate an extremely large amount of energy relative to its size and in comparison, with a traditional capacitor.

Additionally, there is a superconducting inductive energy storage. There are three main components of the storage system of this type: a superconducting coil, an air conditioning subsystem, and a cooling and vacuum subsystem. The storage system stores electrical energy in a magnetic field created by direct current in a superconducting coil, which has been cryogenically cooled to a temperature below its superconducting critical temperature. Typically, when current flows through a coil, electrical energy is dissipated as heat due to the resistance of the wire. However, if a coil is made of a superconducting material, such as mercury or vanadium, in its superconducting state (usually at a very low temperature), zero resistance occurs, and electrical energy can be stored almost without loss.

Moreover, there is a heat accumulator. The first type is heat storage. In general, there are three main types of thermal ESSs: sensitive, hidden, and thermochemical. The basic principle of all thermal ESSs is the conservation of excess energy in the system for use at a later time. Three processes occurring in the heat accumulator are represented by charging, storing, and discharging.

In addition, there are sensitive heat storage systems. This is when the thermal energy is stored as internal energy caused by temperature changes in the material. The energy that is available as heat is stored as a result of its isolation from the environment, which may arise from the presence of physical barriers.

Another type is the hidden heat storage systems. In that case, heat is absorbed or released due to a change in the phase of the material at a constant temperature. For example, the transformation of gaseous, liquid, and solid states.

Yet another type is the thermochemical energy storage. It includes thermochemical reactions and sorption processes that receive energy through the supply of heat. In thermochemical reactions, energy is stored through a reversible reaction and then recovered during the reverse reaction.

The traditional power system has a large number of generating sources, the production of which can be easily changed in accordance with the needs of the load. In addition, most of these systems work interconnected, and the electricity generated by generators in other power systems can be used to balance the load. In such a situation it is very difficult to justify the economic benefits obtained by using energy storage devices.

Deploying drives has three major obstacles that need to be overcome. The first factor is the analysis of the competitiveness of the cost of energy-saving technologies, which also includes the cost of production and integration with the network. To select the best solution, it is necessary to take into account the cost of the life cycle, as well as characteristics for a specific energy storage technology, namely: full cycle efficiency, energy density, and service life.

However, prices for different technologies are constantly changing. The second task is technical. It includes factors that provide user confidence, such as security checks, drive reliability, and performance. Third, it is a nondiscriminatory regulatory and market environment. Value propositions for storage in networks require equal conditions between different service providers (accumulation, price-dependent consumption, generation, network reinforcement, etc.).

Network storage is not schematically compensated, and this constitutes one of its main barriers, since some of the participants associated with the electricity market are part of a regulated market, such as transmission network operators and distribution network operators, while others are part of a deregulated market. This division in organs has reduced the acceptability of energy storage to meet demand and supply.

Space is another barrier associated with the promotion of energy storage. This is one of the key factors that customers consider before installing the system to support power lines for balanced supply and demand at any time. Closer location of ESS to the network enhances the network, increasing its throughput, and increasing revenue.

The network in a remote area may suffer from the integration of RESs, because the network in such places is weak to create a stable voltage due to fluctuations in electricity. This disadvantage cannot be solved economically, but it can be resolved by using energy storage devices to stabilize the power and regulate the network voltage in such remote facilities.

Various articles raising topics of arbitrage use in the power industry, omit the fact that drives affect the equilibrium price, since they consider single use cases. The idea is that drives charge in off-peak hours, at a corresponding lower price. However, by this, additional demand appears, which leads to an increase in the market price. Then the same energy storage replaces inefficient sources with the highest marginal prices during peak hours, thereby reducing the equilibrium price.

Margin pricing establishes the order of generation, so the generation of sources with the lowest marginal value is at the beginning of the supply curve, while production with the highest marginal value is at the end, thereby determining the price of electricity. The marginal price of electricity is the intersection of supply and demand curves. The demand curve is determined by the willingness of all customers to pay for the goods based on their power requirements and supply obligations, forecast, etc. The supply curve is based on all marginal prices of generators, based on generation type, efficiency, risk management, etc.

Previously, accumulator batteries were associated exclusively with small-scale technologies because of their properties and price. However, this situation has been overcome due to the growing market of smartphones, electric vehicles, and other devices that require a longer service life. However, the size of the batteries still depends on their use. Small and large-scale use of energy storage devices varies by site. Thus small-scale technologies, such as batteries, are residential and industrial consumers. At the same time, electricity suppliers mainly use larger storage types which are part of a unified energy system.

As a rule, large-scale use has a greater number of charge and discharge cycles, so the service life is shorter than for small-scale use. Life expectancy is the reason why the number of cycles is more important for large-scale use than for use by the end user for their own purposes. The LIA advantage is the ability to store energy for a longer full cycle time, which is critical at the network level.

Despite this, the rapid development in the electrochemical industry, namely in the LIA market, cannot occur without a change in the legislation and the wholesale market model. The market architecture must be modified and adjusted for more efficient use of fuel and electricity.

Current wholesale electricity market model is not ready for the introduction of batteries of high-power consumption as an independent player in the market. However, some factors that were not taken into account in this work may affect the overall result. There is a large percentage of old generating equipment at TPPs. In the case of replacing the old technology, investing in a new unit or power plant, various options can be compared, in which the batteries of high-power consumption may become preferential because of the market situation, where accumulator battery prices are decreasing, and at the same time the cost of building new power plants increases. However, before making any decisions, it is necessary to provide a feasibility study. In addition, now the availability of inefficient and expensive technologies is required only because of the possible increase in demand that needs to be covered. But if there is a battery at the power plant, this demand may be covered by electricity generated during off-peak hours.

In the future, the price of fossil fuels will continue to grow, so the cost of electricity from TPPs using it will become higher, which will increase the marginal price, which, in turn, is going to make the use of batteries more appropriate and socially and economically acceptable. The cost of power plant equipment is also increasing, so does the price of fuel. Capital expenditures for the construction of a gas power plant currently amount to 800–1000 EUR/kWh. In this case, batteries of high-power consumption make a profitable solution for smoothing peak consumption, as well as to save on the cost of fossil fuels and reduce emissions (Schmidt et al., 2016).

As a rule, the balance of power in the system is maintained by changing the power supply in the network by power plant operators. This mode of control of power plants not only significantly increases the wear of generating equipment, but also leads to additional fuel consumption. The excessive consumption of fuel is especially noticeable when large blocks of power plants participate in the regulation of the variable part of the load schedule. In addition, there is not always a technological ability to quickly start or stop energy production. Moreover, in case of emergency situations when power plants do not have enough power reserves, the load is limited by consumers in order to restore the permissible frequency level. This can lead to significant damage associated with interrupting the supply of electricity to consumers.

The environmental aspect should also be taken into account, since batteries of high-power consumption, first, increases the efficiency of power plants, allowing them to operate in a more efficient way, and, second, replace inefficient generation with accumulators.

The role of the batteries of high-power consumption in the capacity market demonstrates its economic advantages. However, there are other places of possible sharing, the effect of which was not considered in this thesis. In this case, the use of batteries should be considered as a mechanism of various markets, such as a balancing market, where prices at peak hours can reach a much higher cost at certain hours during the day. Other applications are integration with nuclear power plants to increase power factor and frequency control at all levels. In addition, the technical characteristics of the batteries of high-power consumption allow them to participate in the auxiliary market, as they meet all the necessary requirements.

Another development direction for the batteries of high-power consumption is distributed generation. There is a depletion of efficiency potential for centralized power supply systems. Thus the increase in electricity prices, according to estimates, has increased 3.3–3 times over the past 10 years. This leads to the development of distributed energy.

It is also worth noting the integration of RESs. The forecasts that include massive deployment and implementation of the RESs will certainly stimulate the integration of energy storage. The experience of various countries around the world shows that the emerging large-scale wind integration is a factor influencing the deployment of energy storage in power systems.

4.7 Renewable energy source and smart grids

All in all, it appears that RESs would occupy an important position on the future electricity markets governed by smart grids. It has been shown in this chapter, that recently, the global energy industry has faced serious problems: the global energy crisis and the ever-increasing consumption of carbon-driven energy.

The smart grids of the future will likely need large computing power, even though the evolution of information and communication technologies (ICTs) and the possible appearance of artificial intelligence (AI) that would enormously speed up the computations and processes than the algorithms that we are using today.

Thence, this might become a serious problem due to the existence of large geographically distributed data centers that produce "cloud services" and need from tens to hundreds of megawatts of energy and therefore have high electricity bills and may entail negative environmental consequences. The only way out of this situation is the use of RESs.

In order to reduce environmental impacts (such as CO_2 emissions and global warming) caused by the rapid increase in energy consumption, many Internet service providers have begun to take various measures to supply their data centers with renewable energy. Unfortunately, due to the unstable nature of RESs such as wind turbines and solar panels, currently renewable sources are much more expensive than brown energy produced from conventional fossil-based fuels, which can put tangible pressure on Internet budgets and services.

The relevance of this work is to consider the idea of increasing the percentage of RESs used to power data center networks. To implement this idea, one can use a system of intermittent generation of renewable energy, for example, solar energy.

A typical data center smart grids of the future might require a specialized building for hosting server and network equipment and connecting subscribers to Internet channels. The data center performs the functions of processing, storing, and distributing information, as a rule, in the interests of corporate clients—it is focused on solving business problems by providing information services (Al-Fuqaha et al., 2015). Consolidation of computational resources and data storage facilities in the data center reduces the total cost of ownership of the IT infrastructure due to the possibility of efficient use of hardware, for example, redistribution of workloads, as well as by reducing administrative costs.

Data centers are usually located within or in close proximity to a communication center or point of presence of one or more communication operators. The quality and bandwidth of channels affect the level of services provided, since the main criterion for assessing the quality of any data center is server availability time (uptime).

Some data centers offer customers additional services for the use of equipment for automatically avoiding various types of attacks. Teams of qualified specialists around the clock monitor all servers. It should be noted that data center services are very different in price and quantity of services. To ensure the safety of data backup systems are used. To prevent data theft, data centers use various systems to restrict physical access and video surveillance systems. In corporate (departmental) data centers are usually concentrated most of the servers of the organization. The equipment is mounted in specialized racks and cabinets. As a rule, only rack-mounted equipment, that is, standard-sized enclosures adapted for rack mounting, is accepted into a data center. Computers in desktop cases are inconvenient for data centers and are rarely located in them. The main indicator of data center operation is fault tolerance; cost of operation, energy consumption, and temperature control are also of a great importance.

Data center communications are most often based on networks using the Internet protocol. The data center contains several routers and switches that control traffic between servers and the "outside world." For reliability, the data center is sometimes connected to the Internet using a variety of different external channels from different providers.

Some servers in the data center are used to run basic Internet and Internet services that are used internally: mail servers, proxy servers, DNS servers, etc. Network security level is supported by firewalls, VPN gateways, IDS systems, etc. Traffic monitoring systems and some applications are also used.

The intensity of work on the Internet for users from all over the world continues to grow which in turn requires an increasingly developed infrastructure of data centers and additional generating capacity to meet electricity needs. At the same time, data center operators understand that the future belongs to RESs.

In the recent years, there has been a rapid growth of large and geographically wide-spread data centers that require the Internet to support various services such as the so-called "cloud computing." Cloud computing is a method of interaction between a client and a server, in which client information is processed and stored on a remote server; reduces the hardware and software requirements of the client's computer. This arrangement provides fast and reliable method of operating the smart energy systems remotely and is likely to gain wider popularity with the continuing deployment of smart energy grids.

4.8 Conclusion and discussions

RESs, power markets, and smart grids seem to be the pieces of one large mosaic. Solar, wind, and hydro energy sources constitute promising alternatives to the traditional energy sources that are based on carbon fuels. Although being volatile and subjected to seasonal and daily changes, they are becoming more and more ubiquitous as the new technologies for generation and storage of renewable energy appear on the market. Generation of renewable energy is also closely connected with its storage. With regard to this, battery energy storage is also becoming an important piece of the mosaic mentioned above. Traditional battery storage is evolving, and new novel solutions are taking place. However, the new type of batteries might still change the electricity market and the electricity grids as we know them. After the renewable energy is generated and stored there is where the smart grid comes into the picture to manage it and allocate it wisely.

Compared to traditional generation methods, solar energy is one of the cleanest and safest ways to produce energy. But she is not devoid of "pitfalls." For example, the production of modules belongs to the "dirty" industry, which pollutes the environment and is dangerous to human life. Additionally, another negative aspect of solar power system is quite complicated and expensive disposal of waste installations. Due to continuous scientific progress, the cost of production of PV plates is continuously reduced, their service life increases, and production technologies are improved.

The need and prospects for the development of solar energy in different countries of the world is confirmed today by various stimulating legal and economic acts and laws. Many countries adopted a resolution on a mechanism to encourage the use of renewable energy in the wholesale electricity and capacity market, as well as promote changes in the main directions of state policy in energy efficiency improvements based on the use of RESs.

The global wind energy market is growing faster than any other renewable energy. With a total installed capacity of over 60,000 MW and an average annual growth rate of 28%, wind power engineering is booming in many countries around the world and becoming the main source of energy. Subject to political support for the large-scale development of wind energy in combination with energy-saving measures, by 2030, the wind industry will be able to meet 29% of the global electricity demand.

The political decisions that will be made in the coming years will determine the ecological and economic situation in the world for many decades to come. While developed countries need to urgently review their energy strategy, developing countries need to take into account past mistakes and build their economies on the basis of sustainable energy.

According to their potential, wind energy can become the basis for a safe and sustainable energy of the future, based on environment-friendly technologies, creating millions of new jobs and promoting local development. Wind energy can and should play a leading role in the global energy industry of the future.

Despite the fact that hydropower is generally recognized as an important means of ensuring economic development without polluting the atmosphere with CO_2 emissions, the peak of its importance for the global energy supply has been passed a long time ago. Binding to the geographical location and the corresponding gradual exhaustion of the hydropower potential, the need in most cases of withdrawal of land and relocation of residents, the complexity of the design (geographical and geological conditions are unique in each case), large-scale risks of design and construction errors, long construction time— these circumstances led to a decrease in the share of hydropower plants in favor of other forms of energy (thermal and nuclear). Still, hydropower plants constitute one the widely used sources of renewable energy nowadays with some countries (e.g., Norway) heavily relying upon them. Hydropower's share in the global renewable energy mix is likely to decline but its importance is hard to underestimate.

Energy storage is an important part of future energy systems. The accumulated energy will solve many of the problems caused by the ever-changing demand for electricity and provide the ability to easily add smaller systems for generating RESs to the power grid. PSPs have established a steady footing in the current market, but many other storage options are still burdened with high investment costs and low efficiency.

Development in the field of energy storage is carried out every day, and in the future accumulator batteries will most likely be present inside most office buildings, as well as private consumers. Most batteries are likely to be based on lithium-ion batteries. Lead-acid and flow-through batteries will be used additionally for various applications. The world of energy storage is full of new innovations. It is likely that we do not even know the type of technology that will be used in the storage systems of the future. However, the future smart grids are going to need reliable ESSs—either the improved existing ones, or the completely novel ones.

Smart grids of the future are going to employ renewables on the path of transition to the carbon-free economy. In order to accomplish this, they would need an enormous computing power and algorithms that would allow them to optimize, allocate, effectively manage, facilitate exchange of information and data between multiple devices, points, and agents (prosumers). Therefore they would require reliable data centers that would also work autonomously on renewable energy, perform all necessary calculations and operations based on the principles of cloud computing by employing the power and speed of the 5G (or even more advanced) network technologies that would ensure quick and safe transfer of data and information between peers. In order for all this system to work, human decision-making is not going to be enough and the support of powerful AI is going to be needed.

References

Alanne, K., Saari, A., 2006. Distributed energy generation and sustainable development. Renew. Sustain. Energy Rev. 10 (6), 539–558. Available from: https://doi.org/10.1016/j.rser.2004.11.004.

Al-Fuqaha, A., Guizani, M., Mohammadi, M., Aledhari, M., Ayyash, M., 2015. Internet of things: a survey on enabling technologies, protocols, and applications. IEEE Commun. Surv. Tutor. 17 (4), 2347–2376. Available from: https://doi.org/10.1109/COMST.2015.2444095.

Aliyu, A.K., Modu, B., Tan, C.W., 2018. A review of renewable energy development in Africa: a focus in South Africa, Egypt and Nigeria. Renew. Sustain. Energy Rev. 81, 2502–2518. Available from: https://doi.org/10.1016/j.rser.2017.06.055.

Amelang, S., Wehrmann, B., 2019. German onshore wind power – output, business and perspectives. <https://www.cleanenergywire.org/factsheets/german-onshore-wind-power-output-business-and-perspectives> (accessed 19.05.19.).

Andor, M.A., Frondel, M., Vance, C., 2017. Germany's energiewende: a tale of increasing costs and decreasing willingness-to-pay. Energy J. 38, 211–228. Available from: https://doi.org/10.5547/01956574.38.SI1.mand.

Antipov, E.V., Abakumov, A.M., Drozhzhin, O.A., Pogozhev, D.V., 2019. Lithium-ion electrochemical energy storage: the current state, problems, and development trends in Russia. Therm. Eng. 66 (4), 219–224. Available from: https://doi.org/10.1134/S0040601519040013.

Ardizzon, G., Cavazzini, G., Pavesi, G., 2014. A new generation of small hydro and pumped-hydro power plants: advances and future challenges. Renew. Sustain. Energy Rev. 31, 746–761. Available from: https://doi.org/10.1016/j.rser.2013.12.043.

Armaroli, N., Balzani, V., 2007. The future of energy supply: challenges and opportunities. Angew. Chem. Int. Ed. 46 (1-2), 52–66. Available from: https://doi.org/10.1002/anie.200602373.

Barlev, D., Vidu, R., Stroeve, P., 2011. Innovation in concentrated solar power. Sol. Energy Mater. Sol. Cells 95 (10), 2703–2725. Available from: https://doi.org/10.1016/j.solmat.2011.05.020.

Barrios, L., Rodriguez, A., 2004. Behavioural and environmental correlates of soaring-bird mortality at on-shore wind turbines. J. Appl. Ecol. 41 (1), 72–81. Available from: https://doi.org/10.1111/j.1365-2664.2004.00876.x.

Bentley, R.W., 2002. Global oil & gas depletion: an overview. Energy Policy 30 (3), 189–205. Available from: https://doi.org/10.1016/S0301-4215(01)00144-6.

Boluk, G., 2013. Renewable energy: policy issues and economic implications in Turkey. Int. J. Energy Econom. Policy 3 (2), 153–167.

Burroughs, W.J., 2007. Climate Change: A Multidisciplinary Approach, first ed. Cambridge University Press.

Calvert, G., Guguschev, C., Burger, A., Groza, M., Derby, J.J., Feigelson, R.S., 2016. High speed growth of SrI2 scintillator crystals by the EFG process. J. Cryst. Growth 455, 143–151. Available from: https://doi.org/10.1016/j.jcrysgro.2016.10.024.

Chen, H., Cong, T.N., Yang, W., Tan, C., Li, Y., Ding, Y., 2009. Progress in electrical energy storage system: a critical review. Prog. Nat. Sci. 19 (3), 291–312. Available from: https://doi.org/10.1016/j.pnsc.2008.07.014.

Cherp, A., Vinichenko, V., Jewell, J., Suzuki, M., Antal, M., 2017. Comparing electricity transitions: a historical analysis of nuclear, wind and solar power in Germany and Japan. Energy Policy 101, 612–628. Available from: https://doi.org/10.1016/j.enpol.2016.10.044.

Comello, S., Reichelstein, S., 2016. The US investment tax credit for solar energy: alternatives to the anticipated 2017 step-down. Renew. Sustain. Energy Rev. 55, 591–602. Available from: https://doi.org/10.1016/j.rser.2015.10.108.

Constant, E.W., 1983. Scientific theory and technological testability: science, dynamometers, and water turbines in the 19th century. Technol. Culture 24 (2), 183–198. Available from: https://doi.org/10.2307/3104036.

Crampsie, S., 2012. The isle of right [power built environment]. Eng. Technol. 7 (2), 46–49. Available from: https://doi.org/10.1049/et.2012.0204.

Cunha, Á., Martins, J., Rodrigues, N., Brito, F.P., 2015. Vanadium redox flow batteries: a technology review. Int. J. Energy Res. 39 (7), 889–918. Available from: https://doi.org/10.1002/er.3260.

Dai, K., Bergot, A., Liang, C., Xiang, W.N., Huang, Z., 2015. Environmental issues associated with wind energy-a review. Renew. Energy 75, 911–921. Available from: https://doi.org/10.1016/j.renene.2014.10.074.

Dai, J., Yang, X., Wen, L., 2018. Development of wind power industry in China: a comprehensive assessment. Renew. Sustain. Energy Rev. 97, 156–164. Available from: https://doi.org/10.1016/j.rser.2018.08.044.

Dawn, S., Tiwari, P.K., Goswami, A.K., Singh, A.K., Panda, R., 2019. Wind power: existing status, achievements and government's initiative towards renewable power dominating India. Energy Strategy Rev. 23, 178–199. Available from: https://doi.org/10.1016/j.esr.2019.01.002.

De Amorim, W.S., Valduga, I.B., Ribeiro, J.M.P., Williamson, V.G., Krauser, G.E., Magtoto, M.K., et al., 2018. The nexus between water, energy, and food in the context of the global risks: an analysis of the interactions between food, water, and energy security. Environ. Impact Assess. Rev. 72, 1–11. Available from: https://doi.org/10.1016/j.eiar.2018.05.002.

Denisov, S.E., Denisova, M.V., 2017. Analysis of hydropower potential and the prospects of developing hydropower engineering in South Ural of the Russian federation. Proc. Eng. 206, 881–885. Available from: https://doi.org/10.1016/j.proeng.2017.10.567.

Devine, W.D., 1983. From shafts to wires: historical perspective on electrification. J. Econ. Hist. 43 (2), 347–372. Available from: https://doi.org/10.1017/S0022050700029673.

Devine-Wright, P., 2005. Beyond NIMBYism: towards an integrated framework for understanding public perceptions of wind energy. Wind Energy Int. J. Progr. Appl. Wind Power Conver. Technol. 8 (2), 125–139. Available from: https://doi.org/10.1002/we.124.

Dincer, I., 2000. Renewable energy and sustainable development: a crucial review. Renew. Sustain. Energy Rev. 4 (2), 157–175. Available from: https://doi.org/10.1016/S1364-0321(99)00011-8.

Dubey, S., Sarvaiya, J.N., Seshadri, B., 2013. Temperature dependent photovoltaic (PV) efficiency and its effect on PV production in the world-a review. Energy Proc. 33, 311–321. Available from: https://doi.org/10.1016/j.egypro.2013.05.072.

Dusonchet, L., Favuzza, S., Massaro, F., Telaretti, E., Zizzo, G., 2019. Technological and legislative status point of stationary energy storages in the EU. Renew. Sustain. Energy Rev. 101, 158–167. Available from: https://doi.org/10.1016/j.rser.2018.11.004.

EIA, 2016. World energy outlook 2016. < https://webstore.iea.org/world-energy-outlook-2016 > (accessed 29.04.19.).

Ekins, P., Speck, S., 1999. Competitiveness and exemptions from environmental taxes in. Europe. Environ. Res. Econ. 13 (4), 369–396. Available from: https://doi.org/10.1023/A:1008230026880.

Eveloy, V., Ayou, D.S., 2019. Sustainable district cooling systems: status, challenges, and future opportunities, with emphasis on cooling-dominated regions. Energies 12 (2), 235. Available from: https://doi.org/10.3390/en12020235.

EWEA, 2009. The Economics of Wind Energy, first ed. European Wind Energy Association.

Fertig, E., Apt, J., 2011. Economics of compressed air energy storage to integrate wind power: a case study in ERCOT. Energy Policy 39 (5), 2330–2342. Available from: https://doi.org/10.1016/j.enpol.2011.01.049.

Fiandra, V., Sannino, L., Andreozzi, C., Graditi, G., 2019. End-of-life of silicon PV panels: a sustainable materials recovery process. Waste Manage. (Oxford) 84, 91–101. Available from: https://doi.org/10.1016/j.wasman.2018.11.035.

Fontani, D., Sansoni, P., 2015. Renewable energy exploitation for domestic supply. Sustainable Indoor Lighting. Springer, London, pp. 335–355. Available from: https://doi.org/10.1007/978-1-4471-6633-7_17.

Fraunhofer ISE, 2014. Electricity generation from solar and wind energy in 2014. < https://www.ise.fraunhofer.de/content/dam/ise/de/documents/publications/studies/daten-zu-erneuerbaren-energien/stromproduktion-aus-solar-und-windenergie-2014.pdf > (accessed 28.04.19.).

Fu, Y., Liu, X., Yuan, Z., 2015. Life-cycle assessment of multi-crystalline photovoltaic (PV) systems in China. J. Clean. Prod. 86, 180–190. Available from: https://doi.org/10.1016/j.jclepro.2014.07.057.

Fuchs, C., Marquardt, K., Kasten, J., Skau, K., 2019. Wind turbines on German farms-an economic analysis. Energies 12 (9), 1587. Available from: https://doi.org/10.3390/en12091587.

Goudarzi, N., Zhu, W.D., 2013. A review on the development of wind turbine generators across the world. Int. J. Dynam. Control 1 (2), 192–202. Available from: https://doi.org/10.1007/s40435-013-0016-y.

Guo, B., Niu, M., Lai, X., Chen, L., 2018. Application research on large-scale battery energy storage system under global energy interconnection framework. Global Energy Inter. 1 (1), 79–86.

Hau, E., 2013. Wind Turbines: Fundamentals, Technologies, Application, Economics, first ed. Springer Science & Business Media.

He, Z., Xu, S., Shen, W., Long, R., Yang, H., 2016. Overview of the development of the Chinese Jiangsu coastal wind-power industry cluster. Renew. Sustain. Energy Rev. 57, 59–71. Available from: https://doi.org/10.1016/j.rser.2015.12.187.

Hebner, R., Beno, J., Walls, A., 2002. Flywheel batteries come around again. IEEE Spectr. 39 (4), 46–51. Available from: https://doi.org/10.1109/6.993788.

Huang, H., Yan, Z., 2009. Present situation and future prospect of hydropower in China. Renew. Sustain. Energy Rev. 13 (6-7), 1652–1656. Available from: https://doi.org/10.1016/j.rser.2008.08.013.

IRENA, 2018. Renewable power generation costs in 2017. < https://www.irena.org/publications/2018/Jan/Renewable-power-generation-costs-in-2017 > (accessed 18.04.19.).

Jacobson, M.Z., Delucchi, M.A., 2011. Providing all global energy with wind, water, and solar power, Part I: Technologies, energy resources, quantities and areas of infrastructure, and materials. Energy Policy 39 (3), 1154–1169. Available from: https://doi.org/10.1016/j.enpol.2010.11.040.

Jing, J., 1997. Rural resettlement: past lessons for the Three Gorges Project. China J. 38, 65–92. Available from: https://doi.org/10.2307/2950335.

Junge, C.E., 1962. Global ozone budget and exchange between stratosphere and troposphere. Tellus 14 (4), 363–377. Available from: https://doi.org/10.3402/tellusa.v14i4.9563.

Kadro, J.M., Pellet, N., Giordano, F., Ulianov, A., Müntener, O., Maier, J., et al., 2016. Proof-of-concept for facile perovskite solar cell recycling. Energy Environ. Sci. 9 (10), 3172–3179. Available from: https://doi.org/10.1039/C6EE02013E.

Kalantar, M., 2010. Dynamic behavior of a stand-alone hybrid power generation system of wind turbine, micro-turbine, solar array and battery storage. Appl. Energy 87 (10), 3051–3064. Available from: https://doi.org/10.1016/j.apenergy.2010.02.019.

Kalogirou, S.A., 2004. Solar thermal collectors and applications. Prog. Energy Combust. Sci. 30 (3), 231–295. Available from: https://doi.org/10.1016/j.pecs.2004.02.001.

Karltorp, K., Guo, S., Sandén, B.A., 2017. Handling financial resource mobilisation in technological innovation systems-the case of Chinese wind power. J. Clean. Prod. 142, 3872–3882. Available from: https://doi.org/10.1016/j.jclepro.2016.10.075.

Khan, M.J., Bhuyan, G., Iqbal, M.T., Quaicoe, J.E., 2009. Hydrokinetic energy conversion systems and assessment of horizontal and vertical axis turbines for river and tidal applications: a technology status review. Appl. Energy 86 (10), 1823–1835. Available from: https://doi.org/10.1016/j.apenergy.2009.02.017.

Klagge, B., Liu, Z., Silva, P.C., 2012. Constructing China's wind energy innovation system. Energy Policy 50, 370–382. Available from: https://doi.org/10.1016/j.enpol.2012.07.033.

Kouchachvili, L., Yaïci, W., Entchev, E., 2018. Hybrid battery/supercapacitor energy storage system for the electric vehicles. J. Power Sour. 374, 237–248. Available from: https://doi.org/10.1016/j.jpowsour.2017.11.040.

Kumar, M.S., Charanadhar, N., Srikanth, V.V., Rao, K.B.S., Raj, B., 2018. Materials in harnessing solar power. Bull. Mater. Sci. 41 (2), 62. Available from: https://doi.org/10.1007/s12034-018-1554-x.

Kyriakopoulos, G.L., Arabatzis, G., 2016. Electrical energy storage systems in electricity generation: energy policies, innovative technologies, and regulatory regimes. Renew. Sustain. Energy Rev. 56, 1044–1067. Available from: https://doi.org/10.1016/j.rser.2015.12.046.

Latunussa, C.E., Ardente, F., Blengini, G.A., Mancini, L., 2016. Life cycle assessment of an innovative recycling process for crystalline silicon photovoltaic panels. Sol. Energy Mater. Sol. Cells 156, 101–111. Available from: https://doi.org/10.1016/j.solmat.2016.03.020.

Lenzen, M., McBain, B., Trainer, T., Jütte, S., Rey-Lescure, O., Huang, J., 2016. Simulating low-carbon electricity supply for Australia. Appl. Energy 179, 553−564. Available from: https://doi.org/10.1016/j.apenergy.2016.06.151.

Li, Q.A., Kamada, Y., Maeda, T., Murata, J., Iida, K., Okumura, Y., 2016. Fundamental study on aerodynamic force of floating offshore wind turbine with cyclic pitch mechanism. Energy 99, 20−31. Available from: https://doi.org/10.1016/j.energy.2016.01.049.

Li, H., Li, F., Shi, D., Yu, X., Shen, J., 2018. Carbon emission intensity, economic development and energy factors in 19 G20 countries: empirical analysis based on a heterogeneous panel from 1990 to 2015. Sustainability 10 (7), 2330. Available from: https://doi.org/10.3390/su10072330.

Li, S., Zhang, S., Andrews-Speed, P., 2019. Using diverse market-based approaches to integrate renewable energy: experiences from China. Energy Policy 125, 330−337. Available from: https://doi.org/10.1016/j.enpol.2018.11.006.

Liu, B.Y., Jordan, R.C., 1960. The interrelationship and characteristic distribution of direct, diffuse and total solar radiation. Sol. Energy 4 (3), 1−19. Available from: https://doi.org/10.1016/0038-092X(60)90062-1.

Liu, C., Wang, Y., Zhu, R., 2017. Assessment of the economic potential of China's onshore wind electricity. Resour. Conserv. Recycl. 121, 33−39. Available from: https://doi.org/10.1016/j.resconrec.2016.10.001.

Lucas, A., Chondrogiannis, S., 2016. Smart grid energy storage controller for frequency regulation and peak shaving, using a vanadium redox flow battery. Int. J. Electr. Power Energy Syst. 80, 26−36. Available from: https://doi.org/10.1016/j.ijepes.2016.01.025.

Lucena, J.D.A.Y., Lucena, K.A.A., 2019. Wind energy in Brazil: an overview and perspectives under the triple bottom line. Clean Energy 3 (2), 69−84. Available from: https://doi.org/10.1093/ce/zkz001.

Lukic, S.M., Cao, J., Bansal, R.C., Rodriguez, F., Emadi, A., 2008. Energy storage systems for automotive applications. IEEE Trans. Ind. Electron. 55 (6), 2258−2267. Available from: https://doi.org/10.1109/TIE.2008.918390.

Malhotra, R., Ali, A., 2019. 5-Na/ZnO doped mesoporous silica as reusable solid catalyst for biodiesel production via transesterification of virgin cottonseed oil. Renew. Energy 133, 606−619. Available from: https://doi.org/10.1016/j.renene.2018.10.055.

Marchenko, O.V., Solomin, S.V., 2004. Efficiency of wind energy utilization for electricity and heat supply in northern regions of Russia. Renew. Energy 29 (11), 1793−1809. Available from: https://doi.org/10.1016/j.renene.2004.02.006.

Marchenko, O., Solomin, S., 2014. Economic efficiency of renewable energy sources in autonomous energy systems in Russia. Int. J. Renew. Energy Res. 4 (3), 548−554.

Mazur, E., 2013. U.S. Patent No. 8,513,826.U.S. Patent and Trademark Office, Washington, DC.

Minkin, G. (2017). Impacts of large-scale battery energy storage systems on Russian wholesale electricity and capacity market. < http://lutpub.lut.fi/handle/10024/143721 > (accessed 18.05.19.).

Namsaraev, Z.B., Gotovtsev, P.M., Komova, A.V., Vasilov, R.G., 2018. Current status and potential of bioenergy in the Russian Federation. Renew. Sustain. Energy Rev. 81, 625−634. Available from: https://doi.org/10.1016/j.rser.2017.08.045.

Neij, L., 2008. Cost development of future technologies for power generation-a study based on experience curves and complementary bottom-up assessments. Energy Policy 36 (6), 2200−2211. Available from: https://doi.org/10.1016/j.enpol.2008.02.029.

Nelson, V., Starcher, K., 2018. Wind Energy: Renewable Energy and the Environment, first ed. CRC Press.

Neuhoff, K., Barquin, J., Bialek, J.W., Boyd, R., Dent, C.J., Echavarren, F., et al., 2013. Renewable electric energy integration: quantifying the value of design of markets for international transmission capacity. Energy Econ. 40, 760−772. Available from: https://doi.org/10.1016/j.eneco.2013.09.004.

Newbery, D., Pollitt, M.G., Ritz, R.A., Strielkowski, W., 2018. Market design for a high-renewables European electricity system. Renew. Sustain. Energy Rev. 91, 695−707. Available from: https://doi.org/10.1016/j.rser.2018.04.025.

Oh, S.Y., 2015. How China outsmarts WTO rulings in the wind industry. Asian Surv. 55 (6), 1116−1145. Available from: https://doi.org/10.1525/as.2015.55.6.1116.

Ouammi, A., Dagdougui, H., Sacile, R., Mimet, A., 2010. Monthly and seasonal assessment of wind energy characteristics at four monitored locations in Liguria region (Italy). Renew. Sustain. Energy Rev. 14 (7), 1959−1968. Available from: https://doi.org/10.1016/j.rser.2010.04.015.

Palomares, V., Serras, P., Villaluenga, I., Hueso, K.B., Carretero-González, J., Rojo, T., 2012. Na-ion batteries, recent advances and present challenges to become low cost energy storage systems. Energy Environ. Sci. 5 (3), 5884−5901. Available from: https://doi.org/10.1039/c2ee02781j.

Pickard, W.F., Shen, A.Q., Hansing, N.J., 2009. Parking the power: strategies and physical limitations for bulk energy storage in supply-demand matching on a grid whose input power is provided by intermittent sources. Renew. Sustain. Energy Rev. 13 (8), 1934–1945. Available from: https://doi.org/10.1016/j.rser.2009.03.002.

Pires, V.F., Romero-Cadaval, E., Vinnikov, D., Roasto, I., Martins, J.F., 2014. Power converter interfaces for electrochemical energy storage systems-a review. Energy Convers. Manage. 86, 453–475. Available from: https://doi.org/10.1016/j.enconman.2014.05.003.

Ramírez, F.J., Honrubia-Escribano, A., Gómez-Lázaro, E., Pham, D.T., 2018. The role of wind energy production in addressing the European renewable energy targets: the case of Spain. J. Clean. Prod. 196, 1198–1212. Available from: https://doi.org/10.1016/j.jclepro.2018.06.102.

Renn, O., Marshall, J.P., 2016. Coal, nuclear and renewable energy policies in Germany: from the 1950s to the "Energiewende". Energy Policy 99, 224–232. Available from: https://doi.org/10.1016/j.enpol.2016.05.004.

Romano, A.A., Scandurra, G., 2011. The investments in renewable energy sources: do low carbon economies better invest in green technologies? Int. J. Energy Econ. Policy 1 (4), 107–115.

Sahu, B.K., 2015. A study on global solar PV energy developments and policies with special focus on the top ten solar PV power producing countries. Renew. Sustain. Energy Rev. 43, 621–634. Available from: https://doi.org/10.1016/j.rser.2014.11.058.

Sayigh, A., 1999. Renewable energy-the way forward. Appl. Energy 64 (1-4), 15–30. Available from: https://doi.org/10.1016/S0306-2619(99)00117-8.

Schmidt, T.S., Battke, B., Grosspietsch, D., Hoffmann, V.H., 2016. Do deployment policies pick technologies by (not) picking applications? A simulation of investment decisions in technologies with multiple applications. Res. Policy 45 (10), 1965–1983. Available from: https://doi.org/10.1016/j.respol.2016.07.001.

Schmidt, O., Hawkes, A., Gambhir, A., Staffell, I., 2017. The future cost of electrical energy storage based on experience rates. Nat. Energy 2 (8), 17110. Available from: https://doi.org/10.1038/nenergy.2017.110.

Shaaban, M., Scheffran, J., Böhner, J., Elsobki, M., 2018. Sustainability assessment of electricity generation technologies in Egypt using multi-criteria decision analysis. Energies 11 (5), 1117. Available from: https://doi.org/10.3390/en11051117.

Sharma, S., Jain, K.K., Sharma, A., 2015. Solar cells: in research and applications—a review. Mater. Sci. Appl. 6 (12), 1145. Available from: https://doi.org/10.4236/msa.2015.612113.

Tie, S.F., Tan, C.W., 2013. A review of energy sources and energy management system in electric vehicles. Renew. Sustain. Energy Rev. 20, 82–102. Available from: https://doi.org/10.1016/j.rser.2012.11.077.

Tripathi, L., Mishra, A.K., Dubey, A.K., Tripathi, C.B., Baredar, P., 2016. Renewable energy: an overview on its contribution in current energy scenario of India. Renew. Sustain. Energy Rev. 60, 226–233. Available from: https://doi.org/10.1016/j.rser.2016.01.047.

Wagner, H.J., 2013. Introduction to wind energy systems. EPJ Web Conf., 54, 01011. https://doi.org/10.1051/epjconf/20135401011.

Wang, Q., Martinez-Anido, C.B., Wu, H., Florita, A.R., Hodge, B.M., 2016. Quantifying the economic and grid reliability impacts of improved wind power forecasting. IEEE Trans. Sust. Energy 7 (4), 1525–1537. Available from: https://doi.org/10.1109/TSTE.2016.2560628.

Weedy, B.M., Cory, B.J., Jenkins, N., Ekanayake, J.B., Strbac, G., 2012. Electric Power Systems, first ed. John Wiley & Sons, London.

Weisheng, W., Yongning, C., Zhen, W., Yan, L., Ruiming, W., Miller, N., et al., 2016. On the road to wind power: China's experience at managing disturbances with high penetrations of wind generation. IEEE Power Energ. Mag. 14 (6), 24–34. Available from: https://doi.org/10.1109/MPE.2016.2596459.

Wrobel, J.K., Power, R., Toborek, M., 2016. Biological activity of selenium: revisited. IUBMB Life 68 (2), 97–105. Available from: https://doi.org/10.1002/iub.1466.

WWEA, 2015. Wind energy 2050. <http://www.wwindea.org/download/technology/GRID_INTEGRATION_Wind_Energy_2050.pdf> (accessed 27.05.19.).

Xie, S.P., Deser, C., Vecchi, G.A., Ma, J., Teng, H., Wittenberg, A.T., 2010. Global warming pattern formation: sea surface temperature and rainfall. J. Clim. 23 (4), 966–986. Available from: https://doi.org/10.1175/2009JCLI3329.1.

Zavrsnik, B., 1998. The importance of selection and evaluation of the supplier in purchasing management. Management 3 (2), 59–74.

Zeng, S., Liu, Y., Liu, C., Nan, X., 2017. A review of renewable energy investment in the BRICS countries: history, models, problems and solutions. Renew. Sustain. Energy Rev. 74, 860–872. Available from: https://doi.org/10.1016/j.rser.2017.03.016.

Zhang, J., Meng, J., 2006. New information on world uranium resource, production, supply and demand. Uran. Min. Metal. 25 (1), 21–28.

Zhang, S., Andrews-Speed, P., Zhao, X., He, Y., 2013. Interactions between renewable energy policy and renewable energy industrial policy: a critical analysis of China's policy approach to renewable energies. Energy Policy 62, 342–353. Available from: https://doi.org/10.1016/j.enpol.2013.07.063.

Zhao, J., Wen, F., Xue, Y., Dong, Z., Xin, J., 2010. Power system stochastic economic dispatch considering uncertain outputs from plug-in electric vehicles and wind generators. Dianli Xitong Zidonghua (Automat. Electr. Power Syst.) 34 (20), 22–29.

Further reading

European Parliament, 2009a. Directive 2009/28/EC of the European Parliament and of the Council of 23 April 2009 on the promotion of the use of energy from renewable sources and amending and subsequently repealing Directives 2001/77/EC and 2003/30/EC.

European Parliament, 2009b. Directive 2009/72/EC of the European Parliament and of the Council of 13 July 2009 concerning common rules for the internal market in electricity and repealing Directive 2003/54/EC (text with EEA relevance).

Peer-to-peer markets and sharing economy of the smart grids

5.1 Introduction: recent changes on energy markets

Nowadays, the existing electricity markets and grids are undergoing rapid transition that is presupposed and predefined by the novel emerging technological solutions and innovations. The rising popularity and the widespread implementation of information and communication technologies (ICTs) in almost all spheres of our social and public lives made these markets to embrace things like blockchain, social networks, or smartphone apps. These changes inevitably impose notable social impacts on the existing smart grids and pave the road for the creation of the more complex and sophisticated smart grids of the future.

Today's electricity and power grids, some of them centuries-old in design, have been built to house central generators, one-way power stations through high-voltage transmission lines, send consumers via lower voltage power distribution lines and central control centers that collect information from a limited number of power stations called substations. While environmental sustainability is not the main driving force behind the adoption of smart grids, it helps both directly (through energy saving) and indirectly [through promoting the development of electric vehicles (EVs) and renewable energies]. However, as the electricity markets evolve and monopolies become more and more complex, the design of smart grid business models is becoming increasingly complex. Information on electricity production and consumption, wholesale prices, imports and exports, and balances of energy can be found, combined, and downloaded for different periods of time. There is a need to intelligently integrate renewable energy, such as solar and wind, into the grid and ensure that electricity is safe, affordable, and environmentally friendly. Therefore, there is a clear need for an interface between the trade in electricity on the electricity market and the infrastructure of the power grid through which transactions are physically carried out. Dependence on imports on fossil fuels have significant economic consequences, which can leave any economy vulnerable to volatile global energy prices and a significant contribution to the current account deficit.

While existing technologies may stimulate a large part of energy transition system, more innovation, research, and development will be needed to implement new low-carbon technologies to significantly reduce the emissions of harmful greenhouse gases (such as notoriously known CO_2). Priority should be given to the energy regulation frame-work, for example, by increasing the building's energy efficiency standards and complying with the energy efficiency guidelines. International financial institutions and programs such as the World Bank, the Global Environment Center, and the European Union's energy efficiency framework.

Some researchers have tried to model how a particular country can operate a grid with mostly renewable energies, often finding carbon-negative technologies, advanced nuclear energy, and even carbon capture and storage (CCS)-equipped coal-fired power stations, part of the solution kit. Special areas include wholesale design, participation in DER markets, coordination of transmission distribution systems, distribution system operators (DSO) models and distribution markets, microgrids and energy resistance strategies, network architectures for whole systems. There is an order required to facilitate the storage of energy in organized markets, as well as facilitating the storage of energy and renewable energy developers who want to enter the market.

One would probably agree that electricity markets are of a special importance when smart grids and their implications, either technical or social, are concerned, since smart grids generally represent upgraded electricity networks that enable a two-way information and power exchange between suppliers and consumers employing intelligent pathways of communication and complex management systems (Zio and Aven, 2011).

The transition of the traditional electricity markets is indicated by three main trends. First of all, there is a decentralization brought on by a sharp reduction in prices of distributed energy sources (DERs) and digitalization of the smart grids (e.g., smart meters, receivers, detectors, sensors of all types and kinds, as well as other electronic network technologies). Second, it is a growth of Internet of Things (IoT) and the related opportunities it provides (interconnectivity, sharing date in real time, exchanging information). Finally, third, it is the rise of the sharing economy that enables the agents to connect using the social network-type schemes in order to optimize their resources and conduct real-time exchange of their goods or services (or both).

Other important and fundamental changes to the power systems and electricity markets are posed by the decarbonization of market and the energy customers becoming active "prosumers" (proactive, energy-generating and energy-trading consumers operating with DERs). According to the research conducted by the General Electric, the top digital trends for the electricity value network in the next decade or so are the rise of platform economy that is centered around platform-based business models which outreach a technology platform and become the main building block of the digital economy (GE, 2017).

One recent peculiar trend that can be recently observed in the power and electricity markets is the situation when consumers can trade electricity directly with each other rather than with utility companies. As this is starting to happen more and more often, consumers become "prosumers" (i.e., both actual consumers and actual producers of electric energy). This approach that looks upon the network as a platform is market as "peer-to-peer" (P2P) market. P2P markets in electrical energy are already gaining wide popularity all around the world with many people engaging in prosumers communities to generate and trade green energy bypassing large energy producers working carbon fuels.

Recent changes in the energy markets yield many new trends coming up in virtually all sectors of the economy and imposing many important implications. The appearance of the new technologies and the desire to be more competitive is very apparent in the traditionally regulated fields such as power provision developing deregulated approaches to overcome various limitations imposed by the old-fashioned markets. With regard to all these trends described above, P2P markets represent one of the trends that continue to create very interesting innovations and generate novel ideas in the energy sector. For example, this can be the consumers generating their energy through solar panels or wind turbines, storing it on in-house batteries and selling the surplus electricity back to the grid or to other users for a profit. By selling the energy within a network, the producer is becoming a real prosumer when she or he sells her or his excess power often through the existing infrastructure which is commonly referred to as a "system." This emerging issue has led to a new approach of ways in which the transactional amount generated by grid fixed fees can be maintained. A review of pricing methods needs to be considered to keep such prices.

Among the most successful (and widely publicized in the mass media) examples of P2P energy markets are the success stories from the United States (e.g., the Brooklyn Microgrid initiative), United Kingdom (e.g., Piclo project), or the Netherlands (e.g., Vandebron scheme) (see e.g., Khan and Khan, 2018; or López-García et al., 2019).

It becomes very obvious that in a situation when governmental subsidies intended for supporting and subsidizing energy tariffs eventually come to a halt, P2P markets are very likely to replace the traditional "selling-to-the-grid" models. Of course, there are some practical questions arising, for example: How the transactions on the P2P markets would take into account the grid fixed fees? The practice from many P2P electricity market and their success stories from all around the world shows that solutions can be found for all of them. For instance, prosumers can start paying a fixed subscription fee to use network for trade and this fee is used to recover network fees. Apart from that, there are many other approaches and each of them would mean implications for operation of power system. This is becoming specifically important as the popularity of P2P markets increases and the traditional power and electricity markets are becoming two-tier systems (while some prosumers use the grid in the traditional way, the others use it as a platform).

In the few recent decades, the whole process chain of generation, distribution, and selling of energy have morphed and transformed beyond recognition. Conventionally, the key techniques of electricity generation were water and fossil fuel-based sources. However, recently, other methods such as solar, wind, and other environmentally friendly approaches have emerged in response to growing concerns regarding the sustainability and environmental effect of several methods like the use of coal in electricity generation. Similarly, the retail and distribution of power, which was formerly dominated by authorities and associations, has evolved. Across the globe, changes from the electricity utility industry, and the market especially, are asserting pressure on the players from the industry to economize and innovate to stay sustainably profitable and meet the increasingly diverse needs of the electricity buyer (Flaherty et al., 2017). In addition, other elements, such as legal and nonlegal aspects or technological changes, altered the working environment for players in this sector and made it even more complicated and dynamic. Additionally, they also changed the system of pricing and other strategic decisions of the management of firms in the industry.

With regard to all of the above, Lavrijssen and Carrillo Parra (2017) discuss that shifts in the sources and methods of energy generation and distribution are aligned with the new emerging opportunities for the customers who can now generate and distribute electricity among themselves thanks to the actual decentralization of electricity generation. As a result, the distribution and retail selling of electricity are also evolving very rapidly. In addition, Parag and Sovacool (2016) observed that consumer tastes and preferences, such as the preference for green or environmentally friendly electricity, resulted in the emergence of a new category of consumers—i.e., those consumers who produce electricity and sell to the grid or to other consumers (whom we already identified as "prosumers" above), and in so doing affect the larger grid in different ways.

However, the emergence of new renewable energy technologies, such as solar battery systems or wind turbines, has the power of altering consumer perception and attitudes as well as to changing their acceptance of the ways electricity is produced and distributed to the customers. It appears that the connection of the same to the national grid could supply many customers with their power needs which could lead to load and customer defection in a situation when customers move from traditional methods to the new methods of generating their own electricity (Rocky Mountain Institute, 2017). Essentially, this situation implies that customers could reduce their usage of electricity from the main suppliers or retailers as currently constituted in national grid systems, and shift to using the power bought from other customers, who have opted to produce their own electricity using (most typically) renewable energy sources (RESs). These customers could be reacting to pricing structures of current retailers, or acting in line with their conviction, tastes, and preferences.

5.2 From traditional to new electricity markets

Traditional energy markets are often approximated by the electricity markets, since electricity is the most ubiquitous and most notorious form of energy everyone is dealing with in today's world. Traditional electricity markets are based on the classical model drawn in accordance with the conventional principles of supply and demand of the market. It becomes obvious that traditional electricity markets are predestined and shaped up by the main distinguishing features of the main product they are supplying. One would probably agree with the well-known fact that electricity represents a peculiar product that can be consumed continuously and simultaneously by a large number of customers. Electricity represents a commodity that is being sold, purchased, and exchanged on the market, just like any other good or service. In the same time, unlike a typical good or service, it remains a unique product due to the fact that it cannot be effectively stored in large quantities that would allow maintaining a surplus (see e.g., Stoft, 2002; Sovacool, 2009). In economic and business terms, electricity is a stock that is fitted to being sold, purchased, and exchanged. A method that facilitates acquisitions through sales, bids to buy, and short-term trades, by propositions to sell and frequently in the form of an obligation of commercial exchanges is recognized as an electricity market. In electricity, commercial businesses are typically paid or netted by the market administrator or a special-purpose sovereign entity charged exclusively with that function. In order to maintain generation and load balance, market operators do not clear trades but regularly require knowledge of the business. Electricity service involves the delivery of both power (kW), energy (kWh), and power quality (e.g., voltage, frequency, or power failures) at a particular location.

Utilities, public utility commissions, consumer and low-income supporters, and electricity consumers in the industrial sector usually raise concerns about the wholesale market which is very different from a traditional market. State regulators are typically established to change the rules on the retail market to encourage or demand that traditional vertical utilities be integrated into the market to sell their generation equipment and give consumers the opportunity to buy electricity from other generators. In order to stimulate competition, the "deregulation" allowed for the sale of electricity and related products, such as energy management, to go beyond conventional utilities. Management of the wholesale markets for the sale of electricity, electricity and utilities (both in advance and in real time) is often conducted by specially appointed authorities. There are basically three basic products sold in the wholesale market: energy, auxiliary services, and capacity. Today, energy storage is in fact only one of the main factors in the auxiliary service market, and specifically the market for frequency regulation, or the balance between electricity supply and demand over subhours. Energy storage is much more limited in capacity markets and is generally not at stake in the energy markets, as such markets are designed for resources that can deliver energy for hours or days without interruption, that is, power stations. Fully integrated unique functionalities of storage into the software are used to plan resources on the power grid and determine the market price of energy, utilities, and capacity. If the above parameters are required to accurately represent the operational capacity of the available energy storage devices in the system, the storage space can be fully integrated into the economical shipping process, which is used to plan network resources and determine the price of time-sensitive market electricity.

There are also new trends: for example, the world's electricity markets are witnessing a major change as the rapid rise in renewable energy production is taking place. In addition to solar and wind power generators, the Virtual Power Plant (VPP) also includes flexible generators such as biomass and hydro, customers with flexible combined heat and power (CHP) production on site, as well as commercial and industrial customers with flexible requirements (Yu et al., 2019). Since the inception of VPPs, they have expanded their services by offering energy trading on various markets and controlling the reserve, balancing utility and volatile utilities, and (flexible) electricity rates for industrial and commercial energy consumers. Due to the fact that customers buy less power, the market price of electricity is lower in the power supply curve (all other products remain the same). While markets have evolved since their introduction, they are now working to provide reliability and minimize the short-term costs of wholesale electricity, despite the pressure of rising demand, federal and state policy interventions and a huge economic shift in the relative natural gas economy compared to other fuels. The cost of coal and nuclear power stations remained relatively flat, while new and existing natural gas stations, more flexible and relatively low operating, have lowered the wholesale price to such an extent that some of the previously profitable nuclear and coal plants have begun to suffer losses. Changes in energy demand, in particular the apparent disconnection of economic production and electricity demand, have been driven in part by energy efficiency policies. In addition, regional transmission organizations (RTOs) all have some form of price deficiency, where electricity prices can rise above the ceiling during the stress of the system (the prices paid in such markets also have a significant impact on the price of electricity collected by generators from such markets in a bilateral or standard offer contract). Like the wholesalers of RTO capacity markets, the RTO capacity markets depend on a central single clearing price (which allows to achieve profitability for low-cost units), are subject to local prices and impose a maximum ceiling on transactions.

While participating in the wholesale market in RTO can benefit from public energy in terms of cost savings and additional power-of-sale opportunities, there are still some potential problems with markets that require vigilant supervision. In addition to traditional power and 4−6 hours of power demand, new types of power services may be encouraged to manage these intrinsic features of clean-up technology.

For example, the policy makers at the US Federal Energy Regulation Committee (FERC) and PJM Interconnection are focusing on the role of battery energy storage and flexible resources such as distributed resource aggregators in power markets. RES (or more precisely, their abundance) might cause various problems in the network's operation which can vary according to the geographical factor, depending on the mix of RES (e.g., solar and wind), the availability of transmission and distribution capacity (both in terms of accessibility and availability), and a mix of non-renewable energies, among other factors that can be mentioned. The operation of power grids with high intermittent resources is a unique challenge for utility companies and network operators. The restructured markets of wholesale electricity serve the stakeholders by offering three basic services: real-time, cost-effective shipping, resource matching incentives, and long-term recovery.

In light of these issues, one can look at the two key aspects of market design: firstly, whether the energy and utilities markets provide sufficient income to cover all system costs, and secondly, whether the current market designs are sufficient to meet the changing needs of the massive electricity grid. There are also smartphone apps envisaged to reduce reliance on wholesale markets toward flexible suppliers, in particular on demand-side resources which may be able to meet higher energy prices.

Thus, the traditional scheme assumes the existence of energy markets that facilitate unidirectional conformist energy trading with the electricity being transmitted from large-scale generators to consumers over long distances and the payments for electricity being wired all of the way back to the large-scale generators.

Fig. 5.1 describes the well-known traditional electricity market arrangements when the electricity is produced at central facilities and power stations (e.g., coal, hydro, or nuclear). Transformers increase the voltage to higher levels suitable for transmission and gradually step it down to lower levels as the electricity flows through the transmission and distribution networks to be delivered to the end-consumers (e.g., industries, households, and businesses). One can see that the traditional energy market (represented on the scheme here by the conventional electricity grid) creates a link between the generation and the customers (see Fig. 5.1).

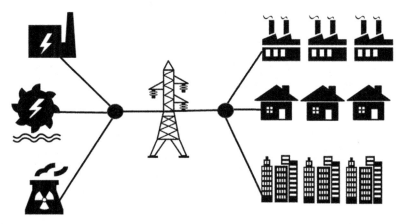

FIGURE 5.1 Traditional energy markets.

According to Lund (2014), the electricity markets consist of key economic features that can be described from the demand side or the supply side. The demand-side features of the electricity market include:

- Inelastic demand: basically, electricity demand does not highly respond to prices of electricity, and how the demand-side response to changes in price over a short period are even lower than that. For instance, many households are usually not aware of their consumption or demand patterns (even though smart metering might result to more flexible consumer demand in future) and users in industries may as well have only inadequate opportunities for shifting their demand to respond to real-time prices;
- Demand volatility: because electricity demand is influenced by factors like seasonal and variations, daily consumption patterns, climate and macroeconomic conditions, it varies considerably over time, possibly leading to spells of too high peak demand;
- Nonexcludability: due to the fact that suppliers usually are not able to connect customers personally, particularly households, the advantages of high system dependability are efficiently shared by most of the users. Thus, this likely challenges the individual customers' incentive for signaling their readiness to pay for augmented supply security.

In addition, the key economic features of the supply side of the electricity markets include the following elements:

- Costs and risks of generation: risks and costs vary largely across technologies used to generate which is caused by variations in capital and costs of operations as well as market risk and technical differences;
- Sunk costs: production capacity investments are generally specific and irreversible, which increases investors' risk of being locked out of the pocket where contracts regulations change of contracts are renegotiated after making investments;
- Intermittent output: various technologies like wind generation vary inherently, possibly transferring the burden to meet the demand for other flexible generators with a controllable output;

- Carbon emissions: given that carbon emissions are by-products of electricity production, they can compose an externality where the impact of climate change is not revealed in prices.

One of the key market failures of the traditional electricity markets is the reduction of final energy production. This can be shown on the example of energy consumption in the European Union which is one of the pioneers in increasing its share of renewables in the total production of electricity and setting up legally binding targets for RESs to cover over 20% of the total energy consumption by 2020. Despite these goals, one can see that energy consumption in the European Union was not consistent with economic output. Moreover, energy production in Europe decreased by 4% between 2010 and 2012 (European Commission, 2014). Production for crude oil and petroleum products reduced by 21% altogether while that of natural decreased by 17%. On the brighter side, the production of renewable energy and solid fuels increased by 9% and 1.3%, respectively. Electricity production also reduced to 3295 TWh in 2012. About 25% of the electricity was generated from renewable sources. To sum this up, the key market failures of EU electricity markets that are distinguished by a high share of RESs and commitment to add even more of those include the following features:

- Inability to provide adequate incentives for reducing carbon emissions by the energy policies governing the electricity markets which threatens the decarbonization goal;
- Absence of investment in generation capacity for meeting demand, especially as higher penetration of recurrent generation like the wind affects the economics of different generation technologies which are more flexible, therefore, threatening the objective of security supply;
- Interruptions of electricity which deprives electricity to users and consumers, causing substantial economic welfare loss;
- Lack of the ability to deliver the desired levels of renewables in the European Union.

There is no doubt that traditional electricity markets can boast a number of advantages. First of all, they yield the economy of scale since large power generating units can produce vast amounts of energy while operated by a relatively low number of employees. Second, interconnected high-voltage transmission network make it possible for the generating plants to be dispatched at any time while the bulk power can be transported over long distances with limited electrical loses and minimum generation reserves. Third, the distribution networks are tailor-made for unidirectional flows of power and accommodating customer loads.

One way or another, with the recent worldwide concerns about the global warming and sustainable economic development, the issues of decarbonization and using renewable energy come to the forefront of electric energy generation. These recent developments make the traditional electricity markets and the traditional electricity models obsolete and call for the new changes and amendments in the form of the RESs and the P2P electricity markets.

5.3 Renewable energy sources and electricity networks

Environmental, sustainability, and energy security concerns have a profound impact on the electricity networks and smart grids worldwide. These concerns spark fundamental changes introducing the decarbonization as the new element into the traditional game of energy generation and energy demand. Decarbonization is understood as the process of replacement of traditional fossil fuels by the RESs as well as advanced technological solutions that prioritize the electrification of transport and heat generation employing EV and electric heat pumps (EHP). The focus on RES that became an important precedence of the governments in many countries, both in the European Union and worldwide, considerably facilitates this process.

It has been already discussed above that the European Commission put forward a legally binding target for RESs to cover 20% of the total energy consumption in the EU Member States by 2020. In the United Kingdom, the Climate Change Act made a step further setting a legally binding target of 50% reduction in greenhouse gases emission by 2030 which would be followed by the next target of 80% reduction of greenhouse gases by 2050. The Paris Agreement signed in December 2015 set up even more demanding goals for. In order to achieve this, the EU Member States set up their own national RES targets fluctuating from 10% in Malta to 49% in Sweden (Calliess and Hey, 2013). Moreover, the EU stakeholders decided that by 2020 more than 10% of transportation fuels used in the EU should be obtained from the renewable sources (Drom, 2014). Thence, all EU Member States have adopted national renewable power goals describing their own measures envisaged to facilitate their renewable targets objectives. This strategy incorporates various separate sectoral targets for heating, transport, and cooling; planned policy measures; the various mix of renewable innovations expected to be implemented as well as the intended mix of cooperation mechanisms. The EU supports public interventions like support systems which remain essential to make renewable energy reforms aggressive. In order to circumvent distorting business and energy rates, support schemes of renewables are meticulously outlined and optimized including strict supportive time frames.

However, looking back in history, it becomes apparent that before the RES started playing a crucial role in generating electricity worldwide, the standard design of electricity markets was marked by the historical development mostly on the Old Continent and in the United States. As a matter of fact, prior to the Industrial Revolution that was launched in the 1760s and the increasing importance of coal that started to be used for powering machine tools and factory systems in the second half of the 19th century, all energy sources were renewable (e.g., water, or biomass).

Generally speaking, RESs represent a clean source of electricity that is an alternative to fossil combustible fuels. These fuels were formed very many years ago when creatures, plants, and animals decayed and were buried in the soil. The electricity mix has continued to be dominated by fossil fuel use with more than a half of the gross power generation being from fossil fuel, but this has been a typical feature for many European countries over the last century. Nowadays, the share of fossil fuel has decreased by 20% since 1990 and this can largely be attributed to the growing popularity of RESs. A sustainable market design is characterized by an approach that can achieve long-term goals and as to accommodate medium term and short-term changes (see Newbery et al., 2018). Moreover, it needs to guarantee a decent supply of power and has to provide an effective utilization of resources and offer simultaneous investment incentives and innovation. Generation of electricity from RESs are commonly located and dispersed in remote regions which are far away from large consumption centers such as cities.

The rise of the RESs paved the way for supporting a more effective electrical energy consumption and metering. The new trend also called for the establishment of the new method for assessing the consumption of electricity using ICTs. Even though stakeholders worldwide are promoting the necessity of implementation and effectiveness of renewable energy for achieving energy security and sustainable development, households and individual consumers are reluctant to embrace these ideas. Lots of consumers are still suspicious when it comes to the RESs and their usage. Moreover, they are against the idea of smart metering when the new generation of electrical and gas meters is able to provide precise information about how much and when the energy was consumed at each given metered household. These attitudes should be changed, since smart meters allow households and businesses to measure electric energy consumption in a more precise way and to reduce the costs of electric energy. All these examples including renewables, energy storage, EVs and energy management systems become the pieces in the mosaic of the increasingly widespread products and services that the smart electricity grids need to embrace in order to stay competitive and up to date. Smart meters become an essential part of smart grids by enabling measurements and analysis of the site-specific information that helps the utility companies to introduce different prices for consumption based on the time of day or the season of the year. Many studies conducted in various countries demonstrate that the installation of the smart meters with in-house displays improved the consumption feedback and significantly reduced residential electricity demand (Carroll et al., 2014). However, many consumers still yield traditional approach and are slow to embrace the new technologies. With regard to this, Rausser et al. (2018) show that the costs of installing and running the smart meters for monitoring the energy uptakes and enhancing savings on the electricity bills might seem higher for consumers than the opportunity costs of ignoring the prices of electricity in their everyday decisions. There might be many opportunities how to cut one's electricity bill, however it appears that all of them require personal discipline, commitment, time, in short efforts many customers no longer possess in our globalized, digitalized, and interconnected world.

The widespread popularity of the RESs has reached its focal point and the world is now adding more capacity for renewable energy annually than all other sources such as natural gas, coal, and oil put together. The share of the renewable electricity output (in percent of the total electricity output) is growing with the most notable increase in Europe, North America, and South East Asia (see Fig. 5.2).

The most significant change occurred in the year 2013 when the 143 GW of clean power was produced compared with 141 GW in modern manufactories that burn fuel. The change will continue to advance and by 2030 more than four times of decent power capability will be appended (Khalid et al., 2016).

According to the International Energy Agency and other data sources, the cost of solar and wind energy proceeds to fall, and this is way cheaper than most in countries of the planet (Thapar et al., 2017). The newest source of power in the mix is solar energy which aggregates 1% of the electricity business now but by 2050 it could be world's biggest individual source. One of the benefits of renewable power is that it can be renewable, and it is hence sustainable and will never run out. The other advantage of renewables is that generation facilities need less maintenance than conventional generators. Renewable energy projects additionally bring economic benefits to many regional areas since most projects are

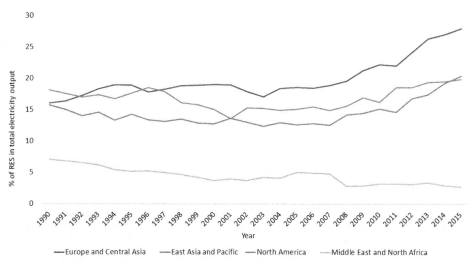

FIGURE 5.2 Renewable electricity output (% of total electricity output). Source: *World Bank, 2019. World development indicators. Available from: <https://data.worldbank.org/indicator/EG.ELC.RNEW.ZS> (accessed 20.1.19.) (World Bank, 2019).*

located away from huge urban centers and outskirts of the capital towns. These economic benefits may be from the intensified use of regional services as well as tourism.

Traditional electricity businesses are broken and can no longer fulfill their essential functions. The power markets are distorted, and disturbing manifestations are emerging. Increasing production expenses and falling commercial rates.

In many EU countries, several factors can be credited to the sinking of commercial electricity costs. These factors involve subsequent slow growth in economic activity after the economic and financial crisis of the 2008 and the subsequent recession but one vital reason as to the fall in price is the adoption and growth of renewable generation. But at times when they are generating system marginal costs are depressed since renewable sources have zero marginal cost and their presence in the market reduces the requirement for higher-cost plants to generate, so there is a fall in market prices. However, this does not confirm a slump in the long-term prices. Renewable power has a tremendous expense than the conventional production systems they replace. If renewable sources were reliant on the market to remunerate cost most would be losing money but since these sources are not reliant on the sources, they go unheeded. In effect, government policies which encourage continued deployment are in conflict with markets and the strains can only increase over time.

Over the centuries, electricity technology has been evolving at a slow pace and it was only in the 21st century that it has made an impressive progress. Underwent a rapid development in the field of energy and energy conservation. At present, humanity has learned to generate energy using RESs. The main sources, as a rule, include solar radiation (solar energy), wind power, energy of rivers and streams (hydropower), tidal energy, wave energy, and geothermal energy.

The transformation of the energy of tides, waves, sun, and wind does not actually require preparation for later use, unlike coal, oil, gas, biomass, hydrogen, as well as all other similar energy sources. If it is reasonable to use only these four main sources of energy, applying the latest developments, it is possible to provide energy to the whole world forever. However, apart from securing the unlimited sources of energy there is another path toward sustainable future—optimization of energy use.

For 130 years, there have been electric light sources, and all this time people have used only two-heat and gas-discharge. Only by the end of the 20th century did a new, improved type of electric light source appeared represented by a semiconductor light sources or light-emitting diodes (LEDs). The mechanism of operation and functioning of LEDs is based on the principle of generating light when an electric current pass through the boundary of semiconductor and conductive materials.

In semiconductors, the current flows when even a small voltage is applied, overcoming the threshold, like water flow, in which energy is released when the threshold is passed, sufficient to rotate the turbine, mill wheels, and so on. It all depends on the height of the threshold and the amount of water flowing. When overcoming the "energy threshold" electrons emit a certain energy. Usually, energy is released in the form of heat, but under certain conditions it can turn into light.

LEDs are the future of lighting, with great efficiency savings, precise control and a variety of practical applications (Yoon et al., 2016). LEDs are semiconductor light sources that are activated when voltage passes through them. Research and development in the field is constantly improving the efficiency, brightness and life of LED bulbs and optimizing them for other exciting applications that are still awaiting their discovery. For example, filament lamps from designers are now reborn and innovative companies apply LED technology to traditional filament lamps to make them both efficient and beautiful. LED technology is advancing rapidly, leaving other energy-efficient lighting technologies in the dust. The "Light Fidelity" (or "Li-Fi") is not a kind of LED bulb, but rather a system of visible light using white LEDs visible to transmit high-speed data (Sharma et al., 2018). Future LEDs are going to be able to save energy by detecting the number of occupants in a room and then directing the lighting to their location. Therefore, as LEDs benefit from the use of automotive lighting, the market for replacement lamps is slowly eroded. As dynamic lighting systems become more grounded and lighting becomes more and more "signature," the collaboration between vehicles and lighting manufacturers is becoming increasingly important. A solid warranty and post-service capabilities are going to be crucial in the automotive lighting market, as car owners may be dissatisfied—or worse—when they are hit by a higher than expected bill for replacing lamps in complex car lighting systems. In the midst of declining revenue, backlight companies should also be able to switch to a new technology: organic diode (called OLED). LEDs can be adjusted to the wavelengths which seem to measurably improve the educational environment. Meanwhile, new home medical devices are entering the market in Europe, using blue LED lights to alleviate back pain. The coloring is the highlight of about 50 years of innovation with LEDs or LEDs, a technology that has been widely developed by Philips and may prove to be one of the ideas that we are always looking for, but rarely identified. So far, early buyers are attracted to the durability of LEDs (which last not only years, but also decades) and their extraordinary energy efficiency (which consumes only 15% of the energy of a bulb) (Montoya et al., 2017).

Invented by Nick Holonyak at General Electric in the early 1960s, the first LEDs were seen as a scientific breakthrough, but not as a consumer product. OLED, which stands for an organic diode, is a kind of LED that uses an organic compound (rather than a traditional synthetic composite) as a film layer that reacts to the electrical currents that generate images (Wang et al., 2017). In addition to technological improvements, the growth of the OLED market is fueled by increasing investment and government support for oil research and development. But for the time being, OLED will dominate future—proof AV solutions, as more products are coming onto the market and research and development is leading. LEDs are currently four to five times more energy efficient than conventional technologies and are expected to be even more efficient—a key advantage, as more than 50% of the total cost of light is due to energy consumption. However, the increase in the use of connected lighting systems in the consumer segment will be slower. This is because two trends in the volume of depress lamps—the increased use of long-term LEDs and the increase in market share of LED luminaires—will be successful. The success will be partly due to the continuous research and investment in order to reduce production costs (which will reduce the price of lamps) and improve the efficiency and life of the lamp. Profitability in the consumer lighting industry will increase in the coming years as a result of the shift to LED lighting. The light is not only a nostalgic object, but also an alternative for those who do not have access to the electricity grid, who previously had to use dangerous liquids to light their lamps at night. As more and more people are realizing the energy-efficient benefits of LED lighting, it seems that everything from headlamps to televisions and wallpapers is now being illuminated by the all-round lighting technology. Unlike light bulbs, LEDs are super compact, do not emit heat and consume only a fraction of the energy—and their flexible, plastic design allows them to be used in a variety of innovative ways other types of lighting simply cannot touch. Holonyak's idea continues to surround the concept of LEDs, as the shape of lighting things to come. Some rushed, backward calculations left both sides undoubtedly at the end of the day when the future of LEDs was in the lighting industry. LEDs could only handle half of the lumen, but by directing the light more efficiently in a much narrower beam, the team proved that it was possible to adapt to the performance of conventional products.

In the early 1960s, the first LEDs appeared on the market, using this effect-indicator element with a weak red, which changed into a green glow after a few years. The light output at that time was no more than 0.1 lm/W (which is 100 times less than that of incandescent bulbs), the service life totaled hundreds of hours, and, of course, they were not even considered as light sources in the conventional sense (Craford et al., 2013).

By the end of the 1980s, fundamentally new semiconductor materials were created, with the help of which the power, brightness, light output and service life of LEDs increased by orders of magnitude. In the new materials, compounds of indium, gallium, and aluminum were used in various combinations. The LEDs on the basis of these materials radiated already quite bright light of red, yellow, green, and orange. In 1996, Japanese specialists from Nichia created the first blue-emitting diodes. With the help of phosphors, they began to transform this light into yellow, giving white light of various shades in combination with blue, as a result, from the end of 1990s, the first lighting devices began to appear simultaneously in different countries. The LED technology played the role of the light source and not the light element. To date, LEDs is the most evolving direction in the field of light sources. Nowadays, there are LEDs of almost all colors of the rainbow are available on the market.

Modern mass-produced LEDs have the following parameters: white light output up to 35 lm/W (higher than that of halogen incandescent lamps), red and green—up to 50 lm/W; service life—50,000 hours. Over a decade ago, Hewlett Packard reported a LED lifetime of 1 million hours (or 120 years of continuous work). This is a pure theory, of course, since LED bulb failure mechanisms tend to be very complex and fundamentally different from the well-known old-fashioned light bulbs, we all got used to. A service life of 50,000 hours is merely an Average Rated Life (ARL) which represents the length of time for 50% of an initial sample of bulbs to fail within a given time span. In spite of all these, the cycle of an ordinary LED still surpasses that of a simple old-fashioned light bulb (Narendran et al., 2001).

All these make everyone think that in the coming decades, LEDs will force out traditional light bulbs and push out obsolete light sources from the market. The bulk of the currently produced LEDs has a dome-shaped case with a diameter of 5 mm. Their rated operating current is 20 mA. Some companies produce LEDs with a diameter of 10 mm with a working current of 40 mA. The highest power of a single LED today is 5 W. For quite a long time, LEDs have been used in light-signaling devices for the road and railway traffic lights, information boards, signs, etc. And it was only recently that they are started to be deployed elsewhere—on certain highways and streets, office and retail premises, as well as in everyday life situations one can see new LED light sources that have created a qualitatively new model of lighting the public and private space.

Although LEDs are considered to be a relatively new technology by many people, they are in most cases superior to traditional light sources. LED lighting products are superior to high-pressure lamps in areas requiring the use of colored light. As seen from today's point of view, LEDs have a number of advantages over traditional systems:

- energy efficiency of LEDs can be up to five times higher than that of halogen and incandescent lamps;
- LED light sources are directional and emit light only in the desired direction. The luminous surface allows the use of more efficient optics and better control of the light;
- the quality of light from white LEDs is now comparable to the quality of light from high-pressure discharge lamps and fluorescent lamps. Modern advances in the LED manufacturing industry provide consistent color and color temperature, equivalent to or even superior to those of traditional light sources;
- significantly increased service life of LEDs compared to traditional light sources. As a result, replacement and maintenance costs are reduced;
- since the advent of LEDs for several decades, there has been an annual decline in the cost of LEDs by an average of 20%;
- compared with fluorescent lamps, LEDs do not emit harmful ultraviolet rays, destructive materials and discoloration paints, making them the ideal lighting solution for installation in shop windows, museums, and art galleries;
- LED lighting devices generate heat, but the beams of light emitted by them are cold. MTR with a well-designed heat sink protects users from excessive and harmful heat;
- light sources using LEDs can operate at low temperatures and withstand the effects of vibrations, which allows them to be used in harsh environments where it is impossible

to install and maintain traditional lamps. LEDs do not have moving parts and filaments that can easily be destroyed and fail;

- operation of LED lighting systems can be controlled by digital controllers, ensuring maximum efficiency and high flexibility;
- LED lighting devices are instantaneous, that is, do not require time to warm up or turn off;
- unlike fluorescent lamps, LEDs do not contain mercury, and are also environmentally friendly;
- most of the manufactured LED lighting fixtures not only meet the requirements of energy efficiency and environmental standards, but also often surpass them. At the moment, standards are being actively developed for testing and measuring parameters that will provide the basis for an accurate comparison of the characteristics of various LED lighting devices among themselves and with traditional light sources.

In addition to a long service life, LEDs have many other advantages: high reliability; small dimensions; high utilization rate of the light flux; easy handling; good safety. A large range of colors and a variety of radiation angles (from 3 degree, that is, a very narrow light beam, up to 180 degree, that is, a uniform glow in the hemisphere) promote the use of LEDs in various lighting devices.

The main disadvantage of lamps of this type is their very high price. As practice shows, a period of 100,000 hours is practically unattainable (Nair and Dhoble, 2015). Quite often, the manufacturer itself gives a guarantee for a period of 3−5 years, and not 10−11 years as envisaged by the technical documentation and the test results. This is attributed to the fact that causes a phenomenon of degradation, that is, a quiet dying of LED crystals. The process is irreversible and is taking two stages. First, the crystals lose their brightness, and then they completely go out. The next drawback of LED lamps is an unpleasant glow spectrum. According to the testimony of psychologists, a considerable amount of consumers negatively respond to the use of such lamps at home.

LEDs provide directional light, even with a lens that expands the angle of illumination. You may need more of these lamps to get the usual uniform illumination. Thus, in this situation, LED lights are inferior to cheaper, more familiar lamps and fixtures. Unfortunately, when choosing LED lamps for quality characteristics, the purchase budget greatly increases, but this problem will disappear with the progress of technical progress, with an increase in the power of LED lamps and a decrease in the cost of their production.

And the last drawback was the low interest of energy companies and the state in energy saving, because it reduces profits. Since there are no real benefits for the use of energy-efficient equipment, all the difficulties and costs are borne by the end-users.

An interesting example of how electric energy can be saved in public spaces is the operation of the LED lighting system using solar power as source of energy. The basis of the LEDs is a semiconductor crystal, surrounded by a reflector, directing light in one direction, and located on a conductive substrate. Voltage is applied to the crystal and the substrate through the inputs. Against external influences, the crystal is protected by a housing made of transparent epoxy or polycarbonate. The upper part of the body, as a rule, is made in the form of a dome with a certain curvature and plays the role of a lens that forms a light beam. Sometimes instead of a dome, "Fresnel lenses" are made, that is, sets of circular concentric microlenses on a common flat base. The internal reflector and the lens housing form the luminous flux emitted by the crystal in an appropriate manner, therefore, in luminaires with LEDs, no additional optical system is required, as with "ordinary" light sources.

To power the LEDs, a constant low-voltage current is needed, the value of which depends on the chromaticity of the radiation: in red LEDs it is 1.9–2.1 V, in green LEDs 2.5–3 V, in blue and white—about 4 V (Pousset et al., 2010). A LED is a semiconductor device with an electron–hole p-n junction, which creates radiation visible to us when an electric current is passed through it in the forward direction. The light spectrum of an LED lamp depends on the chemical composition of the semiconductors used in it. When an electric current passing through the LED circuit in the forward direction, the charge carriers—electrons and holes—recombine with the emission of photons (due to the transfer of electrons from one energy level to another).

The electric current required for the operation of the LED lamp can also be obtained from the solar light energy. The principle of operation of such devices is based on the ability of the junction to convert light into electricity. The battery design is a 200-micron thick silicon wafer with a thin layer of phosphorus deposited on one side and a boron on the other. At the points of contact of silicon with boron, an excess of free electrons occurs, and at the places of its contact with phosphorus, electrons are in short supply, which leads to the appearance of "holes." Light photons hit the surface of the plate and displace excess electrons from the silicon site in contact with boron to the missing electrons in the zone of contact with phosphorus. It creates an orderly movement of electrons, which is the electric current. It is possible to direct current to the LED lamps with the help of metal paths through the plate. During the day, the solar panels charge the batteries in the system. At night, the batteries give electricity to illuminate the street.

At present, the night-time lighting system based on sodium arc lamps and gas-discharge lamps is not considered to be quite convenient, and because of its low efficiency, such a design is refused. The main disadvantages affecting its effectiveness are: short service life (about 1000 hours); circular luminescence of lamps requires the use of lamp-shades, which leads to additional costs; at jumps of a tension of the power supply network the service life of lamps sharply decreases. In addition, the insulation of wires is aging, over time, their replacement is required. Lamps consume a lot of power, which requires the use of more cable with a larger section. But, besides all of the above, the main disadvantage of sodium lamps is low light output, because the ratio of the power of the rays of the visible spectrum to the power of the consumed network is very small and does not exceed 4% (Gutierrez-Escolar et al., 2017).

Street lighting has a big difference from the internal in terms of technical requirements, installation design and safety standards. Of great importance is the power of the lamp, and, consequently, light output. Luminous efficiency is the ratio of the luminous flux to the power consumption, that is, the efficiency of a street lighting lamp. The light output of the DNaT lamps is up to 15 lm/W (Chitnis et al., 2016).

Replacing the old lighting system with a new one (solar-powered MTR) seems to be a rational solution which might increase net savings, shorten the payback period of the project, and help to achieve electricity savings. These indicators indicate the effectiveness, feasibility, viability and investment attractiveness of LEDs over the traditional sources of light.

In addition to the economic effect, the project of replacing the lighting system on the DNaT 150 lamps with a LED solar lighting system has a social effect. Research in the field of psychiatry notes that the soft and even light of LED lamps has a positive effect on the emotional background of a person, they tend to calm and sooth and even support mental health. It is recommended that all office owners switch to LED lighting—this will help improve the efficiency of employees, reduce teamwork, improve mood, and relieve eye fatigue. A number of studies in the field of medicine show that LED lighting accelerates the regeneration of tissues and neurons, therefore it can be successfully used in the treatment and prevention of many diseases. In 2008, researchers from Germany found that when exposed to intense light from LED light sources, every day for several weeks the skin becomes more elastic and younger looking, the complexion improves, and the depth of facial wrinkles significantly decreases (ScienceDaily, 2008). In addition, it is worth emphasizing that in the production of LEDs, hazardous compounds of mercury and other heavy metals are not used, so they are safe for the environment and do not require special disposal rules.

This was just an example how we can achieve massive energy saving without making the consumers to dramatically change their behavior and habits (e.g., streetlamps and streetlights). Prosumers of the future will likely to use LEDs instead of old and obsolete light bulbs and smart grids of the future will favor them and optimize their deployment and operation.

5.4 Basic principles of peer-to-peer markets

P2P platforms create and facilitate trade between a large number "fragmented buyers and sellers" by matching buyers and sellers, using different algorithms inbuilt within the platform. According to Oram (2001), the growth of the Internet and other affiliated technologies, such as the World Wide Web (WWW), has provided individuals or businesses with an ability to create websites on which individuals can list products and make purchases, thereby connecting buyers and sellers. On the designed platforms, sellers enter the particulars of a product good or service they are selling, which are augmented by a multimedia attachment supporting the information provided by the seller (Ert et al., 2016; Einav et al., 2016). On the other hand, prospective buyers, enter the particulars of a product or service they need, which are subsequently run against the information about products goods and services within the database, and a report of possible matches generated for a prospective buyer to select from and an exchange ensues if the terms are agreeable to bother parties (Ert et al., 2016).

The development of Internet, and in particular broadband Internet as well as high-speed mobile networks (such as 4 G or LTE that are now in operation or the long-awaited 5 G that is yet to come and take its place) is a very important feature in the rising popularity of the P2P markets. Fast Internet offers the ease with which information can be exchanged two ways and the transactions can be conducted. The role and importance of broadband Internet access are crucial in the current development of the infrastructure associated with broadband, its use and related activities is in many ways. Government agencies in many developing countries do not track in detail the indicators for broadband development, in contrast to the indicators related to the penetration of traditional telephone and mobile communications. However, the amount of reliable and detailed data is growing. Reliable data is difficult to obtain, since broadband technology is relatively new and widespread only in some developing countries.

Broadband Internet access can have significant economic benefits. The transition to broadband at the company level has a proven positive impact on productivity and job creation (Black and Lynch, 2001). So far, the main benefits associated with the introduction of broadband have been received by large multinational companies, while the main potential for further growth is associated with small- and medium-sized enterprises (SMEs) and can be expressed in increasing the efficiency of operations by expanding the implementation of broadband access to the production, marketing, and distribution.

According to one widely cited World Bank study, with broadband penetration increasing by 10%, the average gross domestic product growth in developing countries is growing by about 1.38% (Minges, 2015). These results became a baseline for economic impact studies related to the introduction of broadband access, as well as a strong incentive for governments to invest in the growth of broadband penetration.

Moreover, there are many areas of economy and social life where broadband Internet is setting up the new standards and helping to perform transition to the smart economy that embeds smart grids. One of these areas is education and learning. The task of implementing broadband in educational institutions, especially in secondary and higher education is directly related to the extremely important task of formation of a technologically literate population and labor force (Frieden, 2005). The deficit of workers with the skills to use computer technology and technology related to the internet hamper overall growth of the ICT sector and the modernization of working methods in all other areas of commercial and government activity.

Among the innovations that are the driving factors of the educational process, it is possible to name such as the use of electronic textbooks, research tools, and training on Internet-based audio-visual materials in presentations, interactive tutorials, digital libraries and teaching materials open access, virtual science labs and museums and all sorts of ways to distance learning and network degree programs.

Social and cultural development of broadband Internet cannot be underestimated. Broadband can have very significant advantages in the social and cultural sphere, and their implementation may be long term. In some cases, the introduction of broadband becomes a complement to the implementation of other development problems and create new opportunities for the development of social inclusion and the expansion of social rights and opportunities.

The potential of ICT in recent years has been most clearly manifested, in particular, in the political sphere. Increasing the ability to communicate certain ideas can contribute to increased political participation.

Last but not least, there is a wide access to information. Across a variety of forms of political participation, ICTs contribute to raising the awareness and effectiveness of such participation. Political parties and other organizations have now much more extensive resources to establish communication with the voters, donors, politicians, and government agencies, as well as to influence them.

In addition, ICT made a significant contribution in the form of e-government programs as well as e-government and public administration. Many governments have realized that there are a number of possibilities of using services and software based on the broadband ICTs for the improvement and development of public services and encouraging citizens to improve their quality of life.

A broadband ICT infrastructure consists of several levels: a data network, access mechanisms and services for end-users, which can be built in different configurations. Broadband-based services that end-users receive are provided by a group of various interconnected and competing providers.

The main elements of the infrastructure and services associated with broadband. First of all, there are trunk networks. Broadband requires high-performance international and national backbone data networks that use fiber optic cables to provide connectivity on the busiest long-distance routes and connections between countries.

All broadband networks and services must ultimately be connected to the global Internet, as well as to national or private networks. Thus, the construction of a fiber-optic network in rural areas, as a rule, requires from 5000 to 8000 dollars per kilometer (Taheri et al., 2017).

One of the main challenges in achieving universal access to broadband is the spread of broadband access beyond urban centers. To this end, it is necessary to create additional capacities for connecting to the nodal points located close to the residential centers of the rural population. Such nodal points of broadband connectivity may include points based on wireless data transmission and/or fiber optic data transmission, and sometimes on the use of satellite technology, all of which data transfer technologies require large costs for construction and maintenance. These factors seriously limit the willingness of commercial operators to expand their backbones far into rural areas where revenue from local broadband services can be low or nonguaranteed.

Broadband Local Area Networks (LAN): the final link of a broadband connection is a wired or wireless connection between the backbone network and end-users. It is often referred to as the "last mile" connection. In many countries, traditional wired telephone networks are widely used to provide individual broadband access for businesses, institutions, and households, including by connecting via an asymmetrical digital subscriber line.

Systems using coaxial television cable and connecting the optical cable to the end-user or FTTH are called ADSL. The cost of such wire connections for homes in cities and villages depends largely on the density of the population in a particular locality, since the growth of the projected demand can significantly reduce the costs of one connection from more than $1000 to less than $100. All these technologies are widely used in the markets of developed countries and are beginning to spread in developing countries, at least in relatively densely populated urban areas with higher income levels.

Data storage and data exchange points are also of a great important here. The storage of a huge amount of digital information and software, as well as the exchange of them, require separate capacities and ever-increasing investments. Companies and governments that deal with terabytes and petabytes of data require access to disproportionately large storage databases—data warehouses—as well as extremely high bandwidth data transfer connections and reliable and secure energy sources, data buildings and protocols. In general, the complex of these technical facilities located throughout the world is called the "cloud." There are the following market growth factors that can be identified in connection to the development of fast and reliable Internet connection:

- Increased coverage and development of data networks. The growth in the coverage of sparsely populated areas by networks of medium and large providers;
- An increase in the number of devices and their diversity (PCs, laptops, tablets, or smartphones) and, as a consequence, an increase in the required quality and speed per household;
- Increasing the attractiveness of tariffs and comprehensive services, offers;
- An increase in the number of home broadband connections. Now providers are guided by the principle "more Internet for the same money" is no longer a question of reducing the cost of tariffs, instead operators increase the speed of access. The tariff policy of operators will continue to focus on special offers on tariffs and package offers, which actually reduce the cost of services separately, but retain the total income of the operator. As a result, the average income from one broadband subscriber will decrease.

With regard to the above, it becomes apparent why the relevance of using optical fiber in broadband Internet connection becomes a must nowadays. The urgency of creating broadband is due to the constant development for the end-users of much more sophisticated equipment compared to what was needed for traditional telephony. As the development of basic networks and data transmission technology, the revolution of broadband has gained momentum due to the continued mass modification of the product in the market for end-user devices that connect to these networks.

The range of consumer and industrial devices, which today is called "intellectual," that is capable of accessing the Internet and connecting to other devices, as well as performing various interactive functions, continues to expand. Smartphones and tablet devices show the most dynamic growth in the equipment market. Personal computers and laptops are still indispensable, especially in business, among other things, the so-called Internet of Things becomes a number of other devices ranging from smart TVs, game consoles and countless other gadgets to security devices, cars and practically all links in the chain of industrial activity.

As the variety of technical devices grows, there is a constant decrease in their cost, which expands the possibilities of diversified use of broadband services.

Most users prefer small and affordable mobile devices, while larger, more robust computers are often used in offices, schools, and access centers, but even with the cost reduction, the cost of such equipment remains a significant barrier to the development of the local broadband service market. communication in developing countries. Rapid change of standards and obsolescence exacerbate this problem, since consumers are not likely to be able to change or upgrade their devices too often.

The value and attractiveness of broadband services are directly dependent on software platforms and operating systems, multimedia applications and the amount of content that is available through broadband. The set of information applications of accessible content continues to expand indefinitely, and in the context of the revolution of social networks, the users themselves become the main source of such content.

The rapid growth of broadband networks and modern devices has led to the emergence of a huge number of new dominant applications ("killer apps"): information and communication applications that are becoming almost ubiquitous. The main part of these applications belongs to the sphere of social networks and has become part of the interactive, user-controlled phase of the development of the Internet, known as "Web 2.0." Facebook (a popular social network), the most common online application, was launched only in 2004, and in less than ten years its audience has exceeded a billion users, and half of them work with this application mainly using mobile devices. Other social networks offer unlimited sharing of YouTube videos, Flickr photos, and Weblogs ideas. Virtually all of these hugely successful Internet applications were created in the mid-2000s, in parallel with the spread of the fast-speed broadband Internet.

The urgency of creating a broadband connection is due to the continuous development for the end-users of much more sophisticated equipment compared to what was needed for traditional telephony. As well as the development of basic networks and data transmission technology, the revolution of broadband communications has received an impetus due to the continued mass modification of the product in the market for end-user devices that are connected to these networks.

All of the above allows the creation of effective P2P markets that is based on Internet communication, high speed of data transfer and a rapid exchange of information. A typical P2P market represents a "platform market" or a "two-sided market" which, according to Weiller and Pollitt (2016) is a multisided market where an intermediary captures the value of the interaction between user groups. Jean Tirole, a Nobel Prize Laureate, popularized this type of markets in his fundamental works (for instance, Rochet and Tirole, 2006) which made it possible to use them in many domains of economic theory, including energy economics, as well as policy.

The common goal for any P2P company is that the lenders and borrowers or the buyers and sellers find a platform to do their transactions. This means that the company will have to come up with a suitable, efficient, and sustainable system that will allow for this exchange. Therefore, the services of P2P platforms are divided into three stages which are Loan Application and credit evaluation, investor funding and last, but not least, loan repayment:

- Loan application and credit evaluation: This is the phase where the person borrowing funds or seeking for a loan fills up the loan application forms and gives all the information, he or she finds credible (Sumit, 2014). The P2P company would then go through all the information that has been presented by the borrower and screens it to ensure that they are true and that there are no discrepancies. They would then check whether the borrower is creditworthy and eligible for the loan;
- Investor funding: At this level, after the application process goes through its time for the investors to choose who to give a loan and who to deny the loan. Before any application is brought to the knowledge of the investors, the P2P company has to come up with an investor base that must all have registered accounts running in unison with the P2P records and systems. This is to provide ease of management of everything that is being processed and those that are yet to be processed. These platforms normally vary depending on the country that the P2P is located (Sumit, 2014). Furthermore, these platforms accommodate investors such as retail lenders, individuals with high net worth, institutions which specialize in lending (such as banks among many others), or other financial institutions and parties which partially fund them. In most cases, the

P2P company offers a disclaimer to the investors that they would not be responsible for any losses they make via their platform. This comes handy with the advice from the company. This makes the investors have a fine and rigorous way of mitigating risks of losing their monies. The platform benefits by having commissions from the transactions that go on between the borrowers and the lenders, and in the end of it all, everyone is happy;

- Loan repayment: This is the final phase of the process. When the agreed time elapses, the platform would begin to collect the amounts that were lent from the borrower's accounts to those of the investors who lent them, inclusive of the interests from which the platform would also have a commission, if applicable (Sumit, 2014). Sometimes to achieve this, platforms usually involve third parties who would be the ones to manage the money movements to the extent of even verifying the borrower's worthiness for a next-time loan. This is achieved by using big data analytics. Apart from these third parties, many P2P companies resort to using debt collection agencies when they realize that there are cases of delayed payments. To make this entire process possible, many companies may get involved from the time of loan application to time of loan repayment.

In general, P2P market can be presented as a decentralized market model where there are no intermediaries or third parties in the market, involving only two individuals who interact in the buying and selling of the said good(s) and/or service(s) (Jayme, 2018). Another definition of this is the market model that allows lenders and borrowers to transact or exchange money directly on an online platform that matches lenders and borrowers (Matofska, 2016). This gives liberty for the buyer to get to interact directly with the seller, and the seller is in a position to know exactly what his market really wants. As a result, being an added opportunity to him or her (the seller) because of instant feedback from clients (Jayme, 2018). In addition to that, P2P is a model that is quite often seen as an alternative to traditional capitalist market, where the business owners have inherent ownership of the finished product and also the means of production (Wales, 2015). This means that the business owners have to hire labor who produce these goods and services.

Even though most people link P2P markets to capitalism, they are quite different. This is because, in capitalism, the workers or those in the business have no ownership to the means of production nor rights to the finished products whereas the opposite is true for a P2P market model. However, this does not mean that P2P cannot exist within capitalism, it does. An example being Open-Source Software which coexists with the retail and commercial software of services such as Uber and Airbnb (Bauwens, 2005).

Uber is a perfect example of an alternative to taxi services while on the other hand, Airbnb is an alternative to hotels and inns. In the real sense, Uber as a company has no cars to their names while Airbnb also has no hotels/restaurants to their name. This makes the P2P model a very important model to the business world in this era of developing technologies, especially in these times of a highly revolving world where everyone is looking for a way to move things, ideas, concepts, and models fast and to make money from the comfort of their beds.

To begin with, technology being an important aspect in the life of every successful business, there are numerous businesses that have adopted the P2P market model who has never grown immune to the impact of technology in their production processes and eventually, finished products. Thus, making people believe that P2P is where money makers meet technology makers because mostly, investors are always out to look for how to make their money work for them which eventually amounts to the lending aspect of the P2P business market model.

P2P is unique in the way it gets capital online from various investors and how it has the capacity to enable borrowers interacts directly with lenders. In all this, the cost efficiency aspect and the methods of innovative underwriting are not excluded in the P2P market model (Mengelkamp et al., 2018). This model is just but a modification of what has always happened before time (before technology began to advance the way it is today) just that back in the days, people never used online platforms to meet up directly with their clients. This means that the clients can be chronically affected when technology remains retarded and refuses to mature as time ticks ahead.

All in all, there are three types of P2P market model platforms. These are represented by the nonprofit platforms, socially responsible platforms, and commercial platforms:

- Nonprofit platforms: these are for investors who are looking for communities who are unable to get loans from banks. Their sources of money are donations, grants, and volunteer works;
- Socially responsible platforms: these are investors and companies who offer their loans for profits though all these loans are directed toward the development of social causes or toward supporting the development of a particular community. They make the profit for their gains and the gains of their investors as they meet social objectives;
- Commercial platforms: they have a larger clientele as the general public are all their clients or customers. They have more resources to provide loans and most P2Ps are of this type.

Accepting the fact that most P2Ps allow a direct operation between borrowers and lenders, there are risks, of course, that the providers may fail to deliver thus making the clients lose money, time, and business as a whole. Therefore, making money from this kind of platform is quite tricky since all the operations are managed by a provider whose service delivery status is unsure (Sumit, 2014). In this case, most investors always seek for P2Ps that have a good reputation and are consistent with their service provision.

Because P2Ps are just intermediaries, most times, the loans are not secured loans which are an advantage on the side of the borrower but a risk to the investor/lender since they are not quite sure of the outcome of the transaction (Milne and Parboteeah, 2016). Technology comes in handy on this because as it advances, there are various measures that can be taken by the P2P companies to ensure that their investors are always safe and are in no position to losing their money.

In addition to these, there is always a liquidity risk on the part of investors since they may not realize that their monies and investments are all locked until maturity of the loans they have given to borrowers (Sumit, 2014). There is, however, a few P2P companies with platforms that gives an option of secondary market functionality that gives investors an eased liquidity. Most platforms should adopt this in order to ensure that investors are secured and that their investments are flexible (Sorin et al., 2019).

The power sector in the postliberalization era is organized in the form of wholesale-network-retail. Most electricity consumers (except self-generators) rely on this hierarchical setting of the market to procure their electricity services (through engaging with retail suppliers who make procurements in the upstream wholesale market). The main reason that, in this model, a small consumer like a typical household cannot transact directly with a wholesaler is "transaction cost." Contrary to the traditional paradigm, P2P is a horizontal market structure in which producers and consumers directly transacts with a "low transaction costs." This horizontal arrangement is partly enabled by technological advances. For example, blockchain-based technology allows for the tracking these transactions at a low cost which is something that was not possible in the past.

P2P platform models can be based on many arrangements: retail, vendor, microgrid, or blockchain. The retail are the P2P energy-trading platforms in retail electricity markers that are used by the value-added service suppliers for their own differentiation. The examples are Piclo (United Kingdom), and Vanderbron (literarily meaning "from the source" translated from Dutch) (the Netherlands). Vendor platforms offer electricity obtained from DER to enhance their product or to help reducing charging prices for EV. The example of this is another platform called "sonnenCommunity" (Germany). Microgrid platforms offer incentives for prosumers in order to gain their support for the creation of microgrids and other community "green" energy projects. The example is Brooklyn Microgrid energy-trading platform (United States). Finally, blockchain platforms use secure decentralized blockchain protocols for supervising smart contract transactions. The example of this platform is Brooklyn Microgrid energy-trading platform (United States).

Overall, one can see that P2P electricity platforms appeared in a range of sectors. P2P platform markets enable the reduction of the transaction costs and allow individual and household suppliers to keep pace and to compete with established traditional suppliers. This can be easily shown using the following scheme: while in the traditional markets, vertically integrated firms control the interactions between producers and consumers, P2P-trading platforms make direct transactions between users possible (see Fig. 5.3).

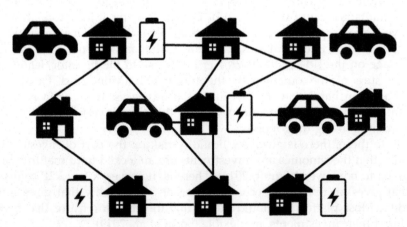

FIGURE 5.3   Peer-to-peer electricity market network.

Overall, it becomes obvious that transactions between prosumers are most efficient when they yield complementary resources preferences. With regard to the above, it becomes obvious that P2P electricity market platforms offer three major advantages: (1) energy matching, (2) reduction of uncertainty, and (3) satisfaction of preferences.

5.5 Peer-to-peer markets and sharing economy

Centuries ago, and in some cases today, commerce and trade took the form of barter trade, where people exchanged goods for goods, and retailer, hypermarkets, corporations, and other similar businesses were virtually nonexistent. In later years, commerce and trade took the shape of businesses in the form of the products, goods, and services providers mentioned above, where if a person wanted a particular good or a service, one would purchase the same from designated providers of the required good or service. However, with technological innovations and development, the introduction and proliferation of computers, mobile telecommunications, the Internet, the WWW, and other related technologies, multiple distribution channels and business models have emerged. One such business model that has developed mainly due to the technological innovation and developments is the P2P model. P2P model is in a way embedded into the "sharing economy" or the economy of direct exchange between people that has evolved into the modern-day phenomenon when transactions are facilitated through the community-based online services. The launch of Uber, a ride sharing service, in San Francisco in 2011 can be used as a starting date for the sharing economy (the word "uberization" is often used to describe the main guiding principles of the sharing economy).

It can be shown that sharing economy and the P2P markets are based on the ICTs as well as the popularity of these technologies amongst the general public. Millions of people in the West use Facebook, Twitter or Instagram daily, while even more people use WeChat or Baidu in the East. Communication and interaction with total strangers over social networks, mobile messengers, and apps is becoming a daily routine. Many people interact more with strangers on social networks than with their own relatives or family members. In the eyes of most people, all these makes ICTs a well-known friend one can rely upon. Therefore, social networks constitute one of the most important steppingstones on the path to the sharing economy (see Fig. 5.4).

There is a clear link between P2P markets and sharing economy. For example, Katz and Krueger (2016) define P2P market as an online platform in which individuals can directly buy and sell goods to each other without any other physical intermediary. Einav et al. (2016) provide a near similar definition, defining P2P markets as online markets in which buyers and sellers can engage in "convenient and trustworthy transactions." As such, drawing from these definitions, it is evident that primarily, P2P markets use the Internet

FIGURE 5.4   Evolution of information technologies into sharing economy.

or the WWW, and facilitate transactions between buyers and sellers without any other physical intermediary or middleman as is the case in the conventional or traditional business model. Examples of P2P markets according to Einav et al. (2016) include eBay, Uber, OLX, Craigslist, and Airbnb, among many others.

Nowadays, online shopping for virtually any item or asset that a person needs, extending beyond the conventional items such as shoes, clothing, groceries, to items such as houses, motor vehicles, and travel destinations, has become the norm. While this may seem normal under modern-day operating environments, barely a half a century ago, in the 1980s and early 1990s, it was not the case. Shah (2011) opined that technological innovation and developments, coupled with demographic and social changes in the society, have quickly transitioned the society from P2P 1.0 to P2P 2.0, demonstrating the breakneck speed in which transitions in the P2P markets are happening.

The 1990s gave rise to technological developments which formed the backbone for the development of P2P markets as they are today. In 1995, Pierre Omidyar started eBay, a consumer auction website, which within a very short period transformed to a website linking buyers and sellers, stocking no items of their own. In the same year, Craig Newmark, started the website Craigslist as is today as an email listing of various types of adverts, which later took on the form of P2P marketplace (Galloway, 2009). Evidently, P2P markets started in the 1990s, with some of the pioneering companies being eBay and Craigslist, which were among the first online platforms to provide users with a medium of connecting individual buyers and sellers directly without another physical intermediary.

While these platforms started as simple websites connecting buyers and sellers or matching demand and supply for different consumables, such as items that people no longer used, the platforms have evolved to include a wide array of goods and services, normally found in different bricks and mortar businesses spread over large geographical areas, giving users a holistic one-stop shopping experience. According to Galloway (2009), the portfolio of products goods and services listed for exchange on the P2P platforms, has grown over the years, taking on the form of different types of new emerging P2P markets, such as P2P lending platforms. Which Katz and Krueger (2016) agree, observing that online lending, transport, and hospitality platforms, have emerged, underlying the fast growth and evolution being experienced in the application of technology and information systems in that particular way.

This example further enumerates the arguments presented by Shah (2011), that technological innovations and developments, augmented by other changes in the society, have led to fast-paced changes in P2P markets, quickly transitioning from P2P 1.0, which is the simple listing of products and services, in their simplest form, for exchange between buyers and sellers, to the emerging platforms offering other services (P2P 2.0).

Uber, Airbnb, and other similar platforms offering transport and hospitality services on a P2P platform are some of the latest entrants in business applying the P2P concept to different products services or goods. There are many popular P2P market platforms such as Indiegogo, Freelancer, oDesk, Prosper, Kickstarter, or Transferwise that enable users to access different types of services and products using direct online interactions between potential sellers and buyers. Further, businesses and online platforms such as Amazon have transitioned to a multiplatform taking the form of the business-to-customer (B2C) and peer-to-peer (P2P) markets because they now allow other businesses and individuals to list their products on the site for sale directly to customers.

5.6 Peer-to-peer networks and electricity trading

Nowadays, electricity markets are continuously changing from different dimensions, including utility company, customer, government, and other stakeholders' perspectives, leading to the emergence of new trends in the industry. Innovations, coupled with technological advancements, have led to the development and use of new and innovative methods or sources of electricity in the electricity generation. Wind, photovoltaic, and parabolic trough solar have emerged as new environmentally friendly and cost-effective methods of generating electricity (Lavrijssen and Carrillo Parra, 2017). It has been observed that the use of RES-based technologies, such as the solar or wind power generation, would create a "win-win" situation in which end-consumers and utility companies yield large savings (see Hirth, 2013). Moreover, most of the technologies based on obtaining electricity from RES do not require massive investments and are affordable for the end-consumers (or prosumers). These factors might have far-reaching effects on all downstream elements of the electrical power industry, such as distribution, pricing, and retailing of electricity to the end-consumers.

Fig. 5.5 outlines the design of the typical P2P platform market. The center point of the market design in the ICT-based online trading services that allow for the effective trading (see Fig. 5.5).

After the generation of electricity by various entities, the same is fed to the national grid by institutional or individual producers and transmitted or distributed to the retailers and end-consumers. In various parts of the world, austerity and other energy conservation measures have been implemented, with an objective of improving power consumption

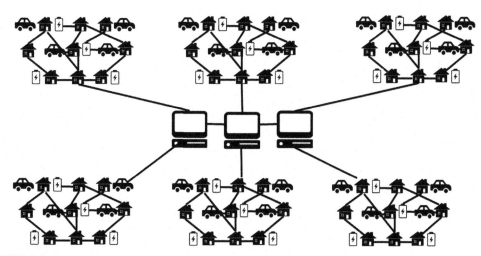

FIGURE 5.5 Peer-to-peer platform electricity market design.

rates, and improving energy efficiency. Probably, attributable to these changes, Flaherty et al. (2017) observed that there have been general declines in the end-user power consumption rates across the globe, with 22 out of 28 countries in the European Union, and other countries from other parts of the world recording similar steep declines in power consumption rates. The report by Flaherty et al (2017) further notes that there is likelihood that there will be continued stagnation of revenues.

With regard to the above, considered together with the argument that electricity utility companies invest a lot of resources in creating and maintaining distribution networks (Klose et al., 2010), it is important for electricity retail companies to come up with innovative methods for shoring up their financial positions. One such strategy, which is a key emerging trend, is the multitier pricing strategy, which includes fixed and variable chargers per kilowatt-hour ($/kW), used in maintaining good financial standing of electricity retail companies (Hirth, 2013). In agreement to that, it can be seen that in a two-tiered pricing system, the variable cost represents the congestion charges, while the fixed element reflects the capacity costs. Combined, these two elements form a price cap, and a retail firm in the industry would need to adjust its pricing of retail electricity accordingly, in order to achieve a trade-off between congestion and capacity to recoup investments and realize profits from operations.

The pricing setting and other related decisions in the electricity utility sector are rather complex, involving multiple players or stakeholders in the sector. Kirschen (2003) price setting decisions for retail electricity are set by the distribution or retailing companies or by a committee of representatives from different stakeholders in the industry, including generators, transmitters, and distributors. Moreover, as observed by Hirth (2013), other factors such as liberalization, increased competition, and inventions leading to alternative sources of energy influence the pricing structures for retail electricity. Evidently, the customer in this case plays a minimal role in price setting. Consequently, with increased options, consumers could become more price conscious, and high fixed prices or parts of the charges to the end-user could increase customer churn.

One of the strategies that can be used involves prosumers paying a fixed subscription fee to use a network for trade. Since the provision of the system is provided by national grid, it implies that the charges will enable the grid to recover an amount of money from prosumers which would have been raised from the traditional market which mainly involves consumer of power that is now met by the prosumers (Benkler, 2013). Additional, by charging a reasonable fee for the network, more prosumers can be encouraged to do the business, and this can generate more revenue. Emphasis on a fee being charged for a system should depend on the amount of power being produced by a prosumer. This is to encourage even the small producers to sell their surplus energy through national networks. The result is to ensure that there is a broad base of revenue generation and this will help in maintaining the amount in the grid fixed fees.

Another approach involves the integration of P2P market within the existing natural power grid. The excess power which consumer generate can be channeled into the national grid. This means that transmitting that power will be a duty of the national grid and this initial fee would still apply. However, such costs can be lowered as the prosumer has helped in reducing the cost of power generation by taking the cost from the national grid. In so doing a lot of customers can be reached and thus maintaining the value of transactions of the power. For example, GridX, a German company specializing in trading on P2P market, practices this by tapping directly into the national grid (see Cohen and Sundararajan, 2015). In return, all the surplus energy is distributed throughout the country.

Thirdly, there is the issue of lowering grid fixed fees. This should result from an added generation of power by prosumers. Since they cater for the cost of power production, the national network is only left with the fewer cost of transmission of energy and thus can lower the fees. Additional, reducing transactional fee can help to encourage more players in this field and thus the market is broadened. By maintaining a broad base with a significant market share, the transaction in the P2P market will help in the maintenance of a steady amount like that in the grid fixed fees (Parag and Sovacool, 2016). This can be further be strengthened if more prosumers are encouraged to integrate with the national grid. Moreover, the subscription fees for the use of the network by prosumers can be regulated in such a way the fees are applied depending on the amount of energy one is generating. This helps to ensure market demands factors play favorably. However, great care should be taken to ensure competition is rewarded to encourage more consumers in P2P market if the base is to be increased.

Another approach that can be used in ensuring P2P market transactional fees takes into account the grid fixed fees is the case when the power is being sold back to the national grid. Once the customers have generated energy, the excess power should be sold to the national grid. This is to ensure the regulation of the market to protect its stability by maintaining the prices of power. The national grid can be in a position to transmit power since there are existing infrastructure and capabilities. By adding to it, the power can be regulated fairly, and thus more stable and fixed fees can be achieved. The only disadvantage of this arrangement is a creation of monopoly market. But considering the importance of the sector toward the whole economy, keeping it in check can help in ensuring the stability of prices. Selling can be encouraged to prosumers by coming up with incentives which aim at allowing more and more people to see to the national grid. This can be issues such as access to more infrastructures for power generation and the likes.

One of the most interesting examples of the ICT-based online P2P energy trading is the use of the blockchain-based technologies and applications at the electricity markets. The technology is an issue of many studies as well as controversies nowadays, so I will not spend much time here describing it and am going to provide just some relevant but concise information and facts.

Blockchain-based technology represents a special form of a distributed database in which all participants in a given network share a consistent copy of the database without the necessity of being connected to the central server. In addition, participants of a given network can conduct P2P transactions which allow their online payments to be transferred directly from one person to another without an intermediary or a central banking authority (Swan, 2015). The technology was originally developed in 2008 as a digital payment system for the cryptocurrency called "bitcoin" but moved well beyond this embedding such sectors as energy trading.

In a blockchain-based distributed database, all participants actually impersonate the role of the intermediary (e.g., the central bank) and share the responsibility of verifying the legibility of the transaction using a prenegotiated consensus mechanism (see e.g. Drescher, 2017). All in all, one can say that the beauty and the functionality of the blockchain technology is its small-scale range that is ideal for the multidirectional trading within the limited geographical area, such as the P2P from the DER and RES. Surely, the blockchain technology has its disadvantages such as limits in throughput (unlike traditional online payment systems such as Visa or Mastercard it cannot cope with tens of thousands transactions per second), growing complexity with each additional node (or agent), or energy-intense operation (Beck et al., 2016). However, its application seems to be very promising for the small-scale P2P prosumer electricity trading.

Generally speaking, the majority of the existing popular P2P electricity trading platforms employ various high-level ICT technologies (e.g., blockchain technology) which allow electricity providers and users (consumers and prosumers alike) to acquire information at a single click of a computer mouse and conduct transactions they would benefit from.

There are some interesting examples of the P2P electricity markets that all use information technologies and ICT solutions for their operation but yield different business models that allow for their operation and smooth functioning. For instance, in Europe there are already mentioned Dutch Vandebron, German sonnenCommunity or UK Piclo. In the United States, the are the Brooklyn Microgrid and Mosaic.

Vandebron in the Netherlands focuses on establishing the local clean energy community (their slogan is "know your clean energy supplier"). The whole idea behind the project is to match the preferences of renewable energy suppliers and consumer. The electricity is marketed at the price preferred by suppliers which helps to reduce consumers' electricity fees. Providers and consumers are charged a monthly subscription fee of about 10 EUR a month.

In Germany, sonnenCommunity offers stable power supply offering in-house batteries and providing the surplus power pool for energy storage. Batteries are sold to the prosumers to store their surplus energy obtained from the renewables. As a result, suppliers get higher revenue than if profiting from FIT and the consumers' electricity fees are declining.

Piclo in the United Kingdom is supported by the government and venture capital. It represents an extensive information service that helps to match the preferences of renewable energy providers, energy consumers, and achieve DUoS (Distribution Use of Service).

In the United States, the Brooklyn Microgrid offers prosumers to trade self-produced energy on microgrid energy markets keeping profits from energy trading within their community. It is built around the blockchain-based microgrid energy market without the need for central intermediaries. The Brooklyn Microgrid provides financial and socioeconomic incentives for the integration and expansion of locally produced renewable energy and sells electricity at a price preferred by suppliers which reduces electricity bills.

Another P2P electricity market from United States is Mosaic, a California-based P2P platform for solar power, that enables everyone to join the scheme and become a prosumer (local solar contractor) for one flat monthly payment. Mosaic provides some upfront capital for the full cost of the solar panels and installation and connects suppliers to their own network of certified local solar contractors.

Sabounchi and Wei (2017) show that when used at P2P electricity markets, blockchain-based technologies help to disintermediate the value chain, redesign and automate energy-trading processes, and help to reduce administrative costs.

One can immediately see the implications for the business model stemming from the usage of blockchain-based technology applied to the P2P electricity markets. The first implication comes from the fact that the use of the blockchain technology enables prosumers to automate the processes and skip intermediaries by triggering payments and energy transactions through smart contracts. Automation and disintermediation reduce administrative costs, which lowers the barrier for implementing P2P electricity-trading systems. As a result, prosumers can profit from lower electricity costs while obtaining higher remuneration.

The second implication originates from the fact that P2P electricity platform markets reduce the burden on the transmission grid which is especially strained by large-scale intermittent renewables (e.g., in the case of Germany). It becomes clear that geographically limited energy markets with integrated balancing mechanisms are needed for increasing the overall efficiency of the grid.

The third implication stems from the fact that prosumers at P2P electricity markets are empowered by participation and can experience a higher sense of belonging to a community benefiting from the existence of the sharing economy. This feature did not exist in the traditional top-down electricity markets of the past. Today's prosumers can opt between supporting their immediate neighboring prosumers and procuring electricity from the utility company or a retailer. In another words, they can independently set their price preferences or negotiate the prices on the P2P electricity platform market.

In short, it becomes clear that the key blockchain-based technology characteristics facilitate the business model of P2P electricity-trading platforms through enabling prosumers to create their own well-functioning thrustless network with a distributed record and the use of the smart contracts. The usage of the smart contracts enables to play with the price preferences, tokenize the generated electricity and to make efficient transactions and payments at the P2P market.

Overall, it is apparent that blockchain-based platforms that are used at P2P electricity-trading markets can provide new approaches for addressing current market challenges by decentralizing the energy supply and balancing the grid through demand response. Blockchain-based technologies foster the decentralization of the electricity market by reducing the entry barrier. Transparency regarding the procured energy mix is a novel transparency and accountability tool to reveal the actual share of green power in energy consumption, influencing consumer behavior and increasing producers' accountability. It has to be made sure that market and pricing mechanisms that use blockchain technologies with the P2P energy-trading platforms need to be properly aligned in order to set the right incentives for the consumers. If these settings are ensured, the blockchain technology can be employed to deal with the market mechanisms using smart contracts.

5.7 Economic and policy implications for the smart grids

All in all, there are economic and policy implications of P2P electricity trading for the smart grids. Most of these implications originate from the nature of the commodity traded at P2P electricity markets. Electricity is a unique good that cannot be stored effectively and must be used or transmitted to a different location immediately after it is produced.

One would probably agree that P2P electricity markets constitute a very similar as the so-called Internet of Energy (IoE) which links them to the gist of the smart grids. In both cases, electricity production and consumption devices are connected to the network and smart meter devices monitor the amounts of electrical energy the prosumers produce and consume while cross-checking for the behavior of other prosumers operating within the same grid (Zhou et al., 2016). However, the flow of electricity obeys the Kirchhoff's voltage law that describes the power flow within an electric system. According to this law, the inputs and outputs in the electrical system should always remain the same at each electric node without any delay (see Wiest et al., 2017). If one goes beyond the case when the consumer is located next door to the green energy producer and is connected to it via local power grid, the description of a typical P2P situation when a particular consumer buys electricity from a particular producer is merely a simplification. In fact, the electricity flowing into the electricity grid is the one that all consumers use together. Thence, it is often not possible for the electricity consumer to identify what type of power is flowing into a plant. The application of the principles of Internet to the energy sector can be used for the optimization of the energy flow but not for automatically applying these concepts to electric energy.

Moreover, when we speak of P2P electricity markets and prosumers, self-generation of electric energy is mostly associated with the RESs. However, the share of energy from renewable sources might be high in the production of electricity for self-consumption but the sectors that are dominant in the world economy, such as transport or heating and cooling, heavily rely upon carbon fuels. This can be shown on the example of the EU Member States.

All in all, P2P electricity trading remains just a small trend in the world economy of electricity trading. New advancements in energy storage technologies such as novel batteries or "power walls" that are currently developed by Tesla and Elon Musk might provide interesting implications and alter the electricity market altogether. However, in its current state, P2P electricity markets remain an interesting experiment that is evolving into something grand but still has a long way to go before it will reach this level.

Let us see into the implications of the P2P electricity trading one by one for the following categories: (1) implications for the retail market, (2) implications for the wholesale market, (3) implications for the transmission grid and the system operator, (4) implication for the DSO.

The main future implications of the P2P electricity trading for all of these categories explained before are multiple.

For the retail market, it is apparent that platforms will be partially taking place of the traditional retail business and that retailers will be extending their business model into platform ownership and operation. For the wholesale market, there will be changes in the market structure (vertical wholesale structure of the market will be dissolving), the new vertical (wholesale-network-retail) and horizontal (P2P) market structure will be created, and the transaction costs will guide the development of the future markets. For the transmission grid and system operator, it might be that operators will not be able to see P2P transactions at horizontal markets but will be able to influence the volume of trade through embedded benefits (i.e., exemption from transmission charges). In addition, it is clear that network transmission charges will need to adapt to this situation. Finally, for the

DSO, the network cost recovery tariff model will need to adapt the new market arrangement, distribution system will need to become smarter and should be able to benefit from P2P transactions to optimize on capital investment and reduce the need for network capacity.

5.8 Conclusions and discussions

Various elements of the traditional electricity markets, including generation; distribution and retailing; and pricing, have evolved over the years. Conventionally, hydro and fossil fuel-based electricity generation methods, controlled by state or institutional actors were the most dominant methods. However, in the recent past, advanced solar electricity generating methods, and wind, among others, have emerged and are gaining increasing popularity, as well as the participation of individuals or households in electric power generating activities. These individuals and households sell the excess power to other users or to the national and regional grids. After power generation, the same is transmitted and distributed to retailers, who working with other stakeholders, price kilowatt-hour usages, using a two-tier system, comprises fixed and variable elements. Price kilowatt-hour fixed charges, coupled with other factors such as personal interests, and new technologies, could contribute to the increased uptake of power from individual and household generators. With the entry of individual power generator and retailers into grid systems, it become important to enter into partnerships for infrastructural development and maintenance, which ensures that each of the parties bears a proportionate part of the costs. In addition, legal, regulatory, and policy frameworks will be developed to ensure quality standards and prices are not exorbitant, and to ensure stability in the sector.

One can see that P2P electricity markets have a profound social impact since they enable prosumers to sell the power in charges depending on the cost of their productions. Thus, at a given place, a prosumer is in a position to sell power at one's prices as market demands and production cost dictate. This in return means that price of power at a given region will be different from that of another region and hence the creation of localized pricing of power. When there are many prosumers, this will increase options of end consumer to who would supply their energy. This in return would give consumer ability to choose their best rate depending on available rates. Furthermore, prosumers when many will help in lowering the cost of power as they try to win large market niche. Additional prosumer will be required to offer great services and better consumer products to gain their market. All of this means specific and distinct local prices in various parts of the market.

In the recent years, P2P market model has been a booming model for businesses globally because of the fact that it is able to link buyers directly to sellers and lenders directly to borrowers. With the help of technology, this market model has been of assistance to many investors who have made money and also helped many sellers to find clients from different parts of the world. P2P markets are, without doubt, the social media of the banking/lending industry, where businesses and individuals meet their financial needs.

Technological developments—computers, Internet, WWW, and their application in different ways—have led to the emergence and growth of new distribution channels and business models. One such development is P2P markets, which are online platforms that connect buyers and sellers of products without any other physical intermediary. Sellers put up adverts for products goods or services, which are then matched with the preferences of a prospective buyer using an algorithm. Starting in the 1990s, with an introduction of eBay and Craigslist among others, P2P markets have since grown to include different platforms providing connections for buyers and sellers of a wide range of products goods and services, supported by technological developments and growth in supporting services. However, if technological innovation and growth does not continue in the same pattern, there is likelihood that further growth of P2P markets will be impeded by the inability of technology to address the key problem areas identified in P2P markets.

To sum it all up, P2P electricity markets contribute to the formation and extension of the small-scale distributed energy resources and the creation of the new platform-type markets. The main advantages of the P2P electricity markets are that the electric energy generated at P2P markets can be linked to the profit, electricity can be generated meeting the requirements of the end-users and the utilization of the resources can be optimized through the creation of the cooperative networks between producers and consumers.

It becomes apparent that P2P electricity markets constitute an integral part of the smart grids. With the current market arrangements at the electricity markets becoming more advanced and moving toward prosumer economy operating via the IoE, P2P market arrangements will become ubiquitous.

There are various issues that need to be considered with regard to the P2P electricity markets and their implications for the future power system. One of these issues is the innovations in energy storage that, once improved and able to deliver better batteries for households and businesses alike, would introduce the significant amount of flexibility for P2P energy trading in practice. If energy storage revolution is going to happen one day, some prosumers might become able to take advantage of the P2P market trading arrangements for maximizing their own profits via scheduling their energy storage in such a way that would interfere with the balance of local generation and consumption.

The integration of P2P electricity markets trading into the local distribution networks will provide a surfeit of technical possibilities and possible benefits. For example, P2P electricity market arrangements would lead to the balance between generation and consumption. Moreover, the use of IT infrastructure (e.g., smart meters, broadband networks, or blockchain technologies) would facilitate P2P energy trading in the distribution networks and make it accessible for small and large prosumers alike on the future energy markets. However, the social acceptance of these innovations might still take some time.

References

Bauwens, M., 2005. The political economy of peer production. CTheory, 12-1. Available from: <https://journals.uvic.ca/index.php/ctheory/article/view/14464/5306&gt> (accessed 18.05.19.).
Beck, R., Czepluch, J.S., Lollike, N., & Malone, S., 2016. Blockchain—the gateway to trust-free cryptographic transactions, 24th European Conference on Information Systems (ECIS), Research paper no. 153, 5—16. Available from: <http://aisel.aisnet.org/ecis2016_rp/153> (accessed 12.11.18).
Benkler, Y., 2013. Practical anarchism: peer mutualism, market power, and the fallible state. Politics Society 41 (2), 213—251. Available from: https://doi.org/10.1177/0032329213483108.

Black, S.E., Lynch, L.M., 2001. How to compete: the impact of workplace practices and information technology on productivity. Rev. Econ. Stat. 83 (3), 434–445. Available from: https://doi.org/10.1162/00346530152480081.

Calliess, C., Hey, C., 2013. Multilevel energy policy in the EU: paving the way for renewables? J. Eur. Environ. Plan. Law 10 (2), 87–131. Available from: https://doi.org/10.1163/18760104-01002002.

Carroll, J., Lyons, S., Denny, E., 2014. Reducing household electricity demand through smart metering: the role of improved information about energy saving. Energy Econ. 45, 234–243. Available from: https://doi.org/10.1016/j.eneco.2014.07.007.

Chitnis, D., Swart, H.C., Dhoble, S.J., 2016. Escalating opportunities in the field of lighting. Renew. Sustain. Energy Rev. 64, 727–748. Available from: https://doi.org/10.1016/j.rser.2016.06.041.

Cohen, M., Sundararajan, A., 2015. Self-regulation and innovation in the peer-to-peer sharing economy. Univ. Chicago Law Rev. 82, 116. Available from: <https://chicagounbound.uchicago.edu/cgi/viewcontent.cgi?referer = https://scholar.google.cz/&httpsredir = 1&article = 1039&context = uclrev_online> (accessed 10.11.18.).

Craford, M.G., Dupuis, R.D., Feng, M., Kish, F.A., Laskar, J., 2013. 50th anniversary of the light-emitting diode (LED): an ultimate lamp [scanning the issue]. Proc. IEEE 101 (10), 2154–2157. Available from: https://doi.org/10.1109/JPROC.2013.2274908.

Drescher, D., 2017. Blockchain Basics: a Non-technical Introduction in 25 Steps. Apress Publishing, Frankfurt am Main. <https://doi.org/10.1007/978-1-4842-2604-9>.

Drom, R., 2014. Electricity capacity markets: a primer. Nat. Gas Electr. 30 (10), 1–8. Available from: https://doi.org/10.1002/gas.21758.

Einav, L., Farronato, C., Levin, J., 2016. Peer-to-peer markets. Ann. Rev. Econ. 8, 615–635. Available from: https://doi.org/10.1146/annurev-economics-080315-015334.

Ert, E., Fleischer, A., Magen, N., 2016. Trust and reputation in the sharing economy: the role of personal photos in Airbnb. Tourism Manage. 55, 62–73. Available from: https://doi.org/10.1016/j.tourman.2016.01.013.

European Commission, 2014. EU Energy Markets in 2014. Publication Office of the European Union, Luxembourg. Available from: <https://ec.europa.eu/energy/sites/ener/files/documents/2014_energy_market_en_0.pdf> (accessed 25.11.18.).

Flaherty, T., Schwieters, N., & Jennings, S., 2017. 2017 power and utilities trends. Available from: <https://www.strategyand.pwc.com/trend/2017-power-and-utilities-industry-trends> (accessed 20.1.18.).

Frieden, R., 2005. Lessons from broadband development in Canada, Japan, Korea and the United States. Telecommun. Policy 29 (8), 595–613. Available from: https://doi.org/10.1016/j.telpol.2005.06.002.

Galloway, I., 2009. Peer-to-peer lending and community development finance. Commun. Devel. Invest. Center Work. Paper, 39, 19-23. Available from: <https://core.ac.uk/download/pdf/6971071.pdf> (accessed 05.12.18.).

GE, 2017. 2017 top digital trends for the electricity value network. Available from: <https://www.ge.com/digital/resource-center?gated-id = 2693> (accessed 10.1.19.).

Gutierrez-Escolar, A., Castillo-Martinez, A., Gomez-Pulido, J.M., Gutierrez-Martinez, J.M., González-Seco, E.P.D., Stapic, Z., 2017. A review of energy efficiency label of street lighting systems. Energy Efficien. 10 (2), 265–282. Available from: https://doi.org/10.1007/s12053-016-9454-7.

Hirth, L., 2013. The market value of variable renewables: the effect of solar wind power variability on their relative price. Energy Econom. 38, 218–236. Available from: https://doi.org/10.1016/j.eneco.2013.02.004.

Jayme, F., 2018. The peer-to-peer lending space: challenges and opportunities for investors Inside Financial & Risk. Inside Financial & Risk. Available from: <https://blogs.thomsonreuters.com/financial-risk/fintech-innovation/the-peer-to-peer-lending-space-challenges-and-opportunities-for-investors/> (accessed 19.11.18.).

Katz, L.F., Krueger, A.B., (2016). The rise and nature of alternative work arrangements in the United States, 1995-2015. National Bureau of Economic Research, NBER paper no. w22667. Available from: <https://www.nber.org/papers/w22667> (accessed 20.09.18.).

Khalid, M., Ahmadi, A., Savkin, A., Agelidis, V., 2016. Minimizing the energy cost for microgrids integrated with renewable energy resources and conventional generation using controlled battery energy storage. Renew. Energy 97, 646–655. Available from: https://doi.org/10.1016/j.renene.2016.05.042.

Khan, S., Khan, R., 2018. Multiple authorities attribute-based verification mechanism for blockchain mircogrid transactions. Energies 11 (5), 1154. Available from: https://doi.org/10.3390/en11051154.

Kirschen, D.S., 2003. Demand-side view of electricity markets. IEEE Trans. Power Syst. 18 (2), 520–527. Available from: https://doi.org/10.1109/TPWRS.2003.810692.

Klose, F., Kofluk, M., Lehrke, S., & Rubner, H., 2010. Toward a distributed-power world, Renewables and smart grids will reshape the energy sector. The Boston Consulting Group Report. Available from: <https://www.bcg.com/documents/file51254.pdf> (accessed 10.08.18.).

Lavrijssen, S., Carrillo Parra, A., 2017. Radical prosumer innovations in the electricity sector and the impact on prosumer regulation. Sustainability 9 (7), 1207. Available from: https://doi.org/10.3390/su9071207.

López-García, D.A., Torreglosa, J.P., Vera, D., 2019. A decentralized P2P control scheme for trading accurate energy fragments in the power grid. Int. J. Electr. Power Energy Syst. 110, 271−282. Available from: https://doi.org/10.1016/j.ijepes.2019.03.013.

Lund, H., 2014. Renewable Energy Systems: A Smart Energy Systems Approach to the Choice and Modeling of 100% Renewable Solutions. Academic Press, Cambridge, MA.

Matofska, B., 2016. What is the sharing economy? Available from: <http://thepeoplewhoshare.com/blog/what-is-the-sharing-economy> (accessed 05.18.18.).

Mengelkamp, E., Gärttner, J., Rock, K., Kessler, S., Orsini, L., Weinhardt, C., 2018. Designing microgrid energy markets: a case study: the Brooklyn Microgrid. Appl. Energy 210, 870−880. Available from: https://doi.org/10.1016/j.apenergy.2017.06.054.

Milne, A., Parboteeah, P., 2016. The business models and economics of peer-to-peer lending. ECRI Res. Report 2016 (17). Available from: https://doi.org/10.2139/ssrn.2763682.

Minges, M., 2015. Exploring the relationship between broadband and economic growth. Available from: <https://openknowledge.worldbank.org/handle/10986/23638> (accessed 21.05.19.).

Montoya, F.G., Peña-García, A., Juaidi, A., Manzano-Agugliaro, F., 2017. Indoor lighting techniques: an overview of evolution and new trends for energy saving. Energy Build. 140, 50−60. Available from: https://doi.org/10.1016/j.enbuild.2017.01.028.

Nair, G.B., Dhoble, S.J., 2015. A perspective perception on the applications of light-emitting diodes. Luminescence 30 (8), 1167−1175. Available from: https://doi.org/10.1002/bio.2919.

Narendran, N., Bullough, J.D., Maliyagoda, N., Bierman, A., 2001. What is useful life for white light LEDs? J. Illumin. Eng. Soc. 30 (1), 57−67. Available from: https://doi.org/10.1080/00994480.2001.10748334.

Newbery, D., Pollitt, M.G., Ritz, R.A., Strielkowski, W., 2018. Market design for a high-renewables European electricity system. Renew. Sust. Energy Rev. 91, 695−707. Available from: https://doi.org/10.1016/j.rser.2018.04.025.

Oram, A., 2001. Peer-to-Peer: Harnessing the Power of Disruptive Technologies. O'Reilly Media, Inc.

Parag, Y., Sovacool, B.K., 2016. Electricity market design for the prosumer era. Nat. Energy 1, 16032. Available from: https://doi.org/10.1038/nenergy.2016.32.

Pousset, N., Rougie, B., Razet, A., 2010. Impact of current supply on LED colour. Light. Res. Technol. 42 (4), 371−383. Available from: https://doi.org/10.1177/1477153510373315.

Rausser, G., Strielkowski, W., Štreimikienė, D., 2018. Smart meters and household electricity consumption: a case study in Ireland. Energy Environ. 29 (1), 131−146. Available from: https://doi.org/10.1177/0958305X17741385.

Rochet, J.C., Tirole, J., 2006. Two-sided markets: a progress report. RAND J. Econom. 37 (3), 645−667. Available from: https://doi.org/10.1111/j.1756-2171.2006.tb00036.x.

Rocky Mountain Institute, 2017. Economics of load defection. Available from: <https://rmi.org/insights/reports/economics-load-defection> (accessed 20.12.18.).

Sabounchi, M., Wei, J., 2017. Towards resilient networked microgrids: blockchain-enabled peer-to-peer electricity trading mechanism. In: 2017 IEEE Conference on Energy Internet and Energy System Integration (EI2), pp. 1−5. Available from: <https://ieeexplore.ieee.org/abstract/document/8245449> (accessed 19.11.18.).

ScienceDaily, 2008. LEDs may help reduce skin wrinkles. Available from: <https://www.sciencedaily.com/releases/2008/10/081020094355.htm> (accessed 18.05.19.).

Shah, S., 2011. The P2P evolution. Available from: <https://techcrunch.com/2011/05/01/P2P-evolution> (accessed 28.12.18.).

Sharma, P.K., Ryu, J.H., Park, K.Y., Park, J.H., Park, J.H., 2018. Li-Fi based on security cloud framework for future IT environment. Human-Centric Comput. Inform. Sci. 8 (1), 23. Available from: https://doi.org/10.1186/s13673-018-0146-5.

Sorin, E., Bobo, L., Pinson, P., 2019. Consensus-based approach to peer-to-peer electricity markets with product differentiation. IEEE Trans. Power Syst. 34 (2), 994−1004. Available from: https://doi.org/10.1109/TPWRS.2018.2872880.

Sovacool, B., 2009. The cultural barriers to renewable energy and energy efficiency in the United States. Technol. Soc. 31 (4), 365−373. Available from: https://doi.org/10.1016/j.techsoc.2009.10.009.

Stoft, S., 2002. Power System Economics: Designing Markets for Electricity, first ed. Wiley., New York.

Sumit, K., 2014. The peer-to-peer (P2P) marketplace, Happiest Minds. Available from: <https://www.happiest-minds.com/wp-content/uploads/2016/10/Peer-to-Peer-Marketplace.pdf> (accessed 21.11.18.).

Swan, M., 2015. Blockchain thinking: The brain as a decentralized autonomous corporation [commentary]. IEEE Technology and Society Magazine 34 (4), 41−52. Available from: https://doi.org/10.1109/MTS.2015.2494358.

Taheri, M., Ansari, N., Feng, J., Rojas-Cessa, R., Zhou, M., 2017. Provisioning internet access using FSO in high-speed rail networks. IEEE Netw. 31 (4), 96−101. Available from: https://doi.org/10.1109/MNET.2017.1600167NM.

Thapar, S., Sharma, S., Verma, A., 2017. Local community as shareholders in clean energy projects: innovative strategy for accelerating renewable energy deployment in India. Renew. Energy 101, 873−885. Available from: https://doi.org/10.1016/j.renene.2016.09.048.

Wales, K., 2015. Internet finance: digital currencies and alternative finance liberating the capital markets. J. Govern. Regul. 4 (4), 10−22495. Available from: https://doi.org/10.22495/jgr_v4_i4_c1_p6.

Wang, Y., Alonso, J.M., Ruan, X., 2017. A review of LED drivers and related technologies. IEEE Trans. Ind. Electr. 64 (7), 5754−5765. Available from: https://doi.org/10.1109/TIE.2017.2677335.

Weiller, C.M., Pollitt, M.G., 2016. Platform markets and energy services. In: Liu, C., McArthur, S., Lee, S. (Eds.), Smart Grid Handbook. John Wiley and Sons, New York.

Wiest, P., Gross, D., Rudion, K., Probst, A., 2017. Rapid identification of worst-case conditions: improved planning of active distribution grids. IET Gen. Trans. Distribut. 11 (9), 2412−2417. Available from: https://doi.org/10.1049/iet-gtd.2017.0148.

World Bank, 2019. World development indicators. Available from: <https://data.worldbank.org/indicator/EG.ELC.RNEW.ZS> (accessed 20.1.19.).

Yoon, H.C., Kang, H., Lee, S., Oh, J.H., Yang, H., Do, Y.R., 2016. Study of perovskite QD down-converted LEDs and six-color white LEDs for future displays with excellent color performance. ACS Appl. Mater. Interf. 8 (28), 18189−18200. Available from: https://doi.org/10.1021/acsami.6b05468.

Yu, S., Fang, F., Liu, Y., Liu, J., 2019. Uncertainties of virtual power plant: problems and countermeasures. Appl. Energy 239, 454−470. Available from: https://doi.org/10.1016/j.apenergy.2019.01.224.

Zhou, K., Yang, S., Shao, Z., 2016. Energy internet: the business perspective. Appl. Energy 178, 212−222. Available from: https://doi.org/10.1016/j.apenergy.2016.06.052.

Zio, E., Aven, T., 2011. Uncertainties in smart grids behaviour and modelling: what are the risks and vulnerabilities? How to analyze them? Energy Policy 39, 6308−6320. Available from: https://doi.org/10.1016/j.enpol.2011.07.030.

Further reading

Morstyn, T., Farrell, N., Darby, S.J., McCulloch, M.D., 2018. Using peer-to-peer energy-trading platforms to incentivize prosumers to form federated power plants. Nat. Energy 3 (2), 94−101. Available from: https://doi.org/10.1038/s41560-017-0075-y.

Schuetz, J., 2016. Where should peer-to-peer markets thrive? Available from: <https://www.federalreserve.gov/econresdata/notes/feds-notes/2016/where-should-online-peer-to-peer-markets-thrive-20161222.html> (accessed 10.12.18.).

World Economic Forum, (2017) The future of electricity new technologies transforming the grid edge. Available from: <http://www3.weforum.org/docs/WEF_Future_of_Electricity_2017.pdf> (accessed 22.12.18.).

6

Consumers, prosumers, and the smart grids

6.1 Introduction: prosumers and smart grids

The recent changes in the energy markets are changing the role not only of the power companies but also of energy consumers. The traditional consumers are becoming "prosumers," or economic agents that both consume and produce energy, and even trade it with each other, in producer and consumer groups, or sell it back to the grid. Prosumers are often organizing themselves for the peer-to-peer (P2P) energy trading and use elaborate mechanisms how to trace and manage complex transactions that involve not only the exchange of information but also that of energy (most frequently electricity).

One can see that at a typical P2P energy market, the size of the generator or consumer is irrelevant. However, in the P2P, any homeowner of a solar system with a capacity of 1.5 kW or higher to the largest coal-fired power plant can freely engage in a solar or wind energy trading system based on which software or technology is used to market (AlSkaif et al., 2017).

In general, P2P technologies enable households and businesses to produce renewable energy for reducing their network demand by using technologies such as storing batteries during peak periods. The popularity of solar and wind energy (e.g., in such "pioneering" countries such as Germany or Australia) means that excess energy is generated during daylight hours which reduces the burden on other conventional energy sources such as natural gas and nuclear power stations.

However, there is a technological gap between what is expected and what we really have and can operate with. Today's smart meters do not have the capacity to support the P2P market and substantial innovations and improvements would be required. The smart meters do not have the ability to control batteries at home, monitor power supplies, and use them for a short period of time or predict when they can supply or consume energy from a house.

The transfer costs could be introduced for each trade via an energy aggregator, which, in combination with a competitive market for network infrastructure, could help to stimulate the transition to a more distributed, more distributed and distributed power grid.

Feed rates pose a major threat to the interests of all actors, consumers and the growth of the microgeneration market.

Therefore in order to achieve the desired financial gain through microgeneration, P2P energy trading will be the ideal platform for market participants. One can see that in the P2P energy trade, it would be possible for consumers to fluctuate an order of sale of their surplus generation in the open market and to generate additional revenue by selling energy to consumers at a better rate (Jogunola et al., 2017).

In deregulated markets with smart meters, smart meters would only keep track of the network usage and reports the data to the independent system operator (ISO) in a given region. If grid customers can predict their energy consumption more accurately than a traditional retailer, in addition to responding to market prices, they will not only save money but also increase the reliability of the network. In the future, there will be smart grid, where there will be more distributed power resources and dynamic distribution costs, which will lead to more localized prices. Although P2P transactions are more economically efficient and will be the basis for a thriving distribution network, it leaves us as a single supplier in a grid of distribution nodes. In this business model, consumers and consumers create energy communities where excess production could be sold to other members. On the P2P trading platform, the traditional supplier is responsible for supplying electricity to consumers, and all transactions are concluded with cryptocurrencies. Therefore traditional suppliers could overtake the trend and propose a solution for themselves before the P2P energy communities emerge (Huang et al., 2010).

In order to implement and realize the benefits of the P2P energy trading model, one needs to have a robust technology platform that meets a number of key requirements. Forecasts using digital network infrastructure and offering predictive analysis that will improve its ability to cope with the growing number of distributed and renewable energy sources (RES) connected to the grid would be needed to tackle this problem effectively.

6.2 Smart grids and renewable energy

RES are sources based on the constantly existing or periodically occurring energy flows in the environment. A typical example of such a source is solar radiation with a characteristic repetition period of 24 hours. Renewable energy is present in the environment in the form of energy, which is not a consequence of purposeful human activity, and this is its peculiarity.

In modern world practice, RES include hydro, solar, wind, geothermal, hydraulic energy, energy of sea currents, waves, tides, temperature gradient of sea water, temperature differences between air mass and ocean, heat of the Earth, and biomass of animal, plant and domestic origin.

There are different opinions about what type of resources should be attributed to nuclear energy. The nuclear waste, given the possibility of its reproduction in the breeder reactors, is enormous and very dangerous. In addition, it can last for thousands of years. Despite this, it is usually referred to as nonrenewable resources. The main argument for this is the high environmental risk associated with the use of nuclear energy (Owen, 2006).

Non-RES are natural reserves of substances and materials that can be used by humans for energy production. An example would be nuclear fuel, coal, oil, and gas. The energy of nonrenewable sources, in contrast to renewable sources, is naturally bound in nature and is released as a result of purposeful actions of humankind.

Coal, oil, natural gas, peat, oil shale, and firewood are reserves of the Sun's radiant energy extracted and transformed by plants. During the reaction of photosynthesis, inorganic elements of the environment—water (H_2O) and carbon dioxide (CO_2)—under the influence of sunlight in plants form organic matter, the main element of which is carbon. In a certain geological epoch, dead plants were undergoing the influence of pressure and temperature regime for millions of years which resulted in a specific amount of solar energy forming organic energy resources, the foundations of carbon which is also accumulated in plants. The energy of water is also obtained by solar energy, evaporating water, and raising steam into the high layers of the atmosphere. The wind arises due to the different temperature of the Sun heating different points of our planet. In addition, the direct radiation of the Sun, falling on the surface of the Earth, has great potential for energy.

Renewable energy represents energy from sources that are inexhaustible on a human scale. The basic principle of using renewable energy is to extract it from the processes constantly occurring in the environment and to provide it for technical use. Renewable energy is obtained from natural resources such as sunlight, wind, rain, tides, and geothermal heat, which are renewable (naturally replenished). In 2013 about 21% of global energy consumption was satisfied from RES. In addition, in 2006, about 18% of global energy consumption was satisfied from RES, with 13% from traditional biomass, such as burning wood (Ellabban et al., 2014).

Hydropower is the next largest source of renewable energy, providing 3.3% of global energy consumption and 15.3% of global electricity generation. In 2010 16.7% of global energy consumption came from renewable sources. The share of renewable energy is decreasing, but this is due to the reduction in the share of traditional biomass, which was only 8.5% in 2010.

The share of modern renewable energy is growing and in 2010 amounted to 8.2%, including hydropower 3.3%, for heating and water heating (biomass, solar and geothermal water heating, and heating) 3.3%; biofuel 0.7%; and electricity production (wind, solar, geothermal power plants and biomass in heat power stations (HPS)) 0.9%. Wind power usage is growing at about 30% a year, worldwide with an installed capacity of 196,600 megawatts (MW) in 2010, and is widely used in Europe, the United States, and China (Popp et al., 2011).

Annual production in the photovoltaic (PV) industry reached 6900 MW in 2008. Solar power plants are popular in Germany and Spain. Solar thermal stations operate in the United States and Spain, and the largest of them is a station in the Mojave Desert with a capacity of 354 MW. The world's largest geothermal installation is placed on geysers in California, with a rated capacity of 750 MW (Hossain et al., 2016).

Brazil has one of the largest renewable energy programs in the world related to the production of ethanol fuel from sugar cane. Ethyl alcohol currently covers 18% of the country's need for automotive fuel. Fuel ethanol is also widely distributed in the United States (Sanchez and Cardona, 2008).

Thermonuclear fusion of the Sun is the source of most types of renewable energy, with the exception of geothermal energy and tidal energy. According to astronomers, the life expectancy of the Sun is about 5 billion years, so that on human scale renewable energy emanating from the Sun, depletion does not present a threat. In a strictly physical sense, energy is not renewed but is permanently withdrawn from the above sources. Of the solar energy arriving on Earth, only a very small part is transformed into other forms of energy, and most simply go into space.

The use of permanent processes is contrasted to the extraction of fossil fuels, such as coal, oil, natural gas, or peat. In a broad sense, they are also renewable, but not by the standards of a person, since their education requires hundreds of millions of years, and their use is much faster.

Therefore the main sources of renewable energy include wind, hydro, tidal, solar, thermal, or biofuel. Let us take a quick look at each of them separately.

Wind energy is an energy industry specializing in converting the kinetic energy of air masses in the atmosphere into electrical, thermal, and any other form of energy for use in the national economy. The transformation takes place with the help of a wind generator, windmills, and many other types of aggregates. Wind energy is the result of solar activity, so it is a renewable energy.

Hydropower plants use the potential energy of the water flow that is used as a source of energy, the primary source of which is the Sun, which evaporates water, which then falls on elevations in the form of precipitation and flows down to form rivers. Hydroelectric power plants are usually built on rivers, building dams and reservoirs. It is also possible to use the kinetic energy of the water flow at the so-called free-flow (non-dammed) hydroelectric power plants.

Tidal energy represents a special type of hydroelectric power station that uses the energy of the tides, and in fact the kinetic energy of rotation of the Earth. Tidal power plants are built on the shores of the seas, where the gravitational forces of the Moon and the Sun change the water level twice a day.

To obtain energy, the bay or the mouth of the river is blocked with a dam, in which hydraulic units are installed that can work both in generator mode and pump mode (for pumping water into the reservoir for subsequent work in the absence of tides). In the latter case, they are called hydro-accumulating power station. The main advantages of power electric stations (PES) are environmental friendliness and low cost of energy production. The disadvantages are the high cost of construction and the varying power during the day, which is why the PES can only work in a single power system with other types of power plants.

Wave power plants use the potential energy of the waves carried on the surface of the ocean. Wave power is estimated in kW/m. Compared with wind and solar energy, wave energy has a higher power density. Despite the similar nature with the energy of tides, ebbs, and ocean currents, wave energy is a different source of renewable energy.

Energy of the Sun is based on the conversion of electromagnetic solar radiation into electrical or thermal energy. Solar power plants use the energy of the Sun both directly (PV solar electric stations (SES) working on the phenomenon of an internal photoelectric effect) and indirectly using the kinetic energy of steam.

Geothermal energy power plants are thermal power plants that use water from hot geothermal sources as a coolant. Due to the absence of the need to heat water, geothermal power plants are to a large extent more environmentally friendly than thermal power plants. Geothermal power plants are built in volcanic regions, where at relatively shallow depths the water overheats above the boiling point and seeps to the surface, sometimes appearing as geysers. Access to underground sources is carried out by drilling wells.

Another popular renewable energy sector specializes in the production of energy from biofuels. It is used in the production of both electrical energy and heat. Biofuels represent fuels from biological raw materials, obtained, as a rule, as a result of processing biological waste. There are also projects of varying degrees of elaboration aimed at obtaining biofuels from cellulose and various types of organic waste, but these technologies are in the early stages of development or commercialization. There are the following ones: (1) solid biofuels (energy forest: firewood, briquettes, pellets, wood chips, straw, husk, and peat); (2) liquid biofuels (for internal combustion engines, e.g., bioethanol, biomethanol, biobutanol, dimethyl ether, and biodiesel); and (3) gaseous (biogas, biohydrogen, and methane).

The experience gained in the world allows us to speak of fixed tariffs as the most successful measures to stimulate the development of RES. The bases of these measures to support renewable energy are three main factors:

- network connection guarantee;
- long-term contract for the purchase of all renewable energy; and
- guarantee of purchase of electricity produced at a fixed price.

Fixed tariffs for renewable energy may differ not only for different sources of renewable energy but also depending on the installed capacity of renewable energy. One of the options for a support system based on fixed tariffs is the use of a fixed premium to the market price of renewable energy. As a rule, the surcharge to the price of electricity produced or a fixed tariff is paid for a sufficiently long period (10–20 years), thereby ensuring the return of investments invested in the project and making a profit.

Worldwide, in 2008 only, $51.8 billion was invested in wind energy, $33.5 billion in solar energy, and $16.9 billion in biofuels. In 2008 European countries invested $50 billion in alternative energy, US countries $30 billion, China $15.6 billion, India $4.1 billion.

In 2009 investments in renewable energy throughout the world amounted to $160 billion, and in 2010—$211 billion. In 2010 $94.7 billion was invested in wind energy, $26.1 billion in solar energy, and $11 billion in energy production technology from biomass and garbage (Sawin et al., 2010).

As in all other cases, there are advantages and disadvantages of alternative energy sources. The first obvious advantage is renewability, environmental friendliness, prevalence, and wide availability. If necessary, these sources can operate autonomously, supplying energy to consumers not connected to centralized power grids. In addition, there is a security of supply. There is a constant rise in prices for traditional fuels and, of course, scientific and technical progress. Modern developments and innovations increase the competitiveness of alternative energy.

Among the disadvantages are the high costs of renewable energy. In the production of electricity through renewable energy on an industrial scale, there are often technical difficulties that are associated with the impossibility of the constant pairing of energy production with its consumption.

Scientists considered that in order to avoid changes in the parameters of the combined energy system, the share of wind generators and solar power stations should not exceed 15% of the total system capacity. Another disadvantage is the high density of energy flows (Chauhan and Saini, 2014). Solar radiation has less than 1 kW per 1 m^2, wind has a speed of 10 m/s, and water flow has a speed of 1 m/s (about 500 W per 1 m^2). While in modern energy devices, we have flows, measured in hundreds of kilowatts, and sometimes megawatts per 1 m^2. The collection, transformation, and management of low-density energy flows, in some cases with daily, seasonal, and weather instability, require significant costs for the creation of receivers, converters, batteries, regulators, etc. High initial capital costs, however, are in most cases offset by low operating costs (Panwar et al., 2011).

Nevertheless, it is important to emphasize that the use of renewable energy is advisable, as a rule, only in an optimal combination with measures to improve energy efficiency: for example, it is pointless to install expensive solar heating systems or heat pumps for homes with high heat losses, it is unwise to use low-efficiency electrical appliances using photoelectric converters such as incandescent lighting systems.

Recently, the world energy sector has faced a serious problem: energy consumption is steadily growing, and the reserves of fossil fuels, on the use of which traditional energy is based, are far from unlimited. The problem lies not only in the fact that natural resources are exhaustible but also in the fact that the available deposits are rapidly depleting, which means that there is a need to develop new ones, which entails huge expenditures. In addition, the inhabitants of the globe have long concluded that the use of fossil fuels harms nature.

Many scientists see a way out of this situation in the development of alternative energy and the promotion of RES. Renewable sources are sources of energy, the reserves of which can be replenished in nature in a natural way. In the foreseeable future, such sources, in contrast to fossil resources, are practically inexhaustible.

RES are divided into traditional and nontraditional (alternative). The traditional sources include the energy of water converted into electricity by hydroelectric power plants, and the energy of biomass traditionally burned to produce heat (this includes firewood, straw, and similar materials). The group of alternative RES includes solar, wind and geothermal energy, the energy of tides, currents, sea waves, water energy converted into electricity by micro-hydro, and also biomass energy used to produce motor fuel, thermal, and electric energy by alternative methods (Demirbas, 2005).

Today, the world community pays great attention to the development of RES, making this area an important area of public policy. Well-funded government programs aimed at developing alternative energy began to appear in different countries of the world. Legislative acts that encourage the use of renewable energy are being widely adopted. However, the contribution of renewable sources to the global energy balance leaves much to be desired. RES provides only about 20% of total global energy consumption, with the overwhelming majority of their contribution coming from traditional RES. Meanwhile, today many scientists of the world hold the opinion that it is necessary to rely on alternative sources.

Until recently, a clear pattern was observed in the development of the energy sector: the development of those areas of the energy sector that ensured a fairly rapid direct economic effect. The social and environmental consequences associated with these areas were considered only as collateral, and their role in decision-making was insignificant (Jefferson, 2006).

With this approach, RES were considered only as energy sources of the future, when traditional sources of energy are exhausted or when their extraction becomes extremely expensive and labor-intensive. Since this future seemed quite distant, the use of RES sources seemed quite interesting, but in modern conditions, it is often an exotic rather than practical task.

The situation has dramatically changed the awareness of humanity of the ecological limits of growth. The rapid exponential growth of negative anthropogenic impacts on the environment leads to a significant deterioration in the human environment. Maintaining this environment in a normal state and the possibility of its self-preservation becomes one of the priority goals of society. Under these conditions, the former, only narrowly economic assessments of various areas of technology, technology, management, are clearly insufficient, because they do not take into account social and environmental aspects.

The impetus for the intensive development of renewable energy for the first time was not promising economic calculations, but social pressure based on environmental requirements. The opinion that the use of renewable energy will significantly improve the environmental situation in the world—this is the basis of this pressure. The economic potential of renewable energy in the world is currently estimated at 20 billion tons of fuel equivalent per year, which is twice the annual production of all types of fossil fuels. And this circumstance indicates the path of energy development in the near future (Ouda et al., 2016). The use of renewable energy does not change the energy balance of the planet. These qualities caused the rapid development of renewable energy abroad and very optimistic forecasts for their development in the coming decade.

In general, one can distinguish the five main reasons for the development of RES: (1) energy security; (2) preservation of the environment and ensuring environmental safety; (3) conquering global renewable energy markets, especially in developing countries; (4) preservation of own energy reserves for future generations; and (5) an increase in consumption of raw materials for nonenergy use of fuel (see, e.g., Fischer et al., 2016; or Pacesila et al., 2016).

The growing interest in renewable energy resources is associated not only with the depleting reserves of fossil fuels and a steady increase in energy consumption but also with an increase in greenhouse gas emissions into the atmosphere. Therefore renewable energy production is becoming more attractive.

Interestingly, not only economically highly developed countries have embarked on the path of intensive development of renewable energy but also developing countries, such as China and India, also show high rates of renewable energy development in the world.

In the EU countries, the share of electricity generated on RES (without hydroelectric power plants) in the total energy balance of the countries increased by four percentage points over the past 10 years, in absolute terms by more than 130 billion kWh of annual output (Bhattacharya et al., 2016). In some other countries around the world, the share of renewable energy in electricity generation exceeds 10%. Such states as Brazil and Mexico are also actively developing domestic renewable energy: the share of RES (without hydropower plants) in the production of electricity exceeds 4%.

In the global structure of electricity production based on renewable energy (with the exception of hydroelectric power plants of more than 25 MW), installations using biomass as an energy carrier tend to prevail. These are plants operating on forestry and agricultural waste, municipal solid waste, biogas and biofuels, landfill gas, etc. Due to the wide variation of biomass types, this resource is to some extent available in every country of the world.

Wind power plants account for about one-quarter of all electricity produced on RES. Wind turbines have become the most widespread in the economically developed countries of Western Europe, individual states of the United States, and in recent years, China and India are among the top five countries in terms of installed wind power plants. In 2007 Germany came out on top in the world in terms of installed capacity of wind installations, overtaking the United States. In 2008 the United States again became the leader in terms of the development of wind power. The third place on this indicator belongs to Spain, then China and India, then the EU countries follow (Premalatha et al., 2014).

The share of geothermal installations in the production of electricity for RES is estimated at 15%. The geography of their use is limited due to the uneven distribution of geothermal resources. The share of solar installations is less than 1%. This is due to the high cost of equipment for the use of solar resources (Lund and Boyd, 2016).

In some countries, installations on various types of RES may prevail: in Denmark— wind installations, in Germany, with the absolute majority of installations on biomass, solar PV installations are in second place; and in Iceland and the Philippines—geothermal stations (Inderberg et al., 2018). Ensuring energy security and environmentally balanced economic growth today are priority areas for development for most of the countries in the world and the development of renewable energy can be one of the ways the country moves in this direction.

The share of geothermal installations in the production of electricity for RES is estimated at 15%. The geography of their use is limited due to the uneven distribution of geothermal resources. The share of solar installations is less than 1%. This is due to the high cost of equipment for the use of solar resources.

In individual countries, installations on various types of RES may prevail: in Denmark there are wind installations, in Germany, the absolute majority of installations are on biomass, solar PV installations come in the second place; and in Iceland and the Philippines, geothermal stations prevail (Rubin et al., 2015).

Ensuring energy security and environmentally balanced economic growth today are priority areas for development for many countries around the world, and the development of renewable energy can be one of the ways the country moves in this direction. The main incentives for the development of renewable sources are the following problems that are exacerbated over time, facing humanity include a number of aspects. First of all, it is the increase in energy needs of the rapidly growing world population. At the beginning of the 21st century, the world energy consumption exceeded 500 EJ/year, or about 12 billion tons of oil equivalent per year. According to various forecasts, already by 2020, the world energy consumption will increase by more than one and a half times, primarily due to developing countries (population growth with a simultaneous increase in the specific per person consumption of energy). With the gradual depletion of cheap stocks of fossil fuels, the possibility of fully satisfying the growing energy needs with reasonable costs raises serious concerns (Court and Fizaine, 2017). Nuclear power after a series of serious accidents at nuclear power plants does not cause public confidence, and its full development is possible only when switching to new types of reactors ensuring the reproduction of nuclear fuel, which is associated with the need to master new technologies and certain additional risks. Thermonuclear energy has not yet left the stage of basic research, and the timing of its possible industrial development is not yet predictable. In this situation, the rate of expanding the use of RES, whose resources are practically unlimited compared to the foreseeable energy needs of mankind, despite the increased costs, seems to be justified.

Ensuring the energy security of countries and regions that are highly dependent on energy imports is another priority. This problem is even more acute and relevant than the previous one. The world is quite strictly divided into countries: exporters and importers of energy resources. The deposits of fossil fuels and uranium are distributed extremely "unfairly" around the world, which causes economic and political crises and creates tension in the world. RES are distributed over the countries of the world more or less evenly and are available in one form or another in any geographic location, which makes them more attractive.

In addition, there is an issue of ensuring environmental safety. The scale of modern energy is still small within the framework of the natural energy balance: the energy consumption of mankind is only about 2/10,000 of the total energy input of solar radiation to the Earth's surface. At the same time, in comparison with the energy spent on photosynthesis processes (about 40 TW), the world energy industry is comparable and, according to estimates, reaches about 20% of it, which indicates the fundamental possibility of a noticeable global influence of energy on the biosphere. Energy is responsible for approximately 50% of all harmful anthropogenic emissions to the environment, including greenhouse gases. There is no doubt that RES are more environmentally friendly than traditional sources (Hussain et al., 2017).

Other important arguments in favor of the development and deployment of the RES are also the concern for future generations. Energy is an extremely inertial sphere of the economy, it takes decades to promote new energy technologies, diversification of primary energy sources is necessary, including through the rational use of RES. Many energy technologies that use the RES have already demonstrated their viability and over the past decade have demonstrated a significant improvement in technical and economic indicators. The specific capital expenditures for the creation of power plants for RES and the cost of the energy generated by them approached that of traditional power plants, and in some cases, the use of renewable energy in some regions and practical applications has become quite competitive.

Let us look at the role and place of RES on the example of a country that is abundant in fossil fuels and natural resources and thus, technically, should be reluctant in spending efforts on developing renewable energy options. A very good example of such a country is Russian Federation. From the point of view of macroeconomic indicators, Russia, it would seem, is abundantly provided with traditional energy resources. Analysis of the energy balance shows that of all the energy resources extracted in the country, about two-thirds are exported abroad. Forty-five percent—in natural form, another 13%—in the form of energy-intensive products of low redistribution (metal, fertilizers, etc.), about 6% are accounted for the energy spent on the transport of energy resources and specified products through Russia abroad. As for oil, today 80% of all oil produced in the country is exported. The approved Energy Strategy of Russia for the period until 2030 actually provides for only a slight relative decrease in the export of energy resources. Export orientation is largely due to the fact that the country's oil and gas sector provides about 17% of Russian GDP and more than 40% of the consolidated budget revenues, and it is extremely difficult to refuse such revenues (Klimenko et al., 2019).

It seems to be calming that, according to the available estimates, Russia ranks first in terms of natural gas reserves (23% of world reserves), second place in coal reserves (19% of world reserves), fifth in oil reserves (4%–5% world reserves). Russia accounts for 8% of world production of natural uranium. However, in Russia, easily accessible deposits of relatively cheap energy resources are rapidly depleted, and the exploration and development of new deposits require huge expenditures (Nezhnikova et al., 2018). It is obvious that the country's energy policy in the near future will require a serious correction toward more efficient use of energy resources. From the point of view of Russia's international obligations on ecology, everything is all right in the country. The sharp decline in production in 1990–2000 led to almost 40% reduction in CO_2 emissions (Klimenko et al., 2019).

Estimates show that even without the adoption of special measures, by 2030 emissions will not reach the 1990 level, and there is no need to be particularly worried about this. These data bring about somewhat pessimistic thoughts: from the point of view of the macroeconomic analysis, RES for Russia seem to be quite irrelevant (Karstensen et al., 2018). However, let us now look at Russia, a bit from other positions: from the position of the country's regions and specific energy consumers. The facts about Russia's energy potential are as follows:

- Two-thirds of the country's territory with a population of about 20 million people is located outside the centralized energy supply networks. These are areas of the country with the highest prices and tariffs for fuel and energy (10–20 rubles/kW and above).
- Most of the country's regions are really energy deficient, needing fuel delivery and energy supply. For them, the solution to the problem of regional energy security is as relevant as for the countries that import energy resources.
- Even though Russia is a gas power, only about 50% of urban and about 35% of rural settlements have access to gas pipes. Instead, they mostly use coal, petroleum products, which are sources of local environmental pollution.
- With the constant growth of tariffs and prices for energy and fuel, the growing costs of connecting to the centralized power supply networks, the autonomous power industry in the country is developing at a faster pace: the introduction of diesel and petrol generators with a unit capacity of up to 100 kW over the past 10 years has exceeded the input of large power plants. Energy consumers seek to provide themselves with their own sources of electricity and heat, which, as a rule, leads to a decrease in fuel efficiency as compared with the combined production of electricity and heat at combination of heat and power (CHP) plants and a decrease in the efficiency of the entire energy sector of the country (see, e.g., Dienes, 2002; or Namsaraev et al., 2018).

Technical and economic assessments show that it is precisely areas with decentralized and autonomous power supply that are most attractive for the effective use of nontraditional RES. Thence, it seems necessary to conduct targeted research and development to substantiate the effectiveness of the practical use of renewable energy in specific conditions, taking into account real climatic conditions and consumer characteristics. It is extremely important, with the support of regional authorities, to create a network of demonstration facilities that vividly show the benefits of using RES and serving as business development centers in this energy sector.

The contribution of nontraditional RES (without large hydropower plants) to the energy balance of Russia does not exceed 1%. Recently adopted government decisions prescribe to bring the contribution of RES to 4.5% by 2020, which will require the commissioning of power plants for renewable energy with a total capacity of 20−25 GW. However, these decisions have not yet been properly supported by legislation and regulations, nor have fundamental decisions been taken to stimulate the development of RES, which makes the implementation of the decisions made problematic (Aalto, 2012).

Russian Federation lags significantly behind the leading countries in the development and mastering the technologies for the use of renewable energy. Nevertheless, there are examples of successful projects in this area. This refers to the creation of several geothermal stations in Kamchatka, the commissioning of which allowed a significant reduction in the delivery of diesel fuel to this region. The private business has made a breakthrough in the development of the production of wood pellets from wood waste. Russia was among the world leaders in terms of pallet production (more than 2 million tons per year). Unfortunately, they are produced mainly for export to European countries, and within the country, their effective use is still constrained by administrative and economic barriers. There are certain successes in the creation of tidal power plants using original domestic developments. A number of companies pay great attention to mastering the technology of large-scale production of photoelectric converters, but, again, focusing primarily on export (Chan and Daim, 2012).

The reasons for the rapid development of renewable energy in the world are not only in the countries' desire to prevent the strengthening of the greenhouse effect but also to optimize the structure of their energy balances and start preparing for a new stage in the development of civilization characterized by the minimum use of carbon fuel in order to create new impulses of industrial development. The world is trying to build a new low carbon economy.

The development of energy based on renewable energy means the production and maintenance of high-tech products—equipment for renewable energy. Thus the formation of the renewable energy industry contributes to the diversification of the economy as a whole and its fuel and the energy complex in particular. In addition to the economic aspects, the social benefits of developing renewable energy in Russia should be considered, among which the fundamental ones are employment growth and improvement in the standard of living of the population.

The official unemployment rate in Russia is quite high. The highest unemployment rate is observed in the republics of the Southern Federal District, as well as in Siberia and the Far East. Unemployment is greatest in rural areas, in which almost half of all the unemployed live in the country (given that the rural population is 26%) (Mishchuk, 2016).

Renewable energy technologies are more labor-intensive than traditional energy (based on the unit of energy produced), thanks to this, their introduction allows you to create additional jobs at all stages: from research and demonstration to production and installation of equipment, operation, and maintenance of the station. Employment growth is maximum for biomass-based technologies that create the preconditions for increasing employment in the agricultural sector and the forest industry.

Income inequality in Russia is one of the most acute social problems. Incomes of rural population are significantly lower than urban ones. At the same time, living conditions in rural areas are much harder than in the city. Due to the fact that electricity supply in many rural areas is unstable, with frequent interruptions (a number of individual settlements do not have access to electricity at all, many households do not have access to running water), ordinary household needs, such as cleaning, washing, washing dishes, and cooking require a significant investment of time and effort.

Renewable energy technologies can improve the quality of life for people in rural remote areas. These technologies are the most effective and often the only means of electrification of remote rural settlements. In addition to lighting, electricity makes it possible to use electrical household appliances (which reduces the time spent on domestic needs) and communications (radio, television, telephone, and Internet), and makes it possible to use modern medical equipment, improve water supply, and increase the efficiency of agriculture.

One of the main environmental benefits of renewable energy is to reduce greenhouse gas emissions, achieved by replacing fossil fuel capacity.

In Russia, the energy sector and its electric power industry make the largest contribution to the total anthropogenic emissions of greenhouse gases in Russia. The main share of emissions is related to the combustion of fossil fuels produced in Russia (oil, natural and associated gas, coal, peat, and oil shale) and their products. Energy also includes volatile emissions from the extraction, storage, transportation, processing and consumption of oil, coal, and gas, as well as emissions from fuel combustion in cases where the energy of combustion is not used (gas flaring in oil fields, combustion of process gases of various productions, etc.). Most renewable energy systems contribute to greenhouse gas emissions only during their production and do not emit CO_2 (or emit a small amount) during their operation.

Biomass and open-cycle geothermal systems are an exception because they emit greenhouse gases during energy production. However, technologies using biomass can be considered as "neutral" in terms of carbon dioxide emissions, since the total amount of greenhouse gases emitted during the combustion of biomass is equal to the amount absorbed during the life of the plants. Greenhouse gas emissions from an open-cycle geothermal plant are tens of times less than emissions from the production of the same amount of energy using traditional fuels.

Concentration in the atmosphere of harmful substances is maximum in large cities with high population density. This adversely affects the health of the population (especially children), most of whom live in large and medium-sized cities.

Thus the use of RES for energy production contributes to the development of its own high-tech engineering base and the creation of new jobs in the regions of Russia. Increasing the use of renewable energy technologies in Russia could help reduce unemployment, improve living conditions, and stop the outflow of population from rural areas, the northern and eastern regions of the country. The development of renewable energy leads to a decrease in the level of environmental degradation and an improvement in the health and well-being of the population.

The need to overcome the lag in the large-scale development of RES is a political task and is dictated by Russia's desire to maintain the status of a world power, to play an important role in solving global energy problems.

Currently, the share of renewable energy in the total electricity generation in the country is extremely small (about 0.9%), despite the fact that Russia has enormous resources of RES. There are no statistics on the production of heat from RES; however, according to expert estimates, heat from RES accounts for 4% of the total heat production in Russia.

In the same time, the technical potential of renewable energy (excluding the potential of large rivers) in Russia is estimated to be more than 20 times higher than the annual internal consumption of its primary energy resources. The economic potential of RES depends on the existing economic conditions; cost, availability, and quality of reserves of fossil fuel and energy resources; electricity and heat prices in the country and regions; and regional distribution of technical capacity and other regional features, etc. This potential varies over time and must be specifically evaluated in the preparation and implementation of specific programs and projects for the development of renewable energy. Today it is about 300 million per year (this is 30% of the annual consumption of primary energy resources in Russia) (Vasileva et al., 2015).

The currently negligible role of RES in the country's energy sector is explained by a number of factors, including: high capital costs for the construction of renewable energy facilities, the lack of specific financial mechanisms of state support, low qualification of personnel, and the lack of reliable information on the availability and economic opportunities of renewable energy experienced by the public, business, and government.

The abundance of reserves of combustible minerals, along with excess generating capacity in the power industry, is often indicated as other deterrents to the development of renewable energy in Russia. Still, currently one finds a sufficient number of applications where renewable energy can be used effectively. The combination of rich Russian renewable energy resources and currently existing advanced technologies in the world provides certain advantages for Russia in expanding the use of RES. One of these applications is off-grid electricity supply and the use of local energy sources for heat generation.

Practically in all regions of Russia, there are at least one type of renewable resources, and in the majority—several types of RES. These are small rivers, waste from the agricultural and forestry complexes, peat reserves, significant wind and solar resources, and low-potential heat of the earth. In some cases, their operation is commercially more attractive compared with the use of fossil fuels, if the latter's supply is unreliable.

Approximately 10 million people in Russia who do not have access to power grids are currently served by autonomous systems running on diesel or gasoline. About half of these diesel and gasoline installations are unreliable, due to interruptions in fuel supplies and/or high prices for imported fuel. In remote areas of the Far North and the Far East, fuel is delivered by rail or road, and sometimes by helicopter, as well as river and sea transport with a limited period of navigation. Such supplies are unreliable and expensive (Martinot, 1999).

Off-grid supplies of electricity based on RES have proven their economic efficiency in many countries; they often avoid the high costs associated with the construction of power lines. In Russia, it would be efficient to use hybrid wind-diesel systems, biomass boiler houses, and small hydroelectric power stations, which may be competitive compared to traditional fossil fuel—based technologies. In order to provide the population with heat and hot water it seems important and relevant to undertake the following measures and steps:

- direct use of geothermal energy for space heating, hot water production, greenhouse heating, grain drying, etc. (in Kamchatka, the North Caucasus, and in other regions with significant geothermal resources);
- transfer of district boilers running on imported fuels to biomass (waste from agriculture and timber industry complex); and
- efficient use of solar collectors in the southern regions of Russia (see, e.g., Martinot, 1998; or Kononov, 2002).

Particularly noteworthy are the widely used heat pump technologies in the world. The conversion factor of the renewable low-potential heat of the source into heat used in heating systems may be 4—6 and higher. Sources of low-grade heat can be: purified water from aeration stations in large cities with a temperature of 16°C—22°C; circulating water from the cooling systems of turbine condensers of thermal power plants, state district power stations, and nuclear power plants, which have a year-round temperature of 12°C—25°C; warm mine waters from decommissioned coal mines; geothermal waters; sea water of the Black Sea coast of the Caucasus and other water bodies; outside air, rocks, and soil; and solar systems and heat accumulators. The strategic task is the development of industrial production of heat pumps at domestic enterprises and regulatory and technical support of their widespread introduction in the coming years.

The world practice shows that renewable energy facilities (solar PV cells, small wind turbines, etc.) have proven to be more cost-effective than traditional power plants, and in some industrial sectors: sea and river navigation, cathodic protection of pipelines and wellheads, power supply of sea gas and oil platforms, power supply of telecommunication devices, etc. The scope of renewable energy in the global industry is constantly expanding, affecting all new areas. In addition to generating electricity at relatively lower costs in specific conditions, the industrial use of RES contributes to the creation of a new renewable energy market, which stimulates the accelerated development of innovative technologies for nonstandard applications.

Generally, one can state that Russian Federation has a significant potential for using renewable energy in industry, but its use is almost close to zero due to mostly economic reasons (already mentioned "Dutch disease"). Thence, it is advisable to use RES in Russia in order to reduce the environmental burden in cities and towns with difficult environmental conditions, as well as in places of mass recreation and treatment of the population, and in specially protected natural areas.

The development of energy technologies based on renewable energy should be one of the key areas of innovative development of the Russian scientific and technical complex and energy.

Nowadays, Russian technologies in the field of renewable energy are comparable in their working and scientific and technical characteristics with foreign technologies. Russia has tremendous experience in the construction and use of small hydropower plants (with a capacity of less than 25 MW); in terms of technology development, the use of tidal energy and geothermal sources is ahead of the EU and the United States. According to the technologies of wind turbines, as well as solar cells and heat pumps, Russia is inferior to the developed countries of the West (Asif and Muneer, 2007). However, the majority of Russian technologies are at the stage of scientific and technical development or demonstration objects, while similar Western technologies are already being used in commercial markets to one degree or another. Due to this, prices for electricity produced on traditional and renewable sources have a huge gap. Therefore if Russia manages to build a viable renewable energy equipment market, based on existing technical and scientific experience, this will give impetus to the development of renewable energy in Russia on a wide scale.

The above example shows that the potential for exploiting the renewable energy does not automatically means that the social acceptance and economic reality are going to be in favor of this endeavor. All in all, among all the factors determining the development of renewable energy, the cost factor is currently fundamental.

6.3 Energy saving and energy efficiency

Energy saving and energy efficiency are one of the most important problems of the modern economy. This is typical both for the facilities of the energy complex and for industrial production consuming fuel and energy resources. The reasons for this are high energy intensity of products, due to the imperfection of the technologies used, the presence of outdated equipment in the production of low energy efficiency, as well as a large amount of losses in the process of working and operating buildings and structures (Sorrell, 2015).

All over the world, lots of attention is paid to the problem of energy saving with the trend starting probably toward the end of the 20th century. Many enterprises began to conduct surveys in order to develop their energy passports and energy-saving management programs that require the implementation of energy-saving measures. They are included in projects aimed at improving the energy efficiency of production, which require certain management methods. The implementation of such projects allows to achieve the targets specified in the energy-saving programs (Gugler and Shi, 2009).

Energy saving and energy efficiency are one of the promising areas of development of the national and global economy. The tasks addressed in these areas are solved through the implementation in practice of complex processes, including analysis, evaluation, long-term planning, and monitoring of implementation, which result in energy-saving programs that are implemented in all sectors of the economy. In particular, at industrial enterprises, energy resources are used to carry out the production process and, in many cases, depending on many factors, can be spent more efficiently. This indicates the possibility of improving the energy efficiency of production and the availability of energy-saving potential, the goal of which is the implementation of programs. At the same time, energy management should create such conditions in the enterprise, in which the consumption of energy resources would be sufficient for the organization's core business and production rhythm but reduced as a whole and did not reduce the company's performance (Hepbasli and Ozalp, 2003).

In modern economic science, energy saving refers to activities aimed at increasing the efficiency of energy use, while associating them with economic, environmental, and social parameters. The purpose of this activity is to increase the efficiency of meeting the needs for energy services; at the same time, as a criterion of efficiency, it is proposed to minimize the costs associated with covering energy needs in the long term, taking into account social and environmental factors of energy saving (Shove, 2018).

The listed definitions reflect a wide range of consideration of energy saving as an activity. Such an approach allows to reveal the universal significance, both national and global, of energy saving in all sectors of the economy. All in all, energy efficiency requires good management.

The concept of "energy efficiency" is originally found in thermodynamics. It consists in determining the efficiency of thermodynamic cycles and obtaining maximum useful work. Thus the energy effect reflects the degree of perfection of the process of converting fuel into energy. Unlike energy conservation, which is mainly aimed at reducing energy consumption, energy efficiency is a useful (efficient) energy use. From an economic point of view, energy efficiency is determined by the use of various resources that accompany and ensure the receipt of this effect, as well as the results of activities. From this we can conclude that energy efficiency shows and is resource efficiency, that is, is a measure of the efficiency of the energy-saving process. It is necessary to agree that energy saving is a synonym for economy, a process mainly directly dependent on activities, that is, process management. Therefore it is necessary first of all to manage the process of energy saving, and the indicator of energy efficiency is an assessment of the efficiency of process management (Patterson, 1996).

With regard to the above, energy saving and resource saving appear to be interdependent concepts. The main goals of energy saving and measures for resource saving at the enterprise are the realized need, the understanding that resource saving is a way to reduce costs, increase product competitiveness, that is, ensuring sustainable operation of the enterprise in the market, and energy saving is a tool to ensure the efficiency of the enterprise. Such interrelation does not contradict the definition of resource saving—as a system of measures to ensure the rational use of resources, to satisfy the increase in the national economy's need for them, mainly at the expense of economy, but specifies, expands, and deepens the economic content of these two concepts. Resource conservation is aimed generally at improving the efficiency of the use of various resources, and at changing individual performance indicators, such as material and resource intensity.

It is necessary to emphasize the importance of energy saving for industrial enterprises, which is the key to economic growth and increased competitiveness due to the impact on the following factors: cost reduction, reduction in the share of payments for energy in the cost of final products, compliance with environmental standards, reduction in the number of fines, impact on image and prestige growth companies, etc. At the same time, apart from the level of industrial enterprises, energy saving is also important in other scales. At each level, its own program, projects, activities, and relevant resource standards for energy-saving measures should be developed depending on the territorial, sectoral, and national energy-saving goals.

Thus given the importance of energy saving as a systemic economic category, it is a process of forming energy efficiency in the use of energy resources, aimed at reducing resource intensity and improving energy resource efficiency, which is a tool for ensuring the economic growth of individual enterprises and industries as a whole with a sufficient level of fuel and energy resources for production processes in industry.

In modern scientific works there are different directions of energy saving, depending on the objects and goals that are considered, which are set in the implementation of this activity. We can distinguish the following areas: technical, socioeconomic, and organizational. Consider each of them in more detail. The technical direction of energy saving is the use of modern energy-efficient technology. In order to achieve the goals of energy saving in this case, measures are taken to modernize, reconstruct production, renovate, etc., which are aimed at saving fuel and energy resources in the process. In addition to the impact directly on the objects of energy consumption, the reduction of their losses also belongs to the technological direction. Energy losses depend on the state of engineering networks, thermal insulation of buildings and structures, etc. In this area, measures on compliance with the production technology, equipment operation modes, assessment of the condition of technological systems and care for them, as well as control over scheduled repairs are also considered. These activities are widely used in industrial enterprises.

The social and economic methods of energy saving include management processes for increasing the energy efficiency of production, improving social awareness of energy efficiency, as well as mastering staff training and human learning processes all for the purpose of reducing the misuse of energy resources. This direction of energy saving depends on the management methods used in organizations.

The organizational component of energy saving includes accounting, rationing, and control over the use of resources. This direction is necessary for analysis, compilation of energy balances and reporting, on the basis of which it is possible to identify savings reserves necessary for planning individual measures and energy-saving programs.

The purpose of energy-saving projects and programs is to increase the efficiency of use of energy resources and reduce their losses using technological measures, as well as organizational and economic tools. At the planning stage, with identified savings reserves, factors affecting the target indicators of projects and programs are determined.

The main indicator of the efficiency of fuel and energy resources consumption by an industrial enterprise is the energy intensity of its products. Despite the progress and success in improving energy efficiency in recent years, many enterprises still have a high share of payments for energy in the cost of their products and can reach up to 40%−50% (Herring, 2006).

At an enterprise, the efficiency of using energy resources is determined by a multitude of factors, while the authors dealing with this topic distinguish many categories of various kinds according to their systematization methods. In the greatest approximation, all factors can be divided into external and internal, which respectively depend on the processes occurring outside the organization and within it.

After conducting the analysis of the external factors, it is possible to identify opportunities and threats for an enterprise that affect the results of operations and energy-saving management. Let me consider external factors in more detail. Energy-saving measures implemented by consumers of fuel and energy resources reduce the demand for energy resources and the load on the fuel and energy complex, and an additional multiplicative effect can be created. These measures, which allow to get the effect of energy saving at industrial enterprises, are divided into two groups: (1) reduction of the volume of direct consumption of fuel and energy resources and the reduction of losses and (2) implementation of the achievements of scientific and technological progress, which have a direct impact on the dynamics of specific energy supply and energy consumption (Abdelaziz et al., 2011).

Providing factors create the conditions necessary for the implementation of energy-saving activities. With their help, it is possible to determine the possibility of combining the achievements of scientific and technological progress with the ever-growing need for energy resources, in order to satisfy it fully.

After analyzing the internal factors that influence the consumption of energy resources, it is possible to determine the reserves of energy saving, on the basis of which its directions in the enterprise are indicated and programs are developed. One of the most important factors influencing the processes of energy saving can be called the imperfection of the methodological base. Different methodologies focus on the assessment of energy efficiency and economic efficiency, which are based on investment analysis methods, but they do not affect the assessment of the complex relationship of resource and energy savings. The mutual influence of these processes and their disclosure is a factor in the growth of the degree of use of energy resources, reducing the level of need for them. The study of this aspect of energy saving will allow creating additional methods of stimulating various kinds (economic, administrative, and legal), as well as improving the management processes of enterprises (Thollander and Ottosson, 2008).

Many organizations are certified to use international ISO standards, in particular for energy management, which is one of the factors affecting the management of energy conservation in enterprises. In the process of implementing these standards, companies face difficulties. Such an obstacle is the incompatibility of methods for managing the costs of energy and resources at enterprises with these standards (Bunse et al., 2011).

This situation is due to the specifics of resource cost accounting and internal specific factors of the organization of work of enterprises. Large organizations use their own methodologies and organize the process of energy management, but in medium-sized and small enterprises this problem is reduced to cost accounting, which requires updating the energy-saving methodology. The basic principles of energy conservation, on which the state policy in this area is based, include the following principles:

- priority of efficient use of energy resources;
- implementation of state supervision over the efficient use of energy resources;
- obligatory accounting by legal entities of energy resources produced or consumed by them, as well as accounting by individuals of energy resources received by them;
- inclusion in state standards for equipment, materials, and structures, vehicles indicators of their energy efficiency;
- combination of the interests of consumers, suppliers, and producers of energy resources; and
- interest of legal entities—producers and suppliers of energy resources in the efficient use of energy resources (Anderson et al., 1996).

Nowadays, energy saving is mostly based on the following principles:

- shifting toward the rational use of energy resources, and not their tough saving (or rationing);
- searching and developing the innovative sources of energy supply, providing comprehensive savings of production factors;
- development of the accounting system and the widespread use of specialized devices (meters) for the implementation of operational and more accurate monitoring; and
- implementation of various measures aimed at reducing energy losses during the operation of buildings and structures (Nguyen and Aiello, 2013).

It is often noted that the principles of energy conservation should correspond to the increase in the efficiency of the entire industrial enterprise. It is possible to draw a parallel between the principles of energy saving and energy-saving measures. Since the implementation of energy-saving potential in enterprises implies the implementation of special projects, they will also be characterized by the following principles:

- identification of positive results from the use of energy-saving measures;
- ensuring compliance with the volume of necessary funds and available resources at the disposal of the enterprise; and
- ensuring minimization of costs for the development and implementation of energy-saving measures in the process (Zhao et al., 2014).

In general, energy-saving projects are based on four key principles of energy saving:

1. Investing in energy saving is justified if its volumes are less than the cost of energy saved during the depreciation period of the results of measures.
2. The reliability of the expected investment indicators decreases with an increase in the period of return on capital investments.
3. The costs of financing capital expenditures have a greater indicator of reliability than current costs in subsequent time periods.
4. The reserve of potential energy savings over time tends to zero, due to the improvement of technology. Thus the potential for saving energy is reduced, and therefore its value expression tends to zero (Masa et al., 2018).

Moreover, the basic principles of energy saving include:

1. Purposefulness—reduction in the total volume of consumption of fuel and energy resources, the magnitude of losses, growth in production, reduction in the energy intensity of products.
2. Systematization—energy saving is a single system consisting of mutually influencing processes (energy-saving projects).
3. Complexity—consists in the development of separate interconnected parts of projects ensuring the achievement of intermediate goals, which should correspond to the main goals of the enterprise.
4. Security—the implementation of energy-saving measures should be provided with the required resources.

5. Consistency—all processes and activities of different levels should be coordinated with each other and ensure the achievement of the main goal of energy saving—reducing the energy intensity of products and services.
6. Timeliness—getting the planned effect in the timeline set by the projects. The implementation of this principle is not always possible under the influence of many factors in the implementation and operation of projects and programs (Enshassi et al., 2018).

Development and implementation of projects and programs in the field of energy conservation and energy efficiency that are carried out in recent years have been one of the leading places in the development of domestic sectors of the economy. The introduction of energy-saving measures allows to reduce the consumption of energy resources due to their more efficient use in the implementation of economic activities and reduction of losses at all stages of the process chain (generation, transportation, and consumption). Among the stimulating factors affecting the implementation of energy conservation measures, it is possible to single out the presence of a common national strategy and an appropriate regulatory framework. Since 2007, the European Union (EU) has been implementing the "Intelligent Energy" program, which aims to reduce the consumption of energy resources, increase energy efficiency, expand the use of RES, and introduce the latest technological solutions in production. Legislative acts of the European Union in the field of energy conservation and energy efficiency can be divided into the following areas:

- energy efficiency and energy services by end users;
- energy efficiency of buildings;
- promotion of new efficient energy production technologies;
- eco-design energy-intensive products;
- marking of household appliances and electronics; and
- harmonization of energy efficiency policies in different countries (Kylili and Fokaides, 2015).

In European Union, energy conservation was formed as a separate activity in the 1970s. The reason was the energy crisis of 1973, which caused economic instability over the next decade. The answer to the problems of the crisis was the implementation of energy-saving principles, so in order to reduce hydrocarbon consumption, modern technologies began to be actively introduced with higher energy efficiency, the use of alternative energy increased, which resulted in the development of economic mechanisms. The urgent need to reduce the consumption of energy resources amid the crisis situation served as a huge impetus for the development of energy conservation (McCalley and Midden, 2002).

Thus taking into account powerful economic incentives and strict requirements and regulations, energy saving began to develop as a strategically planned process implemented in all sectors of the economy. As a result of the implementation of energy-saving programs in Western countries, there was a demand for energy-efficient technologies, which led to the creation of their market. In the public mind, energy saving became the norm, such a lifestyle was prestigious and formed a progressive civic position, based on the idea that saving energy is not only profitable but also fashionable. But economic stimuli remained the main drivers of the development of energy saving, since most of the EU countries are importers of fuel and energy resources, therefore the prices of fuel remain a decisive factor. But even with such fuel prices, the economic motivation to save energy still remains very high and a wide range of economic incentives are applied.

Let us consider the example of a state policy in the field of energy conservation in the experience of some EU Member States. For example, in Germany, financial and economic instruments were used to implement the program for the development of renewable energy. A system of long-term concessional lending was used for organizations and representatives of the private sector who decided to introduce renewable energy. At the same time, at the legislative level, it is determined that the energy supplied to the network from RES must be fully consumed and has priority over energy from sources with traditional types of energy resources. There is also a difference in rates between them. A similar situation allowed to attract citizens and stimulated them to install individual energy sources, such as solar panels. The same changes were made to the material support of the energy metering system since the new provision forced the installation of dual electricity meters, which determine not only the volume of consumption but also how much energy is given to the network. The system of preferential taxation is also practiced for manufacturers of energy-efficient equipment, which makes it possible to reduce their cost in the domestic market, stimulating demand and consumption (Gan et al., 2007).

One of the examples of the state policy of energy saving in the field of heat supply is Denmark. The country has one of the highest rates of centralization of the heating system. More than 60% of the heat load for heating needs is covered by large and small CHP plants. Centralization was achieved by reducing the costs of end users. The achieved indicators of the centralized heating system have reduced the costs of primary fuel by almost two times. At the same time, the state ensured optimal loading of equipment of generating facilities by paying the costs of connecting consumers to the general system. At the same time, consumers, in fact, became the owners of heat supply networks, which forced service companies to strive to increase the efficiency of their work, rather than focus on maximizing profits (Möller and Lund, 2010).

In France, a complex tariff system in the power industry is used, depending on the efficiency class of the equipment used. The more energy-efficient equipment is used, the lower the price for electricity. The tariff system has a wide range of prices. Moreover, for the population there is a stimulating measure for the purchase of high-class household appliances energy efficiency, in which the cost of the acquisition is deducted from the taxable base (Kirschen et al., 2000).

In many countries, the methods of economic incentives for introducing energy-saving projects for both suppliers and consumers of energy resources are widely used. In general, there are many examples of the implementation of large-scale energy-saving projects in various fields of activity. One of the most prominent examples is utilities. In this complex project are implemented to improve the energy efficiency of residential buildings and buildings for various purposes. Organization of energy consumption metering systems to improve the accuracy of operational control projects increases the energy efficiency of generating facilities.

As part of the nationwide projects focusing on the creation of the energy-efficient quarters, there are measures taken by the policymakers and stakeholders providing for the modernization of individual microdistricts and small towns in order to further use this experience in other areas. Such projects have many goals that are interrelated with each other. Thence, the most important of them are increasing the reliability of electricity and heat supply, which leads to an increase in the quality of life of citizens, as well as a reduction in their energy costs.

Despite the recent positive developments, the implementation of energy-saving projects in the framework of municipal and regional programs faces a number of difficulties. One of them is financing, the lack of budget funds does not always allow to cover investment costs of projects. In addition, there are organizational and legal barriers. In most cases, energy-saving measures are accompanied by large expenditures of funds and resources, while the effect of their implementation does not always meet the goals of the projects and the requirements of customers, which causes mistrust on their part to this activity. The reason is the low quality of work performed. In general, management of energy-saving projects is typically carried out in the following main ways:

1. The development, planning, implementation, and management of the project is carried out within the framework of the organizational structure of the entity that initiates it (the enterprise).
2. Project management by an external specialized organization on contractual terms. These organizations are energy service companies.
3. Energy performance contracts used in projects whose purpose is to reduce fuel costs for energy producers and consumers. For the first time, such contracts were used in the United States for implementing energy-saving projects at various facilities and forms of ownership (state, municipal, industrial, etc.) (Bertoldi et al., 2006).

Today, the market for energy-saving services is rapidly gaining momentum around the world. Energy service companies specializing in this type of activity are distributed. According to the legislation, the main task of such companies is to ensure the improvement of energy efficiency and energy saving at the request of the consumer. Consumers in this case are public authorities, municipal level entities, and organizations with regulated types of activities. In practice, energy service companies provide a wide range of services for commercial organizations, including energy consulting, installation of metering devices, installation and maintenance of energy networks, etc. Companies act as contractors, taking responsibility for providing various functions related to the energy use of enterprises that can be moved outside the organization of the customer. In real conditions, both the customer and the contractor do not need actual indicators of resource savings, they require cost reductions. However, the process of implementing energy service contracts has a number of difficulties. The energy service contract implies a certain payment system for the customer to pay for it. Investment costs should be covered by the actual amount of savings in the operation of the object of the contract and paid to the contractor within a certain period of time. In practice, there are practically no methods for monitoring and determining the real value of savings, that is, the effect of project implementation. Therefore the implementation of such a payment mechanism is difficult and the consumer pays the cost of the contract immediately, taking all the risk. Thus the very essence of energy service contracts is depreciated, since the contractor is not responsible for the customer's effects from the implementation of projects and neglect the principles of energy efficiency, discrediting the essence of energy saving.

The situation with energy service contracts differs between countries. Profit is the basis of interest in energy saving. Companies are focused on the consumer, and in the case of the fulfillment of the terms of the contract, they receive from customers freely distributed profits, formed through energy saving. Consumers understand their own potential savings, which they receive without using their own capital at the investment stage of projects, and are not burdened with finding sources of financing, if necessary. Energy service companies are forced to provide a wide range of services and implement packages of activities for each customer, since this option is convenient, and there is no need to involve many narrowly specialized organizations and conduct business with them at the same time.

Thus in the field of energy conservation, it is important to have an interest in its results, the effects obtained, despite the way in which projects are carried out. Since in modern practice, third-party organizations are increasingly being attracted for this, it is necessary to develop the market of energy services and increase the degree of competition, which will ensure an increase in the quality indicators of the work performed. The development of the proposal in this area will increase the energy efficiency indicators of production and involve more enterprises in the process of energy saving.

All in all, it becomes apparent that energy saving covers almost all spheres of life (types of economic and social activities). The processes of energy saving as well as energy efficiency are not limited either in space or in time, the process of improving the methods of production and consumption of energy is endless. Energy-saving measures are aimed at improving technologies, methods of organizing and managing, developing motivational and stimulating mechanisms, and improving the environment. These activities affect the organizational, technical, and economic characteristics of production, the competitive position of the enterprise in the sectoral and product markets, and the financial performance of the enterprise. The variety of effects and objects on which these activities are implemented, the difference in the scale and attracted resources, the timing of implementation, etc. cause the formation of a very complex diverse energy-saving measure.

6.4 Battery storage and smart grids

Energy storage is emerging as a possible way of addressing global electricity system challenges across several different program areas. However, there are specific technical and nontechnical obstacles to the widespread deployment of energy storage apparatus. Therefore it appears crucial to spot innovation processes, systems and mechanics (in a wide sense) that may permit energy storage to help meet grid challenges and deliver industrial expansion from technology growth businesses. The leaders in delivering such new technologies are very few on the market, one of them such as Elon Musk along with his revolutionary projects like Tesla Motors and SolarCity, Powerwall (the Tesla house battery), and only a handful of others.

The values obtained from generating cheap energy obtained from renewable resources throughout off-peak or very low demand periods that might be sold during peak hours that are mainly during the day can be calculated simply by taking the industry cost gap between the time intervals. Therefore the energy stored in batteries by the renewable sources like wind turbines throughout off-peak intervals can be discharged during peak intervals rather than conducting nonrenewable sources like natural gas turbines that are somewhat more expensive. For a comprehensive perspective of the effect on utilizing such technology, it might be required to think about the negative effect of contamination in the environment in acquiring a very clear diagnosis. It can be that despite the ecological benefits caused by energy storage, there might be simultaneous ecological pitfalls. However, it often seems to be quite unproductive to run the wind turbines at night because the generated energy cannot be effectively stored to be used later on. In spite of that, coal-fired power plants may also be left running at night because the energy is also saved thereby leading to an overall impact of a small gain in the entire quantity of contamination as a consequence of the mere presence of the grid (Jacobsen, 2016).

Electric energy may experience transformation to numerous varying kinds for storage. The water is subsequently introduced to flow into the lower reservoirs during peak hours and so triggers water turbines eventually producing energy. There is a proportional relationship between the energy saved, the water quantity at the top reservoir, and the elevation of the waterfall. Pumped hydro storage (PHS) installations generally last for 30 to 50 decades and have a decent round trip efficiency of 65%–75%. Additionally, PHS has a fast response time of under a moment despite its high-power volumes and electricity direction that makes it appropriate for controlling electric network frequency and supplying reserve generation (Díaz-González et al., 2012).

One of the technologies that is most commonly used in energy storage is the battery energy storage system (BESS). BESS is based on a principle when a group or different sets of plentiful cells that are interconnected together, parallel to each other or at both sequences for obtaining a certain value of voltage or ability (Divya and Østergaard, 2009). Then, the electrodes that are typically manufactured from conducting substances are submerged into the electrolyte in a special sealed container which enables to generate an external load (Winter and Brodd, 2004). This exercise enables to obtain electricity battery modules that produce lower voltages. These modules are then attached in a series, parallel or in both modes to the strings and are then capable of producing the desired electrical output and behavior. The numerous kinds of batteries used comprise lead-acid batteries, nickel-cadmium, and lithium-ion batteries. Lead-acid batteries are used for the maximum period of about 140 decades. These batteries are composed of two kinds; flooded batteries that are generally used and valve-regulated batteries now being investigated and developed.

These types of batteries can boast a life cycle ranging from 1200 to 1800 cycles that might vary depending upon the thickness of release with a round trip efficacy ranging from 75% to 80% added with a lifetime of 5 to 15 decades that depends on the operating temperature (Dufo-López, 2015). They are best used for electricity storage over long durations. Nevertheless, they frequently exhibit poor performance at low and high ambient temperatures and also have quite a brief life span. They also need water maintenance with time particularly the flooded type. Effort was led into converting nickel-cadmium and lithium-ion batteries into preferred possibilities for greater power uses especially when it comes to their costs. Nickel-cadmium batteries contain alkaline rechargeable batteries frequently categorized based on its use (Hadjipaschalis et al., 2009). This entails its sealed form frequently used on mobile electric equipment and its own flooded form employed in industrial uses. This form of battery has quite a protracted cycle life of over 3500 bicycles and requires low maintenance. They are nevertheless poisonous because of use of heavy metals that pose environmental and health risks and often suffer from memory effect. Lithium-ion (Li-ion) batteries, on the other hand, have shared program from modern digital gadgets such as mobile phones and electronic devices requiring low power apps. Additionally, they have rapid control and launch capacities along with getting high round trip capacities of 78% more than 3500 cycles. Quite unfortunately, these batteries cannot be used for power backup methods due to their life cycle being depending on the depth of discharge (Young et al., 2013).

Flow BESS is a new kind of technology system depending on the usage of reversible electrochemical responses to get a voltage output signal. This battery kind utilizes two different solutions comprised of two different tanks. Flow batteries are capable of discharging with no harm happening and have a comparatively low self-discharge. In addition, they have extended life spans, are of low care, have readily scalable abilities because of its dependence quantity of the electrolyte preserved, and are capable of storing electricity for extended durations (Chen et al., 2009).

Energy storage receives lots of attention from scientists and practitioners since everyone realizes that an improvement in this technology might forever change the market of electrical energy as we know it. A number of individuals perceive cheap storage since the missing connection between alternative renewable energy, like the solar and wind and everyday dependability. Various companies and institutions worldwide are considering and experimenting with different possibilities of electrical energy storage that would be able to meet different requirements like reducing gridlock and flattening the changes in electricity that take place regardless of renewable energy specifics and generation methods (Urbaniec, 2015). Substantial mechanical companies admit storage an applied engineering which may affect cars, turbines, and client automatics. Others, however, take a dimmer perspective, knowing that storage would not be sensible any time soon. That uncertainty cannot be easily omitted or solved. For a long time now, the prospect of power storage has been in sight but also very far away in practical terms. However, at present, efficient energy battery storage constitutes just a small issue in the whole ocean of other practical problems connected with gathering, storing, and transporting electrical energy over distances (Gallagher, 2009).

6.5 Smart grids and sharing economy

In addition to improving the use of networks, the growing market introduction of decentralized energy resources is creating new markets for justified and transactive two-way energy and information trading. Existing regulatory paradigms and systems are insufficient to allow the same kind of information, payments, and market disruptions that have fueled the sharing economy.

Nowadays, many energy companies are organizing special workshops that have already adapted aspects of the sharing economy in their business model to obtain empirical qualitative insights. As of today, there is a wide whole range of companies operating along the interface between the shared economy and the energy sector.

We have already demonstrated in previous chapters that joint characteristics of the shared economy and the transition of sustainable energy systems, such as the importance of platform organizations, take advantage of digital technologies and increase the potential for interaction between consumers, enabling both areas to learn from each other and open up new paths for researchers from different areas.

Additionally, new developments in the energy industry are taking into consideration new developments in the energy marketplace and give business models created to benefit from these surplus capacities. Moreover, load aggregators and utilities are going to have the ability to benefit from fresh electricity for different clients in real time, even when their distribution and general demand change.

These days, pretty much everybody is able to sell electricity back into the grid; however, the dangers are economically restrictive. Yet another problem here is that some electricity companies are operationally restricted to relying and disseminating electricity that will render some proactive prosumers without this chance. This would, obviously, turned into a discouraging component in the entire new energy economy prosumer-driven design.

According to Li et al. (2018), government subsidies should be used to stimulate investment in energy storage systems if renewable energy is to integrate fully into the industry. The mechanisms for establishing trust, engaging consumers, or managing the sharing economy may be transferable to organizations employing P2P energy trading platforms, energy communities, or local microgrids. In addition, microcredits can be awarded at business-to-customer levels to facilitate the spread of renewable technologies, particularly in developing countries.

It becomes apparent that understanding of energy production may vary from installation and upkeep of facilities, energy trading, energy efficiency, and energy efficiency and the best use of power storage or generation.

Since knowledge of installation, maintenance, and repair of energy technologies and energy efficiency is spreading, there may also be a new market for private services. Goods are economically divided in most business models. Therefore it is another form of economic sharing rather than social sharing. Renewable technologies require a certain amount of space and the right place to operate efficiently, renting space seems to be a viable business model for sharing energy.

While business models in the field are based on the economic sharing of used space, cooperation between social and nonprofit organizations are also a possible form of sharing the potential of unused land. The collaborative consumption, or the shared economy, refers to resource systems that ensure that consumer is not just a passive taker or prices but instead promote the two-sided role of the consumer role, a situation in which a consumer operates as the provider of resources and as their manager in one person. This approach might remind us the "prosumers" at the P2P energy markets. In fact, sharing economy has a lot to do with P2P energy trading models; however, more about that can be found in Chapter 5, Peer-to-peer markets and sharing economy of the smart grids.

It becomes quite clear that sharing economy creates a space for the market for a phenomenon when there appears such a category of financial agreements in which participants remotely discuss their access to services or products besides finding original methods of individual possession. The rising wealth in the developing world, together with population increase is exerting greater pressure on natural resources and has contributed to increasing costs and market volatility. Sharing economy might offer alternative scenarios and solutions how to share this wealth and gain mutual profits and benefits from these interactions. Compared to Uber or even Airbnb, such a sort of cooperation between Uber and Airbnb reduces the surplus capacity in the market without amassing the resources it depends upon.

With agents offering their solutions at the common market, businesses can save money on long-term labor costs and boost marginal earnings in their operations. However, three chief things permit for the mutualization of funds for a vast selection of new products and services and new businesses. To finish this argument, it should be concluded that sharing market platforms and companies are developing a string of issues with the use of competition rules and versions.

6.6 Electric transport

The world is concerned about global climate change and the growing shortage of fossil fuels. It is a common fact that a conventional car is a source of emission of more than 200 different harmful gases and chemical compounds, which are extremely detrimental to the ecology of the environment (Wei and Geng, 2016). Improving the material well-being of the population contributes to the growth of production and the number of cars in the world, which leads to an increase in harmful gases in the atmospheric air, mainly in large cities. In addition, the car is one of the main consumers of petroleum products and these reserves are limited. The introduction of strict environmental requirements for automobiles, accompanied by a rise in oil prices, contributed to the beginning of the search for alternative modes of transport, including the use of innovative technologies (Balcombe et al., 2019).

Currently, the most popular sustainable mode of transportation is electric cars [or electric vehicles (EVs)]. Many countries of the world participate in the development of this innovative sector. Many experts believe that electric transport is the future, as it will improve the environmental situation in large cities, as well as reduce the consumption of fossil fuels. In general, alternative modes of transport include (1) gas vehicle; (2) fuel cell vehicle; and (3) electric transport (Lund and Kempton, 2008).

Gas transport is a competitor in the development of electric transport. Some experts believe in the potential of the compressed natural gas used in transportation. Since the consumption of natural gas in transport in 2010 amounted to 29 billion cubic meters that is less than 1% of the total energy consumption in transport. More than 90% of its use comes from cars. However, there are some restraining factors for using gas instead of oil: (1) reequipment of vehicles is necessary and (2) development of a network of gas stations is required (Singh et al., 2015).

Competition between gas and oil in transport is possible with the use of gas-to-liquid technology, where the final product is similar in quality to fuels from petroleum. However, the costs for the production of such synthetic fuel from gas are equal to 110−140 dollars per barrel, at a gas price not higher than 75 dollars/thousand cubic meters that with projected prices for oil and gas makes these projects unprofitable (Ramberg et al., 2017).

Fuel cell car is another promising but not very frequently encountered technology. The basis of fuel cell vehicles is hydrogen and hydrogen-containing fuels: natural gas, ammonia, methanol, or gasoline. Methanol technology has not yet been tested. The main problem of this type of transport is a high fire and explosion hazard (hydrogen molecules are able to penetrate into the metal structure of the body or tank, seeping out, which can lead to an explosion), which puts the prospects for the development of transport on hydrogen at present under doubt, and the use of fuel cells postponed for the distant future (Liu et al., 2016).

EVs is something we all know thanks to Elon Mask, the founder of Tesla and a globe-trotting entrepreneur who enjoys the spotlight of mass media and social networks and often presents some futuristic visions of projects even large and wealthy countries likes the United States are not comfortable to pursue. EV is basically a vehicle whose drive wheels are driven by an electric motor powered by an electric battery, appeared for the first time in England around mid-18th century. They say Henry Ford was considering EVs which were favored by his wife but later abandoned this idea. The electric car is significantly older than a car with an internal combustion engine. Because of the long range without refueling, as well as the high speed of refueling cars, electric cars were soon superseded. Currently, technological progress has made great strides forward in the development of batteries. Thence, today's electric transport has become a serious competitor to cars with an internal combustion engine (Kalghatgi, 2018).

Electric transport, including electric buses, is considered to be an environmentally friendly mode of transportation. Many experts believe that the wide distribution of electric cars will significantly improve the ecological situation in large cities and densely populated areas, due to the power from electricity from RES. In many countries of the world, national programs for the development of this type of transport and its associated infrastructure have been adopted, which include various measures of motivation for producers and buyers of EVs (Ehsani et al., 2018).

One can make a short overview of national programs of state support for electric transport and methods of implementation. Let us start with the United States. The United States ranks first in EV development investments. The size of the monetary contribution to their development amounted to more than 27 trillion dollars. By 2025, the total amount of tax incentives that encourage the purchase of EVs will be $750 million (Coffman et al., 2017). In the United States, current tax credits are supposed to be improved by

- extending the range of advanced technologies used in modern vehicles in determining their creditworthiness;
- increasing the size of the loan from $7500 to $10,000;
- conversion of the loan in such a way as to allow consumers to benefit when buying a vehicle, and not when registering their taxes; and
- removing existing restrictions on certain vehicles by manufacturers eligible for credit, or, conversely, lowering the limit and, ultimately, eliminating credit at the end of the decade (Palmer et al., 2018).

The purchase of EVs is cofinanced from the federal budget in the amount of 50% of the price, but not more than $7500. In addition, the purchase of EVs is subsidized from the budgets of individual states in the amount of up to $8000. This program implies an increase in electric cars to 1 million by 2025.

In Germany, in the framework of the "Second Program of Economic Incentives," a program was adopted to accelerate the development and implementation of EVs. The Federal Ministry of Transport, Construction and Urban Development (BMVBS) has introduced the Electromobility in the Model Regions program. This program is an addition to the previously launched (in 2006) "National Innovation Program for Technologies for Using Hydrogen and Fuel cells." The programs are aimed at achieving Germany's leading position in the electric car market. From 2009 to 2011, 500 million EUR were allocated to support EVs in Germany (Mazur et al., 2015). During the economic crisis, these programs helped research institutions continue to work on innovative developments. According to some forecasts, German companies are going to present 16 new EVs. In the last 3—4 years, German automakers have invested €12 billion in the development of alternative vehicles. The sales of electric cars in Germany are envisaged to be at least 100,000 per year, and by 2020 the number of EVs will be around 1 million. The total amount of subsidies from the state will be 4 billion EUR by 2025 (Kihm and Trommer, 2014). The objectives of the EV market support program in Germany can be summarized as follows:

- technologically open research and development of EVs on batteries;
- user-oriented testing; and
- transport integration.

Local networks between agents from relevant industries, science, and the public sector, Germany is implementing a unique approach among all projects and programs, which consists of close cooperation on the technological platforms of many companies and research institutes. The results of this work are recommendations that contribute to the further development of EVs in various fields. The program "Electromobility in model regions" can be presented using the following key points:

- Total investment of about 300 million EUR, 220 partners take part in the project, about 150 of them from the car manufacturing sector, component manufacturers and suppliers, power supply companies, logistics, and transport.
- The main focus on the private sector: about 70% of financing was directed to private companies, 43% of which are small and medium-sized enterprises.
- On six content platforms, project partners meet regularly and share experiences. This allowed us to create the largest database of EVs in Germany. The seventh technology platform is responsible for cross-platform communication.
- In the model regions, a total of 2476 EVs were commissioned, including 59 buses, 243 special vehicles, 881 cars, 693 two-wheeled vehicles and scooters, and 600 electric bicycles to analyze the results of project implementation.
- About 70 demonstration projects with various fleets implemented in the field of personal, commercial, and public transport.
- An infrastructure for charging EVs was created, consisting of 1100 charging stations with 1935 charging terminals in public places, semistate, and private households (Rietmann and Lieven, 2019).

China is also a leader in the production and operation of EVs. The total amount of state support by 2025 will be $15 billion. Since 2009, China has been implementing an EV support program in the form of subsidies for the purchase of electric buses over 10 m long and worth $79,000, as well as EVs worth $8800 (Du and Ouyang, 2017).

In 2012 all-EVs registered in Beijing were exempt from annual taxes according to the decision of the State Council of China. In the summer of 2012, China set a goal for its industry to bring the production of electric cars and hybrids to 2 million per year by 2020. In early 2013, the Chinese company Wanxiang Group bought the bankrupt battery manufacturer A123 Systems. In 2015 Nissan and Renault began their production of EVs with its Chinese partner, Dongfeng (Zhou et al., 2015).

Japan is considered the first country to use electric transport. The program for the development of electric transport was developed in 1996. By the end of 2020, sales of electric cars are planned to increase to 15%−20% of all new cars sold, which will be about 1 million electric cars per year. Tax benefits for the purchase of electric cars in Japan account for about 30% of the value of the car, which is the highest rate in the world (Palmer et al., 2018).

France has a national plan for the development of electric transport, according to which by 2020, 2 million EVs and 400,000 charging terminals should appear on the roads.

The total amount of state support is declared in the amount of 2.2 billion by 2025. Since 2012, when buying an electric car, the buyer has the right to receive up to 20% of the value of the car, but not more than 5000 EUR. The AutoPact program provides financing of 250 million EUR for subsidized loans (Rizet et al., 2016).

In Italy, since 2011, there is a system of incentives for the purchase of electric cars (from 5000 EUR in 2012 with a gradual decrease to 1000 EUR in 2015). The total investment in infrastructure development will be 150 million EUR (Comodi et al., 2016).

In Portugal, in accordance with the national energy strategy, by 2020, 10% of all transport will be electrically operated. Portugal has the following sales promotion measures:

- Government subsidies of 5000 EUR for the first 5000 EVs.
- Subsidies in the amount of 1500 EUR to replace an old car with an electric car.
- EVs are not subject to transport and road taxes (Nunes et al., 2019).

In the United Kingdom, in accordance with the state program for the development of electric transport, it is planned to commission 4000 charging terminals for 220,000 EVs by the end of 2020. Since 2011, there has been a program to stimulate demand for electric cars, which is a subsidy from the federal budget for the purchase of an electric car in the amount of 25% of the cost (a limit of 5000 GBP (Pound sterling)). EVs are not subject to transport tax and excise tax. At the same time, legal entities have the right to return the tax, which is equal to 100% depreciation for the first year of service. In addition, free travel and parking are provided for electric transport in London. Additionally, the United Kingdom provided Nissan with a 380 million GBP loan for the development of electric transport production technologies (Bunce et al., 2014).

In Canada, since 2010, in the province of Ontario, part of the amount is reimbursed (up to 8000 CAD for EVs and up to $50,000 for electric buses) for the first 10,000 customers. Compensation is also provided for part of the charging station (50%, but not more than 30,000 CAD). EVs in Canada are issued a "green number" (license place), giving the right to use the fast lane (similar to carpools in the United States) that is typically used by the public transport (Mohamed et al., 2016).

In Norway, the following methods of sales motivation are provided:

- Exemption from one-time fees.
- Exemption from sales tax.
- Exemption from annual road tax.
- Free parking on all places where there is public parking.
- Electric cars are allowed on the bus and taxi lanes.
- EVs have the right to the free use of toll roads.
- Legal entities owning EVs are exempt from property taxation (starting from 2009).

In Sweden and Denmark, there are also many incentives for the development of EVs, for example:

- EVs are exempt from the registration fee (import duty) in Denmark.
- EVs are subjected to free use of toll roads.
- EV owners are eligible for subsidies.
- EV owners are exempted from certain fees and parking fees (Sovacool et al., 2018).

6.7 Vehicle-to-grid

Vehicle-to-grid (V2G) and grid-to-vehicle technologies represent the concept of two-way use of EVs and hybrids, which means connecting the machine to the general grid for recharging the car and returning excess electricity back. Owners of vehicles with V2G-enabled technology are going to have an opportunity to sell electricity to the grid during hours when the car is not in use and charge the car at hours when electricity is cheaper because in many countries the price of electricity depends on the time of day. It will also be possible to connect cars with this technology to one's home power source and then use them as uninterrupted power for the home or office (Habib et al., 2015).

The electric car used for leveling power generation surges should contain a system for controlling the charge of the battery from the electrical network and transfer part of the charge to the vehicle-to-network system. Accumulators of a typical EV allow to transfer from one to several kW/h of electricity to the network without serious consequences. Considering that most of the time the electric car is parked and there may be more than 90% of the lifetime connected to the network, we can say that there is already a solution to the problem of accumulating excess electricity in existing power grids. Owners of cars with V2G technology will not only be able to sell electricity to power engineers during hours when the car is not in use but also charge the car during hours when electricity is cheaper (in many countries the price of electricity depends on the time of day). It will also be possible to connect cars with this technology to one own household and use them as uninterrupted power for the home or office.

The term V2G was coined by AC Propulsion Inc. Now, the technology is widely developed and used by such parties as US government or Google (Mullan et al., 2012). The V2G system requires the construction of a network of charging EVs with the possibility of selecting electricity from the batteries of an EV. This will require standardization of the interface between the network and the EV and the creation of special chargers for EV batteries, which will be able to work as a battery DC inverter into an alternating current grid.

Currently, there is a real eBox EV (Toyota Scion xB conversion into an EV using a lithium-ion rechargeable battery) from AC Propulsion with the V2G system. However, the widespread introduction of V2G systems into reality rests on the absence of a sufficiently large fleet of EVs. When creating a sufficiently large fleet of EVs, the introduction of the V2G systems will make the most of the existing power generation facilities and will open up great prospects for the development of alternative energy. While the widespread introduction of V2G systems into reality rests on the absence of a sufficiently large fleet of EVs. There are three different versions of V2G:

- Hybrid or fuel cell cars, which generate energy from stored fuel and use their own generator to produce electricity. Energy production comes from conventional fossil fuels, biofuels, or hydrogen.
- Cars powered by batteries or hybrid vehicles that use their excess energy from the battery. These vehicles can be recharged during peak periods.
- Cars powered from the solar batteries. The car provides its excess energy power to electrical network. Here, in this case, the car actually acts as a small renewable energy station (Kempton and Tomić, 2005).

Currently, there is already a network of AC charging stations and express-charging complexes for EVs. AC stations are used to fully charge all types of EVs—electric vehicles, electric bicycles, and Segways. Charging an electric car at such a station takes about 6—7 hours. Express-charging complexes imply fast charging of EVs depending on the model of transport and the degree of battery discharge for 10—20 minutes.

In London in December 2006, the first two "gas stations" for EVs were opened, located on the street. Prior to that, 48 free EV charging points operated in parking lots and garages in London. By the end of 2009, about 1000 points were set in the United Kingdom for charging electric cars, of which about 200 in London (Morton et al., 2018).

By mid-May 2010, over 1800 points for charging EVs were built under the ARRA program (American Recovery and Reinvestment Act of 2009) in the United States. At the end of 2011, 500 EV charging points operated in South Korea.

In some large countries such as Russia, relatively recently, the implementation of the first phase of the program for the development of charging infrastructure for electric transport began. The first charging station in the Russian Federation was opened in November 2012. To date, there are 45 stations, including 5 express-charging complexes (Yao et al., 2016).

EVs are widely used at enterprises for the transport of goods within the shops (due to the absence of harmful emissions), at airfields and railway stations. Most electric cars are compact in size with a sufficiently high load capacity (from 0.25 to 5 tons or more) and often with wheels with cast rubber tires (truck tires). By design electric trucks and utility electric cars are close to electric cars. Recently, electric cars (self-propelled electric carts) and tractors designed for the carriage of goods within the enterprise, for medium distances of movement, with frequent change of trailers (trolleys), gained great importance. The most important characteristics for EVs are the amount of tractive effort (up to 25 tons) and the overall dimensions of the EV itself. There are electric cars with lever control in place of the steering wheel and a controller (3 speeds), weight with battery 1500 kg, permitted load 1500 kg. Electric tractors are designed for towing heavy loads, large dimensions, and large lengths. Also, electric tractors are used for towing equipment that does not have an independent course. Electrical tractors are controlled by the operator standing or sitting, depending on the type of work inside or outside the room. Large distribution of electric tractors is acquired in the automotive industry.

The distances to which goods move can vary depending on the area of the warehouse, the layout of the warehouse areas and transportation routes. On short tracks, it is advisable to use self-propelled electric cars, accompanied by a pedestrian operator. With medium transportation distances and greater work intensity, pallet carriers with a platform or platform for the operator are used. To move piece goods or goods in trailed containers, electrically driven tractors and platform trucks are used.

EVs with a fixed platform are also widely used. They are intended for transporting goods in cramped conditions on paved roads at industrial enterprises, warehouses, bases, ports, railway stations, etc. Among them are platform electric cars intended for the mechanization of transport work inside the shops of enterprises and warehouse premises, moving in the interrack space, for bringing cargo directly to workplaces, the dimensions of such electric cars allow it to be moved in freight elevators.

During operation, the driver-operator stands on a spring-loaded platform, while you can adjust the height, which allows operators to work with different body heights. Comfortable side grip ensures safe operation. The platform of the operator is an automatic switch of the tractor, when leaving—the operator's exit occurs automatic braking. At the same time, energy is not wasted but additionally charges the traction battery by returning energy. Usually EVs are widely used in enterprises where horizontal cargo assembly is necessary (automobile industry, airport, railway, or post office).

Warehouse tractors are designed for the transportation of goods in cramped conditions on paved roads at industrial enterprises, warehouses, bases, ports, railway stations, and other objects. It has an increased maneuverability as well as low operating costs. In electric cars, a noncontact static pulse governor of the engine speed is used to control the engine. To control the charge level of the battery in the EVs, an electronic indicator combined with an hour meter is used, while in some types of vehicles there is a switch voltmeter.

The solution of technical problems associated with the introduction of alternative energy sources led to the emergence of cars on electric propulsion. A couple of decades ago it seemed impossible, but innovative technologies in the field of battery and power plant development made EVs a reality. This type of vehicle is finding an increasing number of adherents. In the United States and several European countries, there is already an extensive network of electric "gas stations" where car owners can quickly and relatively inexpensively charge their EVs (Boulanger et al., 2011).

In Russia, despite the fact that such cars have their own specific taxation, which favorably differs for the buyer from taxing cars with internal combustion engines, the demand for electric cars is insignificant. This is partly due to the conservatism of the thinking of the population; there are more weighty reasons that are of a purely practical nature (Belousov et al., 2017).

At the moment, there is no widespread network of recharging stations for EVs in Russia. Thence, development and creation of electric transport in Russian conditions is extremely expensive. However, a scientific and engineering research in the field of construction of battery transport on electric propulsion is possible. Its production has significant differences from the production of cars equipped with gasoline or diesel internal combustion engines. First of all, it is connected with the creation of reliable electric motors with sufficient power and, at the same time, reduced energy consumption and high efficiency, as well as the creation of rechargeable batteries that meet a variety of requirements—high capacity, long service life, the possibility of effective operation at negative temperatures, fast charge cycle, and availability for recharging from a usual household network of 220 V.

Considering that the demand for electric cars in Russia is extremely small, this project will not be able to reach the level of payback even for several years. Even those consumers who decide to transfer to an electric car will turn their eyes toward foreign products. The Russian car industry is significantly lagging behind in development; the reliability of Russian cars with a traditional powertrain causes many complaints, so trust in products with electric motors will be small. For the same reason, few cars will find their customers abroad. All in all, EVs might take a long time to come into popularity and gaining wide acceptance in the case of such an oil-rich state as Russia.

Now, let us take a look at an interesting example that involves the optimization of using EVs in residential households. The data used in this example come from the smart meters readings conducted by the Customer-Led Network Revolution (CLNR) monitoring trial conducted with the 199 households in the Northern England (hereinafter referred to as the CLNR project) that provides the data from the study of household electricity use and tariff behavior between October 2012 and July 2014. The example of Northern England has been selected due to the availability of the data and for the sake of simplicity. Similar results would have been obtained using other hypothetical examples of households using the figures for the renewable energy and electrical vehicles elsewhere.

For the sake of comparison, I also used the data from the trial of 155 households with solar PV collected by the CLNR project between June 2012 and March 2014. The data from the second trial were not so detailed (there were no channels recorded as in the first case) and provided observations on solar power total and the whole home power import (both measured in kW). The data needed to be adjusted to represent the 30-minute data for an average day for an average household with and without solar PV which represents full 17,520 data points for each year (48 half-hour intervals multiplied by the 365 days).

The annual peak kWh export and import charges were the highest day for Tuesday, June 11, 2013 and lowest for Saturday, January 4, 2014. For the sake of comparison, we calculated the annual peak kWh export and import charges for Low Carbon London project (UK Power Networks SmartMeter Energy Consumption Data in London Household) data: the typical day was October 10, 2013 (1.4 kWh), the highest day was June 16, 2013 (3.1 kWh), and the lowest day was December 30, 2012 (0.09 kWh).

The households in our samples were located in the Northeast and were served by the Npower. The company applies four basic methods of electricity payment type: monthly direct debit, prepayment, quarterly variable direct bill, and receipt of bill. Moreover, the electricity tariff can vary depending on the fix and other arrangements and it can assume over 20 different values. The data analysis and the data basic averaging and clustering (from the largest to the smallest) followed the maximum power and the division of equipment (appliances) for the 20 households with solar PV and 20 matching households without solar PV from the total sample of the 199 households of CLNR project. Unfortunately, the IDs and all household information were removed from the dataset by CLNR; hence some irregularities hampered the proper comparison of the data.

The key findings from this particular CLNR study (CLNR, 2013) for the smart meter households can be summarized as the following:

- During the 4 days of greatest network stress, the peak demand average was 0.9 kW between 17:00 and 18:00.
- Electricity demand remained broadly the same at peak hours whether it is on a weekday or at the weekend.
- Households identified as "rural off-gas" and "high-income" customer subgroups' had the demand that could exceed 1 kW at network peak in winter.

The empirical model that is described here was envisaged to demonstrate the differences in network charges for residential consumers in Northern England with and without solar PV installed in their homes and with and without EVs. This empirical model is largely based on the model specified in Simshauser (2016), albeit with some extensions and modifications. According to Simshauser (2016), solar PV consumers in Southeast Queensland have lower metered consumption due to the household own production. This significantly reduces their share of the per kWh costs of the distribution system. However, the existing revenue cap regulation of the distribution charges means that the same revenue has been made as the number of units has fallen, which means that per unit charges have risen and the distribution of their payment between different types of households has dramatically changed.

In order to shed some light on the new arrangements, there are four types of household to be analyzed: households with no PV and no air-conditioning (the poorest group); households with air-conditioning and no PV; households with PV and no air-conditioning; and households with PV and air-conditioning. He looks at how the charging mechanism has shifted the payments and considers a more cost-reflective charging regime where each household pays a fixed charge, a per kW peak charge and a variable per kWh charge which better reflects underlying costs. Simshauser (2016) suggests that households with PV and air-conditioning have only a fractionally lower peak per kW usage compared to those with no PV but air-conditioning. Moreover, households with air-conditioning and no PV pay less than they should toward distribution charges. This creates a disadvantage for the poorest group of households (households without PV and air-conditioning) which currently pays 307 Australian dollars p.a. more (or about £112 (1 AUD = 0.53 GBP as of March 16, 2016), i.e., around 40% more) than those with PV and no air-conditioning.

These findings reveal that the starting point of charging is already unfairly cross-subsidizing peaky users with air-conditioning and that the system has rapidly become much more unfair with the high takeup of PV. Simshauser (2016) suggests that a more cost-reflective three-part tariff schemes would see those with PV and air-conditioning paying 28% more than they are paying now and those without both paying 15% less (with the result that the poorer households pay around 180 AUD (about £95) less). In mathematical terms: the relationship between kWh and kW peak observed prior to the arrival of PV has fundamentally changed, such that kWh is a poor proxy for kW peak demand.

Table 6.1 describes model specifications and outlines its main provisions with the definition of variable and their linkages.

TABLE 6.1 Model specifications.

Variable	Value	Description
t	Year	Time period
N	Unit	Number of electricity users (total number of households)
k		The number of households, $k = [1, N]$
F	£	Payment for the unit of power (kW) per day by one household (k) connected to the electricity network
v	£	Payment for the unit of energy (kWh) by one household (k) connected to the electricity network
M	kWh	Volume of energy used by the household (k) per year (t)
Tariff	£	Tariff (total payment for the electric energy) for the household (k) per year (t)
TR	£	Revenue of the electricity provider for N customers (households)
N_{H1}	Unit	Number of households (no EV, no PV)
N_{H2}	Unit	Number of households (EV, no PV)
N_{H3}	Unit	Number of households (no EV, PV)
N_{H4}	Unit	Number of households (EV, PV)
PV	%	Percent of households with solar photovoltaics (PV)
EV	%	Percent of households with electric vehicles (EV)
P	Probability factor	Probability that households use both PV and EV at a given percent rate of PV and EV deployment
M_{H1}	kWh	Metered import per one household (no EV, no PV) per year (t)
M_{H2}	kWh	Metered import per one household (EV, no PV) per year (t)
M_{H3}	kWh	Metered import per one household (no EV, PV) per year (t)
M_{H4}	kWh	Metered import per one household (EV, PV) per year (t)
P_{EV}	kW	EV consumption
T_{EV}	Hour	Time of EV connection to the network (in hours) per year (t)
P_{PV}	kW	PV battery power
T_{PV}	Hour	Total time of generating electrical energy from the solar PV (in hours) per year (t)
$Tariff_{H1}$	£	Tariff (total payment for electrical energy) by one household (no EV, no PV) per year (t)
$Tariff_{H2}$	£	Tariff (total payment for electrical energy) by one household (EV, no PV) per year (t)
$Tariff_{H3}$	£	Tariff (total payment for electrical energy) by one household (no EV, PV) per year (t)
$Tariff_{H4}$	£	Tariff (total payment for electrical energy) by one household (EV, PV) per year (t)

Own results.

The yearly (t) tariff for one household is calculated as follows:

$$Tariff = F \times 365 + v \times M \tag{6.1}$$

The revenue of the electricity provider obtained by the provider from N customers per year (t) can be calculated as follows:

$$TR = Tariff \times N = (F \times 365 + v \times M)N \tag{6.2}$$

The research objective is to determine the changes in the tariff for the households in case these households would install solar panels (PV) [which would in turn reduce the total metered import (M)] and start using EV [which would increase the total metered import (M)]. Moreover, we will set the total revenue of the electricity providers (TR) as fixed. In each period of time (t), one can identify four basic types of households:

- $H1$—(no EV, no PV)—households without EV and PV.
- $H2$—(EV, no PV)—households without PV but with EV.
- $H3$—(no EV, PV)—households without EV but with PV.
- $H4$—(EV, PV)—households with both PV and EV.

In a given time period (t), the number of these households can be determined using the following formula:

$$N = N_{H1} + N_{H2} + N_{H3} + N_{H4} \tag{6.3}$$

Knowing the percent of PV and EV households and the probability (P) that attributes the deployment of PV and EV to specific households, one can calculate the number of households belonging to each type:

$$N_{H4} = N - Round\left[N\left(1 - P\left(\frac{PV \cdot EV}{10000}\right)\right)\right]$$

$$N_{H3} = N - N_{H4} - Round\left[N\left(1 - \frac{PV}{100}\right)\right] \tag{6.4}$$

$$N_{H2} = N - N_{H4} - Round\left[N\left(1 - \frac{EV}{100}\right)\right]$$

$$N_{H1} = N - N_{H2} - N_{H3} - N_{H4}$$

Each type of household is characterized by the similar pattern of energy consumption per year:

$$M_{H1} = M$$
$$M_{H2} = M + P_{EV}T_{EV}$$
$$M_{H3} = M - P_{PV}T_{PV} \tag{6.5}$$
$$M_{H4} = M_{H2} - (M - M_{H3}) = M + P_{EV}T_{EV} - P_{PV}T_{PV}$$

In its general form, the tariff consists of two parts—payment for the unit of power (fixed tariff, or F) and payment for the unit of consumed electrical energy (variable tariff, or v):

$$Tariff = Tariff(F, v) \tag{6.6}$$

The fixed tariff (F) remains unchanged (not additional power is needed). The only variable part will be (v) given that the volume of consumed electrical energy should remain unchanged after the adaption of PV and EV by some households. The total revenue (TR) should also remain unchanged:

$$TR = TR_{H1}(v) + TR_{H2}(v) + TR_{H3}(v) + TR_{H4}(v) = Tariff_{H1}(v)N_{H1} + Tariff_{H2}(v)N_{H2} + Tariff_{H3}(v)N_{H3} + Tariff_{H4}(v)N_{H4} = const \tag{6.7}$$

As it becomes apparent from (6.7), each household type will use its tariffs that will be based on the changes in the electrical energy consumption (M) due to the EV or PV usage as well as on the changes of the variable tariff (v) caused by the changes in the structure of consumption in the electrical network (caused by the EV and PV deployment). Then, the variable tariff (v) can be calculated as follows:

$$v = \frac{TR - (N_{H1} + N_{H2} + N_{H3} + N_{H4})F \cdot 365}{N_{H1} \cdot M_{H1} + N_{H2} \cdot M_{H2} + N_{H3} \cdot M_{H3} + N_{H4} \cdot M_{H4}} \tag{6.8}$$

For each type of household, the tariffs will be calculated as follows:

$$Tariff_{H1} = F \times 365 + v \times M_{H1}$$
$$Tariff_{H2} = F \times 365 + v \times M_{H2}$$
$$Tariff_{H3} = F \times 365 + v \times M_{H3} \tag{6.9}$$
$$Tariff_{H4} = F \times 365 + v \times M_{H4}$$

Now, let us add an implication of having an EV in each type of a household. With a PV system in place, the EV battery is considered a storage device, as well as a battery that connects the power supply to a distributed PV system.

Let us assume that an EV would add additional 3000 kWh/kW in electricity consumption. We repeat the calculations for the households with and without solar PV but adding some hypothetical EV consumption of 3000 kWh/kW. There are four scenarios to be considered in similarity to a recent study by Strielkowski et al. (2019) that yield a decrease in electricity bill due to the introduced modifications:

- Scenario 1 (no PV, no EV)
- Scenario 2 (PV, no EV)
- Scenario 3 (no PV, EV)
- Scenario 4 (PV, EV)

Table 6.2 reports the results of the model estimations for the case of solar PV and non-PV households in Northern England with and without EV. The calculations presented in Table 6.2 are done as follows:
Nonsolar households: $365 \times £0.0483 + 2540 \times £0.02792 = 17.62 + 70.91 = £88.53$.

TABLE 6.2 Solar PV and non-PV households in Northern England with and without EV.

	Household A	Household B	Household C	Household D
	No solar PV	Solar PV	No solar PV + EV	Solar PV + EV
Maximum demand (kW)	1.69	1.69	1.69	1.69
Metered import (kWh)	2540.08	1800	5540.08	4800
Solar export (kWh)	0	740.08	0	740.08
Gross demand (kWh)	2540.08	2540.08	5540.08	5540.08
Number of customers	3.04 mil.	59,751	3.04 mil.	59,751
Percent of customers (all Npower customers in 2013)	98.1	1.9%	98.1%	1.9%
Base network tariff	£88.51	£67.85	£88.33	£78.92
Difference	B − A		D − C	
	£20.65		£9.4	

Own results.

Solar households: $365 \times £0.0483 + 1800 \times £0.02792 = 17.62 + 50.25 = £67.87$.

Difference: $£88.53 − £67.87 = £20.65$.

In addition, there is also solar households' export: 740.08 kWh $\times$ £0.1292 (Feed-in-tariff (FIT) in 2013) = £95.61.

In addition, one can easily calculate that nonsolar households with EV: $365 \times £0.0483 + (2540 + 3000) \times £0.02792 = £88.33$.

Solar households with EV: $365 \times £0.0483 + (1800 + 3000 − 740.08) \times £0.02792 = £78.92$.

The difference £9.4 is what the solar households save by using what it is generating for charging its EV.

Overall, the issues of solar energy and EV usage bring about the question of whether the current distribution charging mechanism is fair for the consumers (represented by households). In the United Kingdom, a country selected as an example for our case study, there exists a two-part tariff design, with a fixed rate (£/day) and a volumetric rate (£/kWh). When one would start raising a fixed amount of revenue by varying the volumetric charge (as in the case of PV and EV), inequalities in charging would appear.

It becomes apparent that there is a difference of £20.65 (2.792 p/kWh) is what the retail companies pay to the distribution network operators (DNO) for non-PV British households pay in excess per annum compared to the households with solar PV (as in the case of South Queensland). This effect is magnified by the retail tariff (where the unit charge is 14 p/kWh).

This case raises some questions for discussion: first of all, is the current charging methodology efficient and fair (85% per kWh, rest per day)? Second, does the apportionment of charges between fixed, per kW peak and per kWh use of system charges need to be changed? Third, does the advent of a significant new technology at a particular voltage level on the network mean that a new type of charge needs to be introduced at that voltage level (e.g., kW peak export tariff)?

Even though the magnitude is several times less in the case of the United Kingdom than in the case of Australia reported in Simshauser (2016), the core of the problem remains the same: the solar PV households are subsidized by the nonsolar PV households due to the current UK tariff charges.

Let us run a simulation to show how the different type of charging change things in the example described above. We will draw and solve the formulas that follow in order to establish the optimal tariffs and charges in a situation when the total number of customers remains the same but the share of solar PV increases altering the total revenue recovered from the residential customers. We can introduce a fixed element F and a variable element v in order to obtain a new plausible solution. We will put a new factor assuming that the charges were upgraded by the same percentage. Our calculations go as follows: let the total revenue (TR) = 88.51 × 3.04 million + 67.85 × 59,751. The average TR = TR/3.1 million households. We have $F + v$ (fixed and variable tariffs). Then solving the system of equations for x would yield the following results:

$(1 + x)F \times 3.1$ million households+ $(1 + x)v \times (2540 \times 3.04 + 1800 \times 59{,}751) =$ Total revenue

$(1 + x)F \times 3.1$ million households+$(1 + x)v \times (2540 \times 3.1$ mil. $\times 0.78 + 1800 \times 3.1 \times 0.22) =$ Total revenue

$(1 + x)F \times 3.1$ million households+$(1 + x)v \times (2540 \times 3.04 + 1800 \times 59{,}751) =$ Total revenue

$(1 + x)F \times 3.1$ million households+$(1 + x)v \times (2540 \times 3.1$ mil. $\times 0.78 + 1800 \times 3.1 \times 0.22) =$ Total revenue

Furthermore, let us add an implication of having an EV in each type of a household. Let us assume that an EV would add additional 3000 kWh/kW in electricity consumption. We can now repeat the calculations for the households with and without solar PV but adding some hypothetical EV consumption of 3000 kWh/kW.

The two last columns in Table 6.2 show the results for the households with and without solar PV and with EV. The metered import in both types of households increases by 3000 kWh but the solar household has an advantage of using the amount it generates for charging its EV. The difference now is £9.4.

The above results clearly demonstrate that the hypothetical development in variable and fixed tariffs based on the percent of EV used by solar and nonsolar households in order to maintain the same total revenue as per current arrangements.

The results of further simulations for various pecent of PV and EV in solar and nonsolar households are further depicted in Table 6.3 and 6.4.

Overall, the above examples demonstrate that combination of fixed, per unit, and per kWh peaks might yield very different distribution of charges for households with and without solar PV and EV. One can see that the current tariff design applicable on the electricity markets, the attempts to increase energy efficiency using innovative technologies such as PVs and EVs would inevitably lead to the reduction of distribution charges for less wealthier households without any EVs or PVs.

These results and implications might alter the future of the electricity market and the design of network charges, especially in the face of growing decarbonization, the growing use of RES, and the electrification of transport.

TABLE 6.3 F and v tariffs for households with PV.

	Household A		Household A (PV)	
	No solar PV		Solar PV	
Maximum demand (kW)	1.69		1.69	
Metered import (kWh)	2,540.08		1,800	
Solar export (kWh)	0		740.08	
Gross demand (kWh)	2,540.08		2,540.08	
Number of customers	3,040,000		3,040,000	100%
Base network tariff, £	88.5485336		88.5485336	
Total revenue (TR), £	269,187,542.1		269,187,542.1	
$F1$, £	0.0483	$F2$, £	0.0483	100%
$v1$, £ (0% PV)	0.02792	$v2$, £	0.039399463	
		$F3$, £	0.0483	78%
		$v3$, £	0.03613124	
		$F4$, £	0.0483	50%
		$v4$, £	0.03268098	
		$F4$, £	0.0483	20%
		$v4$, £	0.029647632	
		$F4$, £	0.0483	10%
		$v4$, £	0.028757893	
		$F4$, £	0.0483	5%
		$v4$, £	0.028332753	

Own results.

6.8 Conclusion and implications

All in all, it becomes obvious that in the changing global rules of the game, energy consumers are transforming into the new class of energy "prosumers." EVs or V2G systems, advanced battery energy storage, P2P markets, or electric transport are transforming the existing market of electric energy as we know it. Smart meters and smart devices, as well as 5G mobile networks allowing for the high-speed download and transfer or data enable the transition to the Internet of energy being a subset of the Internet of things, a compendium of millions and billions of devices and sensors that are connected together and allow for the two-way transfer of information and data. All these data will feed the artificial intelligence—based algorithms that will become the basis of the smart electricity grids of the future. These new smart grids will know exactly what to do and how to do it in order to save, optimize, and make the most rational choices available.

TABLE 6.4 *F* and *v* tariffs for households with EV.

	Household A		Household A (EV)	
	No solar PV		No solar PV + EV	
Maximum demand (kW)	1.69		3001.69	
Metered import (kWh)	2540.08		5564.08	
Solar export (kWh)	0		0	
Gross demand (kWh)	2,540.08		5,564.08	
Number of customers	3,040,000		3,040,000	100%
Base network tariff, £	88.5485336		88.5485336	
Total revenue (TR), £	269,187,542.1		269,187,542.1	
$F1$, £	0.0483	$F2$, £	0.0483	100%
$v1$, £ (0% EV)	0.02792	$v2$, £	0.012745869	
		$F3$, £	0.0483	78%
		$v3$, £	0.014476818	
		$F4$, £	0.0483	50%
		$v4$, £	0.017501884	
		$F4$, £	0.0483	20%
		$v4$, £	0.022550633	
		$F4$, £	0.0483	10%
		$v4$, £	0.024949704	
		$F4$, £	0.0483	5%
		$v4$, £	0.026351414	
				0%

Own results.

Smart grids are heavily dependent on information that has become an oil (or gold) of the 21st century. Whoever has information has the capability to act logically and to make optimal decisions with low costs. In a way, smart grids have something in common with the provisions of the economic theory that operates with the ideal market distinguished by the perfect information and economically rational agents. However, humans are not these rational agents, even when all information is present and readily available. In 2011 the Nobel Prize in Economics went to Christopher Sims and Thomas Sargent for their "empirical research on cause and effect in the macroeconomy." Christopher Sims is known for his research on "rational inattention." In this framework a general approach is provided for modeling decision-making of economic agents with limited abilities that allows them to obtain and process information (Sims, 2003). The concept of the rational inattention assumes that all pieces of information are freely available, but agents process only the most important of them by using channels of limited information capacity. Humans simply do not pay full attention even to information that is easily attainable with a negligible cost which is very strange given that we live in the era of Internet, Wi-Fi, and smartphones. This is our human nature—we like to go with a flow and let things pass as they are. The computers and algorithms are not like us—they will make decisions in the best possible way. Perhaps this is what scares us in truly smart grids of the future?

The importance of using information can be shown on an example of sharing economy. The concept of sharing economy is not new—the whole thing can be explained by the well-known saying "I do not need a hammer; I just need a nail in the wall." Quite often, only limited resources are required to make something, and the greatest conundrum is to where to find these resources and whom to approach. Internet-based technologies, mostly social networks, allowed the sharing economy to materialize and to become effective. Humans learned how to interact through Internet and Internet-based apps and how to build and verify mutual trust. With all that, sharing things like car drives, apartments, or services, became feasible. There is a two-way flow of information that powers sharing economy: we give "stars" to our Uber driver or AirBnB landlord to "rate" them and they give the same stars to us. As a result, both parties receive a feedback and an opinion and reputation form. The same two-way flow of information will enable the smart grids of the future to be more effective and more productive.

It is easy to predict that things that are more familiar to and requested by an average consumer will find themselves in the center of attention first. One of these things which are also very important for the functioning of the smart grids of tomorrow will be EVs. From today's perspective, the presence of an EV in the product lines of many automobile companies around the world is not economically viable. The companies have their buyers and occupy special niches on the automobile markets which is characterized by low cost of products, parts and spare parts, their widespread prevalence, cars have high maintainability, and their maintenance seem to be possible by independent efforts.

These characteristics, ensuring a stable demand for EVs, cannot be saved in the event of an electric car release—it will be too expensive, while the low levels of reliability and comfort typical of domestic products will not be able to meet the needs of motorists who have the necessary amount. At the moment we do not have the technologies necessary for the production of competitive products, even abroad they are just beginning their mass introduction.

It is necessary to borrow experience that will be accumulated by the world's leading manufacturers of EVs by that time, to develop a wide network of battery charging points, as well as an extensive network of diagnostic and service stations.

The cost of gasoline and diesel fuel is high, in addition, concern for the preservation of the environment is a massive trend in economically developed countries. These reasons cause the rapid growth in the number of EVs abroad, and their market share is growing every year. By some estimates, in 20 years the number of electric cars will exceed the number of cars with internal combustion engines. In the long run, electricity and biofuels may well completely drive out gasoline and diesel engines from the market. Of course, this will happen first in Europe, Japan, and the United States, but over time this trend will reach other countries in Southeast Asia, or Latin America, and then Russia and Africa. Then one can talk about the production of this type of car virtually everywhere around the world. Mass technologies involving EVs are becoming available for consumers and, over time, environmentally friendly transport would also start traveling the roads and highways, preserving our oil reserves and helping to turn from the carbon economy to the one based on RES which are not only exhausted but also play an important role in the economies of many countries (the notoriously known "Dutch disease").

All in all, consumers and prosumers are to become important players in the smart grids of the future. They will shape up the way how the future market of energy will function and will provide feedback and information necessary for the successful operation of complex power networks.

References

Aalto, P. (Ed.), 2012. Russia's Energy Policies: National, Interregional and Global Levels. first ed. Edward Elgar Publishing.

Abdelaziz, E.A., Saidur, R., Mekhilef, S., 2011. A review on energy saving strategies in industrial sector. Renew. Sustain. Energy Rev. 15 (1), 150–168. Available from: https://doi.org/10.1016/j.rser.2010.09.003.

AlSkaif, T., Luna, A.C., Zapata, M.G., Guerrero, J.M., Bellalta, B., 2017. Reputation-based joint scheduling of household appliances and storage in a microgrid with a shared battery. Energy Build. 138, 228–239. Available from: https://doi.org/10.1016/j.enbuild.2016.12.050.

Anderson, W.P., Kanaroglou, P.S., Miller, E.J., 1996. Urban form, energy and the environment: a review of issues, evidence and policy. Urban Stud. 33 (1), 7–35. Available from: https://doi.org/10.1080/00420989650012095.

Asif, M., Muneer, T., 2007. Energy supply, its demand and security issues for developed and emerging economies. Renew. Sustain. Energy Rev. 11 (7), 1388–1413. Available from: https://doi.org/10.1016/j.rser.2005.12.004.

Balcombe, P., Brierley, J., Lewis, C., Skatvedt, L., Speirs, J., Hawkes, A., et al., 2019. How to decarbonise international shipping: options for fuels, technologies and policies. Energy Convers. Manage. 182, 72–88. Available from: https://doi.org/10.1016/j.enconman.2018.12.080.

Belousov, E.V., Grigor'ev, M.A., Gryzlov, A.A., 2017. An electric traction drive for electric vehicles. Russian Electr. Eng. 88 (4), 185–188. Available from: https://doi.org/10.3103/S1068371217040034.

Bertoldi, P., Rezessy, S., Vine, E., 2006. Energy service companies in European countries: current status and a strategy to foster their development. Energy Policy 34 (14), 1818–1832. Available from: https://doi.org/10.1016/j.enpol.2005.01.010.

Bhattacharya, M., Paramati, S.R., Ozturk, I., Bhattacharya, S., 2016. The effect of renewable energy consumption on economic growth: evidence from top 38 countries. Appl. Energy 162, 733–741. Available from: https://doi.org/10.1016/j.apenergy.2015.10.104.

Boulanger, A.G., Chu, A.C., Maxx, S., Waltz, D.L., 2011. Vehicle electrification: status and issues. Proc. IEEE 99 (6), 1116–1138. Available from: https://doi.org/10.1109/JPROC.2011.2112750.

Bunce, L., Harris, M., Burgess, M., 2014. Charge up then charge out? Drivers' perceptions and experiences of electric vehicles in the UK. Transp. Res. Part A: Policy Pract. 59, 278–287. Available from: https://doi.org/10.1016/j.tra.2013.12.001.

Bunse, K., Vodicka, M., Schönsleben, P., Brülhart, M., Ernst, F.O., 2011. Integrating energy efficiency performance in production management-gap analysis between industrial needs and scientific literature. J. Clean. Prod. 19 (6–7), 667–679. Available from: https://doi.org/10.1016/j.jclepro.2010.11.011.

Chan, L., Daim, T., 2012. Exploring the impact of technology foresight studies on innovation: case of BRIC countries. Futures 44 (6), 618–630. Available from: https://doi.org/10.1016/j.futures.2012.03.002.

Chauhan, A., Saini, R.P., 2014. A review on integrated renewable energy system based power generation for stand-alone applications: configurations, storage options, sizing methodologies and control. Renew. Sustain. Energy Rev. 38, 99–120. Available from: https://doi.org/10.1016/j.rser.2014.05.079.

Chen, H., Cong, T.N., Yang, W., Tan, C., Li, Y., Ding, Y., 2009. Progress in electrical energy storage system: a critical review. Prog. Nat. Sci. 19 (3), 291–312. Available from: https://doi.org/10.1016/j.pnsc.2008.07.014.

CLNR, 2013. Insight report: domestic solar PV customers. <http://www.networkrevolution.co.uk/wp-content/uploads/2015/01/CLNR-L090-Insight-Report-Domestic-Solar-PV.pdf> (accessed 23.04.19.).

Coffman, M., Bernstein, P., Wee, S., 2017. Electric vehicles revisited: a review of factors that affect adoption. Transp. Rev. 37 (1), 79–93. Available from: https://doi.org/10.1080/01441647.2016.1217282.

Comodi, G., Caresana, F., Salvi, D., Pelagalli, L., Lorenzetti, M., 2016. Local promotion of electric mobility in cities: guidelines and real application case in Italy. Energy 95, 494–503. Available from: https://doi.org/10.1016/j.energy.2015.12.038.

Court, V., Fizaine, F., 2017. Long-term estimates of the energy-return-on-investment (EROI) of coal, oil, and gas global productions. Ecol. Econ. 138, 145–159. Available from: https://doi.org/10.1016/j.ecolecon.2017.03.015.

Demirbas, A., 2005. Potential applications of renewable energy sources, biomass combustion problems in boiler power systems and combustion related environmental issues. Prog. Energy Combust. Sci. 31 (2), 171–192. Available from: https://doi.org/10.1016/j.pecs.2005.02.002.

Díaz-González, F., Sumper, A., Gomis-Bellmunt, O., Villafáfila-Robles, R., 2012. A review of energy storage technologies for wind power applications. Renew. Sustain. Energy Rev. 16 (4), 2154–2171. Available from: https://doi.org/10.1016/j.rser.2012.01.029.

Dienes, L., 2002. Reflections on a geographic dichotomy: archipelago Russia. Eur. Geogr. Econ. 43 (6), 443–458. Available from: https://doi.org/10.2747/1538-7216.43.6.443.

Divya, K.C., Østergaard, J., 2009. Battery energy storage technology for power systems—an overview. Electr. Power Syst. Res. 79 (4), 511–520. Available from: https://doi.org/10.1016/j.epsr.2008.09.017.

Du, J., Ouyang, D., 2017. Progress of Chinese electric vehicles industrialization in 2015: a review. Appl. Energy 188, 529–546. Available from: https://doi.org/10.1016/j.apenergy.2016.11.129.

Dufo-López, R., 2015. Optimisation of size and control of grid-connected storage under real time electricity pricing conditions. Appl. Energy 140, 395–408. Available from: https://doi.org/10.1016/j.apenergy.2014.12.012.

Ehsani, M., Gao, Y., Longo, S., Ebrahimi, K., 2018. Modern Electric, Hybrid Electric, and Fuel Cell Vehicles, first ed. CRC Press.

Ellabban, O., Abu-Rub, H., Blaabjerg, F., 2014. Renewable energy resources: current status, future prospects and their enabling technology. Renew. Sustain. Energy Rev. 39, 748–764. Available from: https://doi.org/10.1016/j.rser.2014.07.113.

Enshassi, A., Ayash, A., Mohamed, S., 2018. Key barriers to the implementation of energy-management strategies in building construction projects. Int. J. Build. Pathol. Adapt. 36 (1), 15–40. Available from: https://doi.org/10.1108/IJBPA-09-2017-0043.

Fischer, W., Hake, J.F., Kuckshinrichs, W., Schröder, T., Venghaus, S., 2016. German energy policy and the way to sustainability: five controversial issues in the debate on the "Energiewende. Energy 115, 1580–1591. Available from: https://doi.org/10.1016/j.energy.2016.05.069.

Gallagher, K., 2009. Acting in Time on Energy Policy. Brookings Institution Press, Washington, DC.

Gan, L., Eskeland, G.S., Kolshus, H.H., 2007. Green electricity market development: lessons from Europe and the US. Energy Policy 35 (1), 144–155. Available from: https://doi.org/10.1016/j.enpol.2005.10.008.

Gugler, P., Shi, J.Y., 2009. Corporate social responsibility for developing country multinational corporations: lost war in pertaining global competitiveness? J. Bus. Ethics 87 (1), 3–24. Available from: https://doi.org/10.1007/s10551-008-9801-5.

Habib, S., Kamran, M., Rashid, U., 2015. Impact analysis of vehicle-to-grid technology and charging strategies of electric vehicles on distribution networks-a review. J. Power Sources 277, 205–214. Available from: https://doi.org/10.1016/j.jpowsour.2014.12.020.

Hadjipaschalis, I., Poullikkas, A., Efthimiou, V., 2009. Overview of current and future energy storage technologies for electric power applications. Renew. Sustain. Energy Rev. 13 (6), 1513–1522. Available from: https://doi.org/10.1016/j.rser.2008.09.028.

Hepbasli, A., Ozalp, N., 2003. Development of energy efficiency and management implementation in the Turkish industrial sector. Energy Convers. Manage. 44 (2), 231–249. Available from: https://doi.org/10.1016/S0196-8904(02)00051-1.

Herring, H., 2006. Energy efficiency: a critical view. Energy 31 (1), 10–20. Available from: https://doi.org/10.1016/j.energy.2004.04.055.

Hossain, M.S., Madlool, N.A., Rahim, N.A., Selvaraj, J., Pandey, A.K., Khan, A.F., 2016. Role of smart grid in renewable energy: an overview. Renew. Sustain. Energy Rev. 60, 1168–1184. Available from: https://doi.org/10.1016/j.rser.2015.09.098.

Huang, A.Q., Crow, M.L., Heydt, G.T., Zheng, J.P., Dale, S.J., 2010. The future renewable electric energy delivery and management (FREEDM) system: the energy internet. Proc. IEEE 99 (1), 133–148. Available from: https://doi.org/10.1109/JPROC.2010.2081330.

Hussain, A., Arif, S.M., Aslam, M., 2017. Emerging renewable and sustainable energy technologies: state of the art. Renew. Sustain. Energy Rev. 71, 12–28. Available from: https://doi.org/10.1016/j.rser.2016.12.033.

Inderberg, T.H.J., Tews, K., Turner, B., 2018. Is there a prosumer pathway? Exploring household solar energy development in Germany, Norway, and the United Kingdom. Energy Res. Soc. Sci. 42, 258–269. Available from: https://doi.org/10.1016/j.erss.2018.04.006.

Jacobsen, M., 2016. Economies and policy of large-scale battery storage. Kleinman Center for Energy Policy, University of Pennsylvania. <http://kleinmanenergy.upenn.edu/policy-digests/economics-and-policy-large-scale-battery-storage> (accessed 29.01.19.).

Jefferson, M., 2006. Sustainable energy development: performance and prospects. Renewable Energy 31 (5), 571–582. Available from: https://doi.org/10.1016/j.renene.2005.09.002.

Jogunola, O., Ikpehai, A., Anoh, K., Adebisi, B., Hammoudeh, M., Son, S.Y., et al., 2017. State-of-the-art and prospects for peer-to-peer transaction-based energy system. Energies 10 (12), 2106. Available from: https://doi.org/10.3390/en10122106.

Kalghatgi, G., 2018. Is it really the end of internal combustion engines and petroleum in transport? Appl. Energy 225, 965–974. Available from: https://doi.org/10.1016/j.apenergy.2018.05.076.

Karstensen, J., Peters, G.P., Andrew, R.M., 2018. Trends of the EU's territorial and consumption-based emissions from 1990 to 2016. Clim. Change 151 (2), 131–142. Available from: https://doi.org/10.1007/s10584-018-2296-x.

Kempton, W., Tomić, J., 2005. Vehicle-to-grid power implementation: from stabilizing the grid to supporting large-scale renewable energy. J. Power Sources 144 (1), 280–294. Available from: https://doi.org/10.1016/j.jpowsour.2004.12.022.

Kihm, A., Trommer, S., 2014. The new car market for electric vehicles and the potential for fuel substitution. Energy Policy 73, 147–157. Available from: https://doi.org/10.1016/j.enpol.2014.05.021.

Kirschen, D.S., Strbac, G., Cumperayot, P., de Paiva Mendes, D., 2000. Factoring the elasticity of demand in electricity prices. IEEE Trans. Power Syst. 15 (2), 612–617. Available from: https://doi.org/10.1109/59.867149.

Klimenko, V.V., Klimenko, A.V., Tereshin, A.G., Mitrova, T.A., 2019. Impact of climate changes on the regional energy balances and energy exports from Russia. Therm. Eng. 66 (1), 3–15. Available from: https://doi.org/10.1134/S004060151901004X.

Kononov, V.I., 2002. Geothermal resources of Russia and their utilization. Lithol. Min. Resour. 37 (2), 97–106. Available from: https://doi.org/10.1023/A:1014893531132.

Kylili, A., Fokaides, P.A., 2015. European smart cities: the role of zero energy buildings. Sustain. Cities Soc. 15, 86–95. Available from: https://doi.org/10.1016/j.scs.2014.12.003.

Li, X., Chalvatzis, K., Stephanides, P., 2018. Innovative energy islands: life-cycle cost-benefit analysis for battery energy storage. Sustainability 10 (10), 3371. Available from: https://doi.org/10.3390/su10103371.

Liu, Y., Lehnert, W., Janßen, H., Samsun, R.C., Stolten, D., 2016. A review of high-temperature polymer electrolyte membrane fuel-cell (HT-PEMFC)-based auxiliary power units for diesel-powered road vehicles. J. Power Sources 311, 91–102. Available from: https://doi.org/10.1016/j.jpowsour.2016.02.033.

Lund, H., Kempton, W., 2008. Integration of renewable energy into the transport and electricity sectors through V2G. Energy Policy 36 (9), 3578–3587. Available from: https://doi.org/10.1016/j.enpol.2008.06.007.

Lund, J.W., Boyd, T.L., 2016. Direct utilization of geothermal energy 2015 worldwide review. Geothermics 60, 66–93. Available from: https://doi.org/10.1016/j.geothermics.2015.11.004.

Martinot, E., 1998. Energy efficiency and renewable energy in Russia: transaction barriers, market intermediation, and capacity building. Energy Policy 26 (11), 905–915. Available from: https://doi.org/10.1016/S0301-4215(98)00022-6.

Martinot, E., 1999. Renewable energy in Russia: markets, development and technology transfer. Renew. Sustain. Energy Rev. 3 (1), 49–75. Available from: https://doi.org/10.1016/S1364-0321(99)00002-7.

Masa, V., Stehlík, P., Tous, M., Vondra, M., 2018. Key pillars of successful energy saving projects in small and medium industrial enterprises. Energy 158, 293–304. Available from: https://doi.org/10.1016/j.energy.2018.06.018.

Mazur, C., Contestabile, M., Offer, G.J., Brandon, N.P., 2015. Assessing and comparing German and UK transition policies for electric mobility. Environ. Innov. Soc. Trans. 14, 84−100. Available from: https://doi.org/10.1016/j.eist.2014.04.005.

McCalley, L.T., Midden, C.J., 2002. Energy conservation through product-integrated feedback: the roles of goal-setting and social orientation. J. Econ. Psychol. 23 (5), 589−603. Available from: https://doi.org/10.1016/S0167-4870(02)00119-8.

Mishchuk, S.N., 2016. Russian−Chinese agricultural cooperation in the Russian far east. Reg. Res. Russia 6 (1), 59−69. Available from: https://doi.org/10.1134/S2079970516010093.

Mohamed, M., Higgins, C., Ferguson, M., Kanaroglou, P., 2016. Identifying and characterizing potential electric vehicle adopters in Canada: a two-stage modelling approach. Transp. Policy 52, 100−112. Available from: https://doi.org/10.1016/j.tranpol.2016.07.006.

Möller, B., Lund, H., 2010. Conversion of individual natural gas to district heating: geographical studies of supply costs and consequences for the Danish energy system. Appl. Energy 87 (6), 1846−1857. Available from: https://doi.org/10.1016/j.apenergy.2009.12.001.

Morton, C., Anable, J., Yeboah, G., Cottrill, C., 2018. The spatial pattern of demand in the early market for electric vehicles: evidence from the United Kingdom. J. Transp. Geogr. 72 (C), 119−130. Available from: https://doi.10.1016/j.jtrangeo.2018.08.020.

Mullan, J., Harries, D., Bräunl, T., Whitely, S., 2012. The technical, economic and commercial viability of the vehicle-to-grid concept. Energy Policy 48, 394−406. Available from: https://doi.org/10.1016/j.enpol.2012.05.042.

Namsaraev, Z.B., Gotovtsev, P.M., Komova, A.V., Vasilov, R.G., 2018. Current status and potential of bioenergy in the Russian Federation. Renew. Sustain. Energy Rev. 81, 625−634. Available from: https://doi.org/10.1016/j.rser.2017.08.045.

Nezhnikova, E., Papelniuk, O., Gorokhova, A.E., 2018. Russia-china energy dialogue: research of the most promising energy areas for interrelation. Int. J. Energy Econ. Policy 8 (1), 203−211.

Nguyen, T.A., Aiello, M., 2013. Energy intelligent buildings based on user activity: a survey. Energy Build. 56, 244−257. Available from: https://doi.org/10.1016/j.enbuild.2012.09.005.

Nunes, P., Pinheiro, F., Brito, M.C., 2019. The effects of environmental transport policies on the environment, economy and employment in Portugal. J. Clean. Prod. 213, 428−439. Available from: https://doi.org/10.1016/j.jclepro.2018.12.166.

Ouda, O.K.M., Raza, S.A., Nizami, A.S., Rehan, M., Al-Waked, R., Korres, N.E., 2016. Waste to energy potential: a case study of Saudi Arabia. Renew. Sustain. Energy Rev. 61, 328−340. Available from: https://doi.org/10.1016/j.rser.2016.04.005.

Owen, A.D., 2006. Renewable energy: externality costs as market barriers. Energy Policy 34 (5), 632−642. Available from: https://doi.org/10.1016/j.enpol.2005.11.017.

Pacesila, M., Burcea, S.G., Colesca, S.E., 2016. Analysis of renewable energies in European Union. Renew. Sustain. Energy Rev. 56, 156−170. Available from: https://doi.org/10.1016/j.rser.2015.10.152.

Palmer, K., Tate, J.E., Wadud, Z., Nellthorp, J., 2018. Total cost of ownership and market share for hybrid and electric vehicles in the UK, US and Japan. Appl. Energy 209, 108−119. Available from: https://doi.org/10.1016/j.apenergy.2017.10.089.

Panwar, N.L., Kaushik, S.C., Kothari, S., 2011. Role of renewable energy sources in environmental protection: a review. Renew. Sustain. Energy Rev. 15 (3), 1513−1524. Available from: https://doi.org/10.1016/j.rser.2010.11.037.

Patterson, M.G., 1996. What is energy efficiency? Concepts, indicators and methodological issues. Energy Policy 24 (5), 377−390. Available from: https://doi.org/10.1016/0301-4215(96)00017-1.

Popp, D., Hascic, I., Medhi, N., 2011. Technology and the diffusion of renewable energy. Energy Econ. 33 (4), 648−662. Available from: https://doi.org/10.1016/j.eneco.2010.08.007.

Premalatha, M., Abbasi, T., Abbasi, S.A., 2014. Wind energy: increasing deployment, rising environmental concerns. Renew. Sustain. Energy Rev. 31, 270−288. Available from: https://doi.org/10.1016/j.rser.2013.11.019.

Ramberg, D.J., Chen, Y.H., Paltsev, S., Parsons, J.E., 2017. The economic viability of gas-to-liquids technology and the crude oil-natural gas price relationship. Energy Econ. 63, 13−21. Available from: https://doi.org/10.1016/j.eneco.2017.01.017.

Rietmann, N., Lieven, T., 2019. How policy measures succeeded to promote electric mobility—worldwide review and outlook. J. Clean. Prod. 206, 66−75. Available from: https://doi.org/10.1016/j.jclepro.2018.09.121.

Rizet, C., Cruz, C., Vromant, M., 2016. The constraints of vehicle range and congestion for the use of electric vehicles for urban freight in France. Transport. Res. Proc. 12, 500–507. Available from: https://doi.org/10.1016/j.trpro.2016.02.005.

Rubin, E.S., Azevedo, I.M., Jaramillo, P., Yeh, S., 2015. A review of learning rates for electricity supply technologies. Energy Policy 86, 198–218. Available from: https://doi.org/10.1016/j.enpol.2015.06.011.

Sanchez, O.J., Cardona, C.A., 2008. Trends in biotechnological production of fuel ethanol from different feedstocks. Bioresour. Technol. 99 (13), 5270–5295. Available from: https://doi.org/10.1016/j.biortech.2007.11.013.

Sawin, J.L., Martinot, E., Sonntag-O'Brien, V., McCrone, A., Roussell, J., Barnes, D., et al., 2010. Renewables 2010-global status report. <https://inis.iaea.org/search/search.aspx?orig_q = RN:46105565> (accessed 19.05.19.).

Shove, E., 2018. What is wrong with energy efficiency? Build. Res. Inform. 46 (7), 779–789. Available from: https://doi.org/10.1080/09613218.2017.1361746.

Sims, C.A., 2003. Implications of rational inattention. J. Mon. Econ. 50 (3), 665–690. Available from: https://doi.org/10.1016/S0304-3932(03)00029-1.

Simshauser, P., 2016. Distribution network prices and solar PV: resolving rate instability and wealth transfers through demand tariffs. Energy Econ. 54, 108–122. Available from: https://doi.org/10.1016/j.eneco.2015.11.011.

Singh, S., Jain, S., Venkateswaran, P.S., Tiwari, A.K., Nouni, M.R., Pandey, J.K., et al., 2015. Hydrogen: a sustainable fuel for future of the transport sector. Renew. Sustain. Energy Rev. 51, 623–633. Available from: https://doi.org/10.1016/j.rser.2015.06.040.

Sorrell, S., 2015. Reducing energy demand: a review of issues, challenges and approaches. Renew. Sustain. Energy Rev. 47, 74–82. Available from: https://doi.org/10.1016/j.rser.2015.03.002.

Sovacool, B.K., Noel, L., Kester, J., de Rubens, G.Z., 2018. Reviewing Nordic transport challenges and climate policy priorities: expert perceptions of decarbonisation in Denmark, Finland, Iceland, Norway, Sweden. Energy 165, 532–542. Available from: https://doi.org/10.1016/j.energy.2018.09.110.

Strielkowski, W., Volkova, E., Pushkareva, L., Streimikiene, D., 2019. Innovative policies for energy efficiency and the use of renewables in households. Energies 12 (7), 1392. Available from: https://doi.org/10.3390/en12071392.

Thollander, P., Ottosson, M., 2008. An energy efficient Swedish pulp and paper industry-exploring barriers to and driving forces for cost-effective energy efficiency investments. Energ. Effic. 1 (1), 21–34. Available from: https://doi.org/10.1007/s12053-007-9001-7.

Urbaniec, M., 2015. Towards sustainable development through ecoinnovations: drivers and barriers in Poland. Econ. Sociol. 8 (4), 179–190. Available from: https://doi.org/10.14254/2071-789X.2015/8-4/13.

Vasileva, E., Viljainen, S., Sulamaa, P., Kuleshov, D., 2015. RES support in Russia: impact on capacity and electricity market prices. Renew. Energy 76, 82–90. Available from: https://doi.org/10.1016/j.renene.2014.11.003.

Wei, L., Geng, P., 2016. A review on natural gas/diesel dual fuel combustion, emissions and performance. Fuel Process. Technol. 142, 264–278. Available from: https://doi.org/10.1016/j.fuproc.2015.09.018.

Winter, M., Brodd, R.J., 2004. What are batteries, fuel cells, and supercapacitors? Chem. Rev. 104 (10), 4245–4270. Available from: https://doi.org/10.1021/cr020730k.

Yao, L., Lim, W.H., Tsai, T.S., 2016. A real-time charging scheme for demand response in electric vehicle parking station. IEEE Trans. Smart Grid 8 (1), 52–62. Available from: https://doi.org/10.1109/TSG.2016.2582749.

Young, K., Wang, C., Wang, L.Y., Strunz, K., 2013. Electric vehicle battery technologies. Electric Vehicle Integration Into Modern Power Networks. Springer, New York, pp. 15–56.

Zhao, X., Li, H., Wu, L., Qi, Y., 2014. Implementation of energy-saving policies in China: how local governments assisted industrial enterprises in achieving energy-saving targets. Energy Policy 66, 170–184. Available from: https://doi.org/10.1016/j.enpol.2013.10.063.

Zhou, Y., Wang, M., Hao, H., Johnson, L., Wang, H., 2015. Plug-in electric vehicle market penetration and incentives: a global review. Mitig. Adapt. Strat. Global Change 20 (5), 777–795. Available from: https://doi.org/10.1007/s11027-014-9611-2.

Further reading

Suberu, M.Y., Mustafa, M.W., Bashir, N., 2014. Energy storage systems for renewable energy power sector integration and mitigation of intermittency. Renew. Sustain. Energy Rev. 35, 499–514. Available from: https://doi.org/10.1016/j.rser.2014.04.009.

The role and perception of energy through the eyes of the society

7.1 Introduction: the importance of energy for the society

One would probably agree with me that every society needs reliable, stable, cost-effective, socially, and environmentally acceptable supply of energy resources for satisfying its economic and social needs. This is predetermined by the creation of conditions for the formation and implementation of state policies as well as by the positive attitude to energy-related issues.

Energy is important for the society and we all realize it on way of another. However, energy is also a topic that raises many controversies among people. Some feel that humankind is wasting too much energy, others believe that there are still potentials and new sources of energy to be discovered and explored. And of course, there are also people who do not care about energy-related issues at all. For them, energy is a given thing that has always been there and always will. Why to bother about the source of the electricity that comes from the socket on one's wall as soon as it charges one's smartphone?

Probably everyone now realizes that in the face of global climate change, renewable energy represents a promise of hope for reducing the carbon footprint of the energy sector, but there are many challenges to be tackled before it is integrated into our society. Simply, there are too many issues related to the social acceptance of these technologies that, together with various technical "bugs," make them look suspicious or even unacceptable for many members of the society. For example, energy storage systems—a way to capture and store energy for later use—are just emerging to meet the challenge of increasing renewable energies but still require many improvements for making them an item of everyday use for each and every one of us. Nevertheless, energy storage is on the verge of widespread distribution, and many stakeholders have already demonstrated their dedication to providing a more reliable and cleaner future. A more effective energy storage would surely unleash the potential of renewable energy, which would have a huge impact on the reduction of dependence on fossil fuels and the reduction of harmful carbon emissions. However, nowadays not everyone agrees that climate change should be a major focus in our society, not even numerous energy-intensive companies, which would benefit from installations helping to reduce energy demand during peak periods (batteries would be able to deliver energy when demand is high and store energy when demand is low), thereby reducing the cost of electricity and making them more sustainable and more independent.

Energy policy of every country on our planet should facilitate massive deployment of clean energy solutions, support research and development of new clean energy technologies, and maintain a large part of our coal, oil, and gas reserves in the earth. Emerging technologies and new trends are opening the way for our energy system to be 100% clean and renewable. The experience with the current wave of renewable technologies suggests that support for public policy may accelerate the arrival of such low-carbon instruments. It appears that solar and wind power will play an important role in every vision of a 100% renewable energy system. However, it would require strengthening the international cooperation and support for developing countries to expand infrastructure and improve technology for modern deliveries and sustainable energy services as a means of mitigating climate change and its effects. Only joint efforts to increase the share of renewable energy and clean technology with fossil fuels in the global energy portfolio will help to reduce climate change and its impact.

Energy efficiency programs should be introduced globally, giving tax exemptions to companies that are proving to be energy efficiency initiatives (e.g., energy-efficient housing), product design (e.g., energy-efficient equipment), and services (e.g., industrial combined heat and energy). Governments and nongovernmental organizations should combine the energy interests of the public and private sector to create a reliable, affordable, and environmentally friendly "energy mix." Clean and sustainable coal and nuclear power stations are also the driving force behind the growth of local jobs and keep companies' spending predictably low for the life of the plant. Clean coal technology is a low-cost supplier of reliable US electricity and contributes to the diversification of fuel in a highly dependent natural gas state, such as Florida.

However, the future seems to be in favor of the renewable energy sources. Compared to fossil fuels, which are typically mechanized and capital-intensive, the renewable energy industry is more laborious. Renewable energy currently provides affordable electricity throughout the country and can stabilize energy prices in the future. Although renewables require an initial investment to be built, they can then run at a very low cost (for most clean energy technologies, the "fuel" is free of charge). The more intensive use of larger volumes of renewable energy can reduce the price and demand for natural gas and coal, increasing competition and diversifying our energy supply.

For example, China has put energy efficiency at the forefront of its policy of improving energy security, reducing the pressure on domestic resources (especially coal and water for heat generation) and reducing the impact on the environment as the economy grows. Energy efficiency and conservation are currently at the forefront of its industrial planning and development strategies, which are reflected in its objectives to reduce energy intensity (energy consumption per unit of GDP) by 20% since 2005 until 2020 (see, e.g., McNeil et al., 2016). Chinese authorities already recognized and acknowledged the increase in energy efficiency at home and business levels, which has led to financial savings over time.

Future development is crucial for long-term availability in increasing amounts of reliable, safe and environmentally friendly sources. The concern for a reliable future of energy is only natural because energy provides "essential services" for human life—heat, cooking, and production, or electricity for transport and mechanical work.

At present, the energy provided to such services is derived from fuels—oil, gas, coal, coal, nuclear, wood, and other primary sources (solar, wind, or water)—all of which are not used until they are converted into the necessary, mechanical, or other types of energy.

In many countries around the world, much primary energy is wasted because of the inefficient design or operation of the equipment used to convert it into the services it needs, although awareness of energy saving and efficiency is encouraging.

Smart technologies can track and manage energy consumption patterns, deliver flexible energy tracking day by day, take advantage of better storage options and combine group manufacturers to create virtual power stations. In addition, renewable energies in their current supply are not profitable without high government subsidies, or they use huge amounts of land or damage the environment in some way. Thence, large-scale governmental support and funding should be provided to the renewable energy until it becomes economically viable and profitable. This support should not be limited to monetary sources but should also include promotion campaigns, public relations agenda and marketing targeted at the general public, and promoting the benefits of the renewables and their importance for the sustainable development of human society and balanced economic growth.

Governments worldwide understand the importance of the energy-related agenda and would like to inform and persuade their citizens of its absolute importance for achieving and preserving their respective countries' security, sustainability, and competitiveness.

One of the pioneers in doing just this is the European Union (EU). Within the EU, changes in the aspect of its energy strategy and security are discussed as the new approach that can be summarized as follows: "energy transition—structural change in energy system—change of energy mix" (Rausser et al., 2018). The perspective of European Supergrid and of the European energy market is not currently just a political idea and intentions. One can see that the privatization of European electricity markets would inevitably cause higher cross-border trade of electricity: led to higher power exchange and that the benefits are more significant if privatization measures be in place for a longer period (Torriti, 2014).

If we will stay a little longer and keep using the example of the EU, it becomes apparent that energy transition occurring in the EU is a state of things when environment and economy come together hand in hand. Indeed, there is proved scientific and policy position of existing the relationship between economic development and the consumption of the energy. Balitskiy et al. (2014), Buchanan et al. (2014), or Pepermans (2014) show on the empirical data analyses that relationship between economic development and the consumption of energy sources, that is, natural gas, is positive, while the relationship in the causality direction from the consumption of the energy (i.e., natural gas) to economic development in the EU appears to be negative.

Since the inception of electricity deregulation and market-driven pricing throughout the world, the electricity consumption in the EU countries have been looking for a means to match consumption with generation. Currently, there is a strong policy commitment in EU and Organization for Economic Co-operation and Development (OECD) countries to increase the energy efficiency of residential buildings, and it is widely assumed that this will naturally and automatically reduce domestic energy consumption (Galvin, 2014). The existence of broad-brush rebound effects based on changes in energy efficiency and energy consumption in each of the 28 EU countries plus Norway was proved in calculative researches (Inderberg, 2015). Thence, most old EU states show rebound effects in the expected range of 0%−50%. However, the range for newer EU countries is 100%−550%, suggesting that energy efficiency increases are not a good predictor of energy consumption.

New trends in the renewable energy sources helped to support energy consumption in more effective way. However, with new trends appeared the necessity of new method to assess the consumption as well as the need for new technologies. Declaring on the government level the necessity and effectiveness of renewable energy for state's energy secure and sustainable development sometimes did not diffuse speedily in households consuming mind. Therefore most consumers having low anticipation to new energy sources had seen just policy recommendation to shift to the renewable energy. Traditional electrical and gas meters only measure total consumption, and so provide no information of when the energy was consumed at each metered site. The demand to overcome this gap was covered in some way by means of new technologies, particularly by appearing of smart meters that are gaining growing popularity and widely used by households and businesses for measuring power consumption and helping to reduce the costs of energy.

There are many environmental and social problems that can be reduced if people would participate in smart energy systems; however, little is known about which factors motivate people to actually participate in these systems (Van der Werff and Steg, 2016; Kowalska-Pyzalska, 2018). The factors that influence individuals' interest and actual participation in smart energy systems were tested (Koirala et al., 2018). The results of many academic studies showed that there is a big variance in interest and actual participation in smart energy systems; however, the reason for these differences was not completely revealed. Other recent articles analyzed practices and perceptions of stakeholders on including users in smart grids experiments in the Netherlands (Verbong et al., 2013). Also, the interaction between new "smart" energy feedback technologies and households were investigated in Norway (Skjølsvold et al., 2017). These case studies highlighted the need for social reconfigurations in all discussions on domestication of feedback technologies and in-depth analysis of public preferences and acceptance of smart grids. Other issues, such as scripting, control, and privacy in domestic smart grid technologies were analyzed in Danish pilot study. Comparison of two residential Smart Grid pilots in the Netherlands and in the United States was performed focusing on energy performance and user experiences (Obinna et al., 2017). These studies showed that smart grid setups influence energy performance and that end users prefer technologies that automatically shift their energy use. Verbong et al. (2013) analyzed the difference and relationship between smart grids and smart users and presented several ideas for involving users in developing smart grid technologies. Benders et al. (2006) proposed new approaches for household energy conservation based on personal household energy budgets and energy reduction options. This approach needs to be investigated further by defining the role of energy saving in accepting smart grid technologies.

All in all, both the analysis of the existing policy reforms and strategies and the review of the academic research literature yield that energy is and should be an important item of the everyday agenda for the society. It is just a question how to make it really appealing and acceptable for all members of this society. Many people are opportunistic and would not eagerly embrace new energy-saving or energy-efficient technologies without proper incentives provided either by the governments or by the markets. The commitments to the low-carbon future or the energy generation based on renewables that most governments of the developed countries agreed upon make this topic a very relevant issue. However, one needs to realize that the behavioral energy consumption patterns of the end users are hard to break since they represent informal institutions (such as habits) that typically requite lots of time to be changed. As they say, old habits always die hard.

This chapter focuses on the discussion how the society perceives the role and the importance of energy and energy-related issues (including power markets). The chapter contemplates over the debate how society perceives the important energy-related issues and how (or whether at all) consumer behavior changes with the introduction of smart meters, smart houses, and electric vehicles (EVs).

7.2 How does society perceive energy-related issues?

In recent years, environmentally friendly energy sources, also known as "green" or "renewable" energy sources have received great attention and acceptance as governments, organizations, and people around the world take on environmental responsibility. Energy suppliers in such countries have seen the enormous impact that renewable energy can have on the bottom line. Investing in renewable energy can also have a huge impact on government spending.

However, the dependence on fossil fuels continues to undermine the energy market which has led to a large number of people not being able to access cheap electricity. Extraction, transport, and use of energy can have a negative impact on the health, environment, and economy of society. Nowadays, more and more people start to think how energy can affect the economy, security, environment, and health in their respective societies. Energy supplies are traded all over the world, and the impact of energy consumption has a global impact, so anyone can appreciate the breadth of energy issues. Discussions on the scale, historical, socio-environmental, and geographical variations of such data are conducted with a purpose to draw the implications for future energy consumption.

On the other hand, some countries or even regions experiment with "going green"—that is, switching to renewable energy for providing their electricity. As the new-generation solar and wind technologies continue to improve, the cost of fossil fuels becomes integrated into the cost of electricity, so it is possible that renewable technology can produce electricity at kilowatt-hour's competitive cost with fossil fuels. Such experiments occurred in Germany or the United Kingdom, as well as in Central America, and allowed to generate electricity from renewables (mostly solar and wind) for a period from several days to a week. The absolute champion is which Costa Rica managed to generate energy through a combination of renewable sources—heavy rain, solar, wind, and geothermal energy—for the period of the whole 75 days (Barbosa et al., 2017).

Even though there are significant economic and national security concerns about the availability of fossil fuels, as well as important environmental issues related to their use—including, but not limited to, climate change, the times are changing. In addition, there is a changing nature of demand—nowadays most economies are moving away from production to a service and information economy and the imports of energy-intensive products and materials are increasing.

While the implementation of current technologies has the potential to reduce energy consumption and CO_2 and other greenhouse gases (GHG) missions, new technological and scientific advances are likely to bring long-term benefits. These technologies help to reduce the amount of GHG generated by the production of usable energy include renewable energies such as solar, wind, bioenergy, geothermal, hydropower, nuclear and carbon capture, and storage (CCS) used in fossil fuels or biomass.

Moreover, there are attempts to strengthen international cooperation and support the developing countries in expanding their infrastructure and improving technology for modern deliveries and sustainable energy services as a means of mitigating climate change and its effects. Renewable energy sources are perceived as the viable means of mitigating climate change. It becomes apparent that efforts to increase the share of renewable energy and clean technology with fossil fuels in the global energy portfolio will help to reduce climate change and its impact. Climate objectives and the consequences of low-carbon paths of development could have good outcomes for the global energy intensity, carbon intensity of the global energy mix, and the global demand for different energy sources. The Earth's population is currently 7 billion people with projections of reaching 10 billion by 2050; therefore global energy system directly affects the main drivers of energy demand. Providing reliable and affordable energy to support prosperity and improve living standards is linked to the need to do so in a way that reduces environmental impact, including the risk of climate change. Due to the versatility, comfort, and lack of emissions, electricity consumption as part of the total energy consumption has increased with modernization and prosperity. Many experts believe that climate change and other long-term concerns will require the transition to natural gas and then to a hydrogen-based economy which relies on the introduction of noncarbon sources and the sustainable use of biomass. Such policies could further broaden the contribution of nonfossil primary energy sources to the world's power mix in the coming decades.

However, many people still have a negative attitude toward renewable energy. They think it is either too complicated or too expensive to generate. Some of them also think that there is no purpose of developing new technologies for promoting renewable energy, since the return of these investments will take too long and would not probably happen during their respective lifetime. Moreover, many people are not keen on supporting renewable energy projects if they concern them directly or would bring the intrusion into their personal space of giving up some comfort for the sake of environment. This is known as the "not-in-my-backyard" (NIMBY) approach (Liu et al., 2018; Kashintseva et al., 2018; Liebe and Dobers, 2019).

The NIMBY approach can be best shown on an example of the consumer attitudes toward industrial carbon dioxide (CO_2) capture and storage (ICCS) products and technologies. It is apparent that carbon dioxide (CO_2) emissions from the industries and the utilization of the resources constitute a source of environmental pollution and severe global warming based on the evidence from the United States, China, India, Latin America, Russia, and the EU. According to the United States Department of Energy (2012), the manufacturing sector, including cement plants, chemical plants, refineries, paper mills, and other manufacturing facilities, contribute on average, more than 25% of CO_2 emissions, the equivalent of 5.5 million metric tons of CO_2 emissions.

Various initiatives have been rolled out over the years to ensure that the CO_2 emissions are not harmful to the environment or are put to better use. One of these approaches is the industrial carbon capture and storage (ICCS) technologies which facilitate the decarbonization of the manufacturing and other sectors of the economy that contribute to the increase of the global CO_2 emissions. Reiner (2016) points out that CCS technologies might constitute an essential route to meet climate mitigation targets in the power and industrial sectors. The ICCS process is aimed at ensuring that the CO_2 emissions, which have been growing at a rate commensurate with the industrialization and globalization levels, are minimized in order to utilize the available renewable and nonrenewable sources of energy sustainably and protect the environment (Fernández et al., 2016). According to some studies, widespread deployment of carbon capture and storage could account for up to one-fifth of the needed global reduction in CO_2 emissions by 2050 (Bowen, 2011).

It becomes apparent that ICCS technologies constitute a very effective channel for decarbonizing energy-intensive industries, including the steel, cement, refineries, as well as chemical industries which have reached to the maximum theoretical efficiency. Just to give an example: nowadays EU has its industrial CO_2 emissions dominated by iron and steel production (19%), chemicals industry (15%), petroleum refining (14%), as well as cement and lime production (11%) (Global CCS Institute, 2017). Moreover, ICCS deals with a variety of procedures whereby the emissions are captured at the source of production, and they are transported through the most suitable means such as pipeline and finally stored on a permanent basis (Bhatta et al., 2015). However, ICCS technologies also consume large amounts of energy and lead to a loss of efficiency: the reduction of the emitted CO_2 amounts to a maximum of 65%–90% provided that the storage is secure and permanent (Krüger, 2017).

However, the problem is that the general feelings that consumers might experience toward all the issues concerning ICCS and the challenges that are posed toward meeting the satisfaction of the consumers might differ. The threat of global warming that is becoming irreversible and dangerous for the further development of human civilization, calls for the global and viable decarbonizing solutions that ICCS technology presents despite all its shortcomings. The new energy balance for the 21st century is likely to include large CCS plants, either state-owned of private deployed across various locations around many countries. Understanding the potentials of the ICCS and its importance for the energy sector should be delivered to citizens and presented as a favorable outcome backed up by the transparent data and success stories. Nevertheless, it does not always happen this way even though the consumers of the 21st century are more knowledgeable and demanding than those before them. Additionally, they are more educated and have immediate access to diverse information from various sources which is enabled by the widespread use of Internet and social networks (Coyle, 2016). With the increased levels of awareness created about the risks that the increased levels of CO_2 emissions could pose to the well-being of the human race, and the increased levels of sensitization on the need for environmental management, one can observe that the end consumers are increasing their levels of environmental stewardship (Rodrigues et al., 2016). Furthermore, at the individual level, end consumers are making purchasing decisions based on the reputation of companies in relation to environmental management strategies, implying that early adopters in this field, are poised to enjoy increased business from the modern-day end consumer (Lubin and Esty, 2010). Finally, the global consumer is increasingly playing an activist role, where they actively engage each other and corporates in the manufacturing sector to agitate for increased awareness and implementation of strategies aimed at managing the CO_2 emissions from productive activities (Chen et al., 2018). Evidently, from the foregoing, the modern-day end consumer has a positive attitude and engages in activities that foster environmental stewardship and will support initiatives such as the implementation of ICCS and other related projects that mitigate the growing CO_2 emissions.

Nevertheless, there are also a large number of issues related to the consumers' attitudes toward ICCS technologies and their deployment that reflect on the NIMBY perspective described earlier. There are lots of cases where the clear explanation and communication between the stakeholders and industries and the end consumers are required. The general public needs to be made aware of all costs and benefits of the ICCS, as well as about the advantages it presents and the outcomes in terms of halting the CO_2 emissions in the short run by the large-scale application of CCS technologies. Quite often the communication goes wrong, and the pros and cons are not explained correctly. For example, Broecks et al. (2016) show that people find arguments about climate protection less appealing and persuasive than normative arguments or arguments about benefits of CCS for energy production and economic growth. In addition, Kraeusel and Möst (2012) investigate the level and influencing factors of social acceptance of CCS on the example of Germany and find out that the attitude toward CCS is neutral and the level of willingness to pay for CCS technology is much lower than for renewable energy.

Moreover, it might be that the debate on ICCS technologies with the general public and end consumers should be done on the microlevel: some findings demonstrate that small-scale engagement processes might present a viable alternative to standard community consultation techniques for engagement around the siting of CCS facilities (Baker et al., 2009).

According to Wennersten et al. (2015), the major barriers for implementation of ICCS on a large scale are not technical, but economic and social. The key challenge for ICCS is to gain wide public acclaim, which is likely to shape up the future political attitude to it (Baker et al., 2009). Such an approach requires a transparent communication about safety aspects early in the planning phase and dealing with potential disasters and hazards such as major leaks of CO_2.

With the growth in industrialization realized so far in the global society and further projected growth moving forward, it has become important for various stakeholders to come together and implement initiatives that mitigate this trend. Consequently, there has been a lot of sensitization and creation of awareness, coupled with the formulation and implementation of laws, regulations, and policies aimed at supporting the achievement of a reduction in, and management of CO_2 emissions. Because of the increased levels of awareness, the costs incurred in managing the CO_2 emissions notwithstanding, companies, resellers, and end consumers have a positive attitude toward these initiatives.

Thence, one can see how negative or reluctant attitude to ICCS technologies that seem to exist in many countries, including the world's developed economies, might present threats for the whole sustainability concept. Consumer attitudes toward these novel technologies might be influenced by a plethora of factors in which behavioral ones are likely to play the key role. It appears that consumer attitudes toward industrial CO_2 capture and storage products and technologies require further investigation for gaining a deeper understanding of this problem and finding ways for shifting the negative attitudes for achieving the positive energy balance.

The above example explains how in spite of their importance and relevance, many people perceive energy-related issues in the everyday lives. "NIMBY" attitude becomes "not-in-my-lifetime" or "not-me-paying-for-this" approaches that unfortunately can hamper all that formidable process of transition to the low-carbon economy and sustainable growth and development.

7.3 Smart meters and their social impacts

Smart meters have a number of benefits and favorable impacts on sustainable energy development. For example, they are increasingly used in the spheres where products and services of the electricity grids must be accommodated safely and efficiently in renewable generation, energy storage, EVs, and energy management systems. Smart meters may be part of a smart grid, but alone, they do not constitute a smart grid. In other turn, smart meters provided a way of measuring the site-specific information, allowing utility companies to introduce different prices for consumption based on the time of day and the season. The improved consumption feedback provided, and in particular, the installation of in-house displays, has been shown to significantly reduce residential electricity demand in some international trials (Carroll et al., 2014).

Smart meters represent the devices that are widely used nowadays for monitoring, recording, and reporting back the energy consumption patterns of the end consumers. With smart meters gaining wider popularity both in the EU countries and abroad, a question arises whether these devices would actually be capable of changing the consumer behavior and lead to the significant results in shifting the demand and supply of energy.

One of the recent relevant studies describing smart meters trial included 6 digital meters installed in 69 broadly representative Queensland households at the customer switchboard circuit level separately measuring half-hour load for "general power" (e.g., fridge, tumble dryer, washing machine, toaster, kettle, clothes iron, computers, televisions, game consoles, etc.), air-conditioning, electric hot water systems, household lighting, oven, and solar photovoltaic (PV) units (Simshauser, 2016).

The analyses of the representative household's load of a household involved in a Southeast Queensland trial revealed that even though the consumers were using the smart meters and could scrutinize their electricity bills and monitor their consumption patterns and the resulting costs (or even follow their generation and import of solar energy back to the grid), their electricity consumption behavior did not change. In a Southeast Queensland trial study, reported general power load accounted for 52% of household final demand, electric hot water represented 18%, air-conditioning constituted 17%, lighting consumed additional 10%, and oven use was at 3% (Simshauser, 2016).

When it comes to other countries, for example, United Kingdom, the British government announced its intention to mandate smart meters for all UK households by 2020 which is largely presupposed by the balancing requirements that can be lowered by

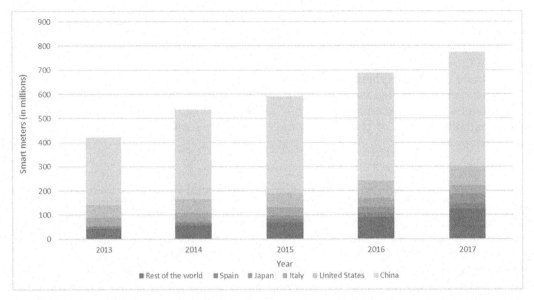

FIGURE 7.1 Global cumulative smart meter installations around the world (in millions). *Source: IEA, 2019. Smart grids: tracking clean energy progress. <https://www.iea.org/tcep/energyintegration/smartgrids> (accessed 20.05.19.).*

allowing the flexibility of demand for electricity due to the increased usage of renewables (Roscoe and Ault, 2010). One very useful source of data that is based on smart meters readings in the United Kingdom is provided by the Customer-Led Network Revolution monitoring trial that offers an overview of household electricity use and tariff behavior (Bulkeley et al., 2015).

The numbers of smart meters around the world are growing at the unprecedented speed. According to IEA (2019), progress in smart meter installment is due to the massive rollout in China, EU, and the United States with South Asia and India still lagging behind slightly. Nevertheless, smart meters might be the blessing for these economies since digital energy networks tend to reduce the need to build new power lines or invest in physical network assets. Fig. 7.1 provides an overview of that growth.

There are various approaches to studying the smart grids and smart metering approach which can be classified along the following subject areas:

- home energy management systems;
- infrastructure;
- communication media—protocols, variables managed by the system and software;
- role of the end user.

Utility companies propose that from a consumer perspective, smart metering offers potential benefits to householders. Undoubted argument in regard to smart meter deployment in the EU is the potential for households to save energy. There are sound results on the empirical effect of feedback on energy consumption. Carroll et al. (2014) showed that feedback and other information provided in the context of smart metering are mainly effective in reducing and shifting demand because they act as a reminder and motivator. Buchanan et al. (2014) studied the empirical data about the usage and response to energy monitors among consumers. Their findings suggested that positive feedback can be achieved due to the physical and conscious visibility of consumption as well as to the complete knowledge about consumption. It appears that the visibility of energy consumption for private households might fall into the sustainable development concept by substantial reduced consumption in the long run. However, there is still a concern how feedback might work and the processes it involves (Joachain and Klopfert, 2014). Nachreiner et al. (2015) argue that feedback by itself is hardly sufficient. Therefore in order to achieve the most effective design of smart meters, the opportunities which these offer to deliver feedback and to supplement this with other information should also be considered.

Hence, despite the ground strand of researches in this field that there are the effects on household energy consumption of the feedback provided by smart meters, there is still a deep concern in the motivation for households to use the feedback to save energy. Awareness campaigns, education and training programs, label schemes and smart metering, and other similar initiatives have great potential to encourage consumers to use less energy and realize efficiency savings of up to 30% (Wilson, 2014 or Schultz et al., 2015). In addition, Joachain and Klopfert (2014) explore how the emerging trend of using complementary currencies for sustainability policies could translate into new interventions adapted to the smart meter deployment and capable of promoting more autonomous forms of motivation compared to interventions using official currencies.

Table 7.1 draws an overview of several information-based electricity demand studies involving smart meters and their impact on human behavior.

Many scientists describe their concerns regarding the cost, health, fire risk, security, and privacy effects of smart meters and the remote controllable "kill switch" that is included with most of them. Many of these concerns are focused on the wireless smart meters (e.g., those that report their measurements over the Internet or Bluetooth) with no home energy monitoring, control, or any safety features. Hence, there is a need in more in-depth look at the specific energy consumption determinants in each country in order to find more precisely what is driving consumption levels with or without smart meters (Galvin, 2014; Inderberg, 2015).

The variability and limited predictability of these sources have brought many technical challenges to grids. Many regulatory communities and system operators in the power sector have established grid codes to ensure proper connection of these renewable sources (Eltigani and Masri, 2015).

However, there are already attempts to build systemic treatments of energy system transitions to characterize the coevolution of value capture and structural incentives in the electricity distribution system, that is, drawing on semistructured interviews and focus groups undertaken with smart grid stakeholders in the United Kingdom (Hall and Foxon, 2014).

Leiva et al. (2016) claim the development of smart grids and smart meters requires regulations that take into account the technological capabilities and the needs of users both in the present and in the immediate future. Their review of the energy policies aimed at the implementation of smart metering infrastructures in Spain, Europe, and around the world concluded in the necessity of developing the concept of informational governance. Naus et al. (2014) conducted interviews with Dutch stakeholders and held observations at

TABLE 7.1 Information-based electricity demand studies on smart meters.

Research	Type of information	Results	Remarks
Darby (2006)	Direct feedback: – Self-meter reading – Direct displays – Interactive feedback via PC – Prepayment meter – Energy advice with meter reading – Cost plugs on appliances	5%–15% savings	Range of international studies with different types of direct feedback
Darby (2006)	Indirect feedback: – More frequent bills – Electricity bills based on readings plus other detailed information	0%–10% savings	Range of international studies with different types of indirect feedback
Wood and Newborough (2003)	Electronic feedback via consumption indicator on electric cookers	15% reduction	44 UK households; focus on electricity for cooking
	Paper-based information pack on electricity consumption of cooking appliances and electricity savings tips	3%	
Dulleck and Kaufmann (2004)	Leaflets describing energy efficiency; introduction of energy efficiency appliance certifications	7% reduction in electricity demand	Impact on long-run rather than short-run demand
Darby (2010)	Durable effects from the adoption and use of energy displays	5%–15%	Smaller sample and lower proportion of people who are not interested or lose interest during the trial
Commission for Energy Regulation (2011)	Electronic feedback: – Smart meter reading – Regular detailed electricity bills Direct feedback: – Electricity-saving educational campaign via administered surveys	Self-declared reduction in electricity demand, low response to renewable	4232 households in pretrial and 3423 households in posttrial
Faruqui et al. (2010)	Direct feedback: – Data on real-time and historic usage – Prepayment meter	7% (14%)	Customers buying on credit (combined with prepayment)
Simshauser (2016)	Indirect feedbackStudying the effect of tariffs on solar PV	Poorest group pays about 30 AUD more than the richest group	Both households with own generation (solar PV panels) and without were involved
Rausser et al. (2018)	Indirect feedback Studying the effects of smart meter trial in Ireland	Owning accommodation negatively effects energy-saving behavior	Households keen on changing their electricity consumption are those interested in the positive impact on environment

Own results.

workshops to examine three information flows (between household-members, between households and energy service providers, and between local and distant households). They reached the conclusion that this approach is convincible to create the changes in domestic energy practices and the social relations. They also concluded that these changes are contextual and emergent.

The feasibility of smart meters implementation heavily depends on residential electricity customers' willingness to pay for it and, furthermore, an understanding of factors that have an impact on this willingness.

Metering-only solutions are popular with utilities because they fit existing business models and have cheap up-front capital costs, often result in such "backlash." The political economy of the UK electricity system has coevolved such that there is a mismatch between where benefits accrue and where costs are incurred, leading to a problem of value capture and redeployment.

Gerpott and Paukert (2013) tested the hypotheses in a sample of 453 German-speaking residential electricity customers revealed that trust in the protection of personal smart meters data and the intention to change one's electricity consumption behaviors after new technology deployment are the constructs most strongly related to residential willingness to invest in smart meter. It appears that in many cases expectations regarding electricity volume saving and environmental awareness contributed less toward explaining the willingness to join smart grip and to install smart meters in residential households.

With regard to the debate outlined above, I can draw up an interesting experimental study on the social implications of smart meters carried out by Rausser et al. (2018) which I presided and which I prepared and executed. The data from the Irish Commission for Energy Regulation (CER) encompassing the findings from the Smart Metering Electricity Customer Behavior Trials (CBTs) that involved from 4232 to 3423 households and businesses (small and medium-sized enterprises) and was held in between 2009 and 2010 were employed for the study. The unique approach of the trial was that it included two survey questionnaires (pretrail and posttrial) administered over the telephone and also contained some interesting supplementing quantitative data (e.g., Likert scales, dummies, selections, etc.) as well as open-ended questions. A total of 4232 respondents (individual households) were involved in pretrial and 3423 households were involved in posttrial. The second questionnaire was administered about 12 months after the first one at the same households that previously participated in the trial.

The purpose of the questionnaire survey was to assess the impact on consumer's electricity consumption in order to inform the cost—benefit analysis for a national rollout. Electric Ireland customers and BordGáis Energy (a utility that supplies gas and energy to consumers in the Republic of Ireland) customers who participated in the trials had an electricity smart meter installed in their homes (or on their business premises) and voluntarily agreed to take part in the data collection in order to contribute to establishing how smart metering might influence energy usage taking into account demographic and lifestyle factors, as we home sizes and consumer behavior. The numbers of customer installations for the field trials were as follows:

- metering system with GPRS communications—5800 single-phase and 500 three-phase meters throughout Ireland;

- metering systems with power line communication (PLC)—1100 single-phase meters for customers in Limerick and Ennis (eight urban and three village locations);
- metering systems with 2.4 GHz wireless mesh—1591 m installed in Cork and 690 m installed in the rural area of County Cork.

In addition to the questionnaire surveys, the trials collected information on the electricity (in kWh) consumed during 30-minute intervals by households. The basic data analysis reveals that out of 4232 respondents in pretrial, only 24 reported that they used renewable energy (e.g., solar PV) to heat their homes and 62 reported that they used renewable energy (e.g., solar PV) to heat water in their homes.

Moreover, out of 3423 respondents in posttrial only 13 reported that they used renewable energy (e.g., solar PV) to heat their homes and 51 reported that they used renewable energy (e.g., solar PV) to heat water in their homes.

The first step was to apply the one-way analysis of variance (ANOVA) to the Irish data. ANOVA is typically used in a situation such as we have—one categorical independent variable and one continuous variable. Our independent variable consists of a number of groups (levels) represented by the possible answers on the Likert scale (from 1—strongly agree to 5—strongly disagree and (Table 7.2) also an additional possibility 6—DK (declined to answers), hence from 5 to 6 possibilities to choose from depending on the type of question).

The basic analysis of frequencies and the one-way ANOVA [one control group (households interested in using smart meters and understanding their importance) and one test group (households failing to see the importance of smart meters and unable to adjust or change their electricity consumption based on their usage)] yields the following results with the mean = 1.39, standard deviation = 0.899, $N = 4232$ (Table 7.2).

Table 7.3 reports the results of the one-way ANOVA for the question mean = 1.23, standard deviation = 0.624, $N = 4232$ for the question "I/we am/are interested in changing the way I/we use electricity if it reduces the bill."

TABLE 7.2   One-way ANOVA results for the question "Our society needs to reduce the amount of energy we use."

		Likert scale	Frequency	Percent	Cumulative percent
Valid	1	Strongly agree	2687	78.5	78.5
	2		430	12.6	91.1
	3		175	5.1	96.2
	4		39	1.1	97.3
	5	Strongly disagree	58	1.7	99.0
	6	DK	34	1.0	100.0
	Total		3423	100.0	

Own results.

TABLE 7.3 One-way ANOVA results for the question "I/we am/are interested in changing the way I/we use electricity if it reduces the bill."

		Likert scale	Frequency	Percent	Cumulative percent
Valid	1	Strongly agree	3572	84.4	84.4
	2		450	10.6	95.0
	3		133	3.1	98.2
	4		41	1.0	99.1
	5	Strongly disagree	36	0.9	100.0
	Total		4232	100.0	

Own results.

TABLE 7.4 One-way ANOVA results for the question "I/we am/are interested in changing the way I/we use electricity if it helps the environment."

		Likert scale	Frequency	Percent	Cumulative percent
Valid	1	Strongly agree	3238	76.5	76.5
	2		697	16.5	93.0
	3		199	4.7	97.7
	4		59	1.4	99.1
	5	Strongly disagree	39	0.9	100.0
	Total		4232	100.0	

Own results.

Table 7.4 reports the results of the one-way ANOVA for the question mean = 1.34, standard deviation = 0.716, $N = 4232$ "I/we am/are interested in changing the way I/we use electricity if it helps the environment."

Summing up the results demonstrated in Tables 7.2–7.4, one can clearly see that the consumers in the Irish trial were, on average, more interested in maximizing their own profit and are only concerned about the changing the way of using electricity if it means direct monetary benefits and personal gains.

In addition to the above statistical analysis of the data and for the purpose of illustrating the consumers' attitude toward smart grids and the new technologies associated with them (e.g., autonomic power systems), it was decided to run an empirical model with limiting the scope of the analyzed data to the household data from the posttrial survey. The empirical model was constructed and analyzed for two main reasons: (1) the research project concerned individual consumers who are able to influence their electricity bills (if only they wanted to) by using smart meters and new technologies that allow them to monitor their energy uptake over various times of the day on different days; (2) the posttrial survey was analyzed due to the fact that the respondents were able to get themselves familiar with smart meters and technologies and made their opinions about them. They also learned how to reduce their energy uptake and save the money (if the benefits of doing so are higher than their opportunity costs).

In general, it was envisaged that households took into account the provision of information on their usage of electricity and the prices they have to pay by using less electricity, even when classified for the type of home, heat, household characteristics, etc. Nevertheless, it was not clear whether they are keen to pay the price for the advancing technologies if the economic outcomes are less clear for them.

All in all, one can pose the following research question to tackle this subject: Does the willingness to change the way people use electricity (with the help of smarter grids and new technologies) depend on their economic and social status? And if so, does this status influence the self-selection of people in favor of smarter grids and technologies?

A simple binary response model was employed for testing the research question articulated above. The binary response model can be presented in the following general form (βs represent coefficients to be estimated):

$$P\left(y = \frac{1}{x}\right) = G(\beta_0 + \beta_i x_i) \tag{7.1}$$

where y is the binary variable that has two outcomes, x is the vector of regressors, P is the probability function, G is a link function (with cdf being a separate case for probit as presented below), and β_0 and $\beta_i x_i$ are the model parameters estimated by the maximum likelihood.

For the purposes of this chapter, the probit model was selected. Probit model using standard cumulative distribution function which is considered to be more realistic in the majority of situations and that can be specified as follows:

$$G(x) = \Phi(x) = \int_{-\infty}^{x} \phi(x)dv \tag{7.2}$$

where $G(x)$ is the cumulative distribution function representing a special case of probit from the link function specified above and $\phi(x)$ is the standard normal density ($\phi(x) = (2\pi)^{-1/2} exp(-x^2/2)$) with π being is the mathematical constant, and exp being the base of the natural logarithm (also a constant).

Thence, the formula specified in Eq. (7.2) means that our probit model can be derived from the latent variable formulation where the term "e" has a standard normal distribution.

Overall, the empirical model based on binary response model employed the following coded variables:

Changing the use of electricity to reduce bills—a binary variable taking up the value of 1 for a "yes" or 0 for a "no" (the original statement in the questionnaire was "I am interested in changing the way I use electricity if it reduces the electricity bill").

Changing the use of electricity to help environment—a binary variable taking up the value of 1 for a "yes" or 0 for a "no" (the original statement in the questionnaire was "I am interested in changing the way I use electricity if it helps the environment").

TABLE 7.5 Results of probit model estimation for Ireland smart meter trial.

Probability of changing the use of electricity to reduce costs	
Helping environment	1.998*** (0.107)
Being employed	0.410*** (0.083)
Own accommodation	−0.046 (0.171)
Constant	0.260 (0.191)
N	3423
Prob> chi^2	0.000
Pseudo R^2	0.268

Note: *P*-values in parentheses *** $P < .001$.
Own results.

Being employed (in employment)—employment status of the chief income earner in the household [taking 1 for employed people (all types of paid employment) and 0 otherwise].

Own accommodation—a binary variable indicating whether the respondent in question owned her/his accommodation in which she/he resided when the survey was conducted.

The results of the empirical model estimation are presented in Table 7.5.

In general, one can see that the results are quite straightforward and came through as expected. It appears that respondents who were keen on changing their electricity consumption with the help of smart meters (and perhaps the new, even smarter grids of the future) are also those who would change their consumption patterns in order to have the positive impact on environment. The fact that the respondent was employed and did not depend on anyone also had a positive impact on changing the electricity consumption (with or without the help of novel technologies). However, the magnitude of impact was much smaller than in the previous case.

As for the fact of owning an accommodation (the most expensive item in the budget of the majority of people in most of the EU countries), it came through as negative and insignificant. This result indicates that saving on energy might be more relevant for people who do not own their own accommodation and therefore try to save on all possible items.

When discussing and contemplating over the conclusions of the reported study with the electricity consumers in Ireland, one can make in the context of the carried-out surveys, the main point is that under the current circumstances, smart electricity systems might be irrelevant for the majority of average consumers. This is happening because the costs of installing and running the smart meters to monitor the energy uptakes and to enhance saving on electricity bills might seem higher than to those consumers than the opportunity costs of not caring about the prices of electricity at all. There are many ways how to save on electricity but all of them are quite cumbersome and involve personal commitments and adjustments in habits and behavioral patterns. In order to be attractive for an average consumer, autonomic systems and smart grids of the future need to be more appealing in economic sense. Not only should they help people to save electricity and to appeal to their environment-friendly nature but they should be able to motivate and to promote ways how to make money on becoming an integral part of smart energy systems by playing the role of proactive prosumers.

Furthermore, the findings of the Irish electricity consumers' trial study are likely to support the idea that household consumers are likely to remain rational economic agents in the way they consume the power and pay for energy by maximizing, optimizing, and behaving economically. The acceptance of the smart metering by the general public and the policy-makers is likely to happen if and only if it will mean some economic advantages. In case their further development and implementation would mean additional costs added to the consumers' electricity bills with no real profit or visible advantage for the end users in sight, the majority of people will be against such innovations and will rather support traditional was of delivering and controlling power and energy. Thence, it might seem to be necessary to either reduce the cost of the smart meters, provide smart meters for free or offer subsidies on their installation, or introduce the requirement to install smart meters in households by law.

There are also some negative words that can be said about the smart meters. In many cases they have also proved to be a dangerous fire hazard items: due to a common hardware failure, volatile devices can easily explode and ignite houses. On the other hand, microwaves typically produce between 600 and 1200 W of power for heating food, or 6000 to 12,000 times the power of a router or a smart meter. Gas-powered smart meters work at lower frequencies (450−470 MHz) and emit at a lower wattage (0, 820 W). In other words, in a typical household, this type of smart meters can receive the data from larger distances.

Typically, public service commission requires independent certification to demonstrate the safety of smart meters and withstand heat, fire, voltage, surge, and self-heating.

Smart meter measures and transmits energy consumption directly to the electricity grid, eliminating the practice of estimated bills, which means no more surprises for electricity bills.

Intelligent meters work as part of the smart grid and improve the detection and notification of power failure.

The smart meters, therefore, do not emit a wireless signal, but introduce some form of electric contaminant to one's home wiring known as "dirty electricity." It is difficult to understand why utilities are so inclined to install wireless smart meters when smart meters can do the same job but much safer by connecting to a normal telephone line. Theoretically, if a smart meter is read by a metering man, it is not a wireless smart meter, but the only way to get a definitive answer is to test it with a radio frequency (RF) meter. Smart meter technology works well below the Federal Communications Commission's (FCC)s maximum Electromagnetic field (EMF) level and is used safely by many utilities. Smart meters are one of the latest targets for people who love to be afraid of electromagnetic radiation and "dirty electricity."

Smart meters are the next step in technology to measure electricity consumption in residential and commercial buildings. Although exposure to electromagnetic field (EMF) by using a mobile phone is more extensive than living with a smart meter, the evidence does not yet show any health risks associated with the use of mobile phones.

When it comes to smart meters, people fear that they will be used to violate our privacy and collect all kinds of information about us. In addition, there have been concerns about the safety of smart meters, mainly because they emit the same RF waves as mobile phones and Wi-Fi devices. The amount of radio energy that people are exposed to from the smart meter depends on how far they are from the smart meter antenna and how the smart meter emits their signal. In addition, the walls between the person and the smart meter antenna further reduce the exposure to the RF. Since radiotherapy is a potential carcinogen, and smart meters emit RF radiation (RFR), intelligent meters can potentially increase the risk of cancer.

Since smart meter systems are installed in cities, there has been no evidence of such devices and no privacy or cyber security issues.

Many residents attest to Detroit Edison (DTE) Energy Corporation's intimidation, with many people witnessing a lack of customer service, a lack of communication, an unjustified elimination of analogue meters with a smart meter, and a sudden shutdown of people. Milham (2012), a doctor, an epidemiologist, and the author of the book *Dirty Electricity*, explains the smart metering equipment, dirty electricity, and harmful health effects of smart meters on human health. Some estimates suggest that the construction of the grid could lead to a 4% reduction in energy consumption by 2030, but real evidence shows that smart meters do not have any energy savings.

In fact, real evidence shows that the smart grid has artificially increased energy consumption, which has led to thousands of homeowners complaining about ridiculous charges. Utilities say that smart meters are only "switched on" for 45−60 seconds a day, when sending and receiving information—barely enough to worry.

David Carpenter, director of the Institute for Health and the Environment, professor of environmental sciences in Albany's School of Public Health in New York city, and an expert in the field of the effects of environmental and energy pollution on human health, states that the exposure to RFR at elevated levels for long periods of time increases the risk of cancer, damages the nervous system, and adversely affects the reproductive organs (Pope, 2019). However, the same is said about the smartphones we all have with us at any moment of every day.

Smart meters that we have in operation today use the same technology as smartphones. And there have been lots of studies proving the health risks of smartphones followed by a comparable amount of studies proving that there is no risk in using smartphones. With the advancement of technology, it is planned that each appliance and device should have an intelligent network device built into it, which will behave like a mobile phone. Utility companies all around the world are replacing decades-old, but still perfectly functioning analogue meters with the smart meters in people's homes with the aim of wireless electricity consumption by sending radio frequency signals to nearby homes, which are then sent to local nodes and then to local utilities. The location of intelligent meters seems to be a great convenience, but the meters use 2.4 GHz microwaves (and 900 MHz) for communication, which is what our microwave ovens, Wi-Fi routers, and other wireless devices typically use.

One way or another, smart meters constitute an inevitable part in the envisaged Internet of energy; therefore they will soon become ubiquitous and common. It might be that they will stay in a form or sensors and small sensor devices located everyone around one's home or at the premises of a large industrial enterprise, or it also might be that they will migrate into our smartphones (or the devices we will be using in the place of today's smartphones) in the future. With regard to this, I can recall an interesting paper by Angrisani et al. (2018) that described how smartphones and tablets can be used to control the power consumption of some typical household appliances by means of technologies that are already used and spread in a simple and effective way. Smartphones are equipped in all kinds of sensors and cameras that allow accurate measurements. In addition, virtual reality and the cloud computing are applied to assess every electrical appliance and take measurements of its energy intensity and usage. It turns out that using one's smartphone as a smart meter makes a person to become aware of the actual impact, both economic and environmental, of energy spending and monitoring.

7.4 Smart houses

In EU Member States, North America, and Russia, buildings account for about 40%−50% of the total energy consumption. In global energy costs, the share of real estate is also impressive—almost a third, 31% of all energy used on Earth is spent on heating and power supply to buildings. The volume of accumulated scientific, engineering and construction knowledge, and technologies allows to significantly reduce this enormous share and today, in mass order, to "serially" build "energy-neutral" ("climate-neutral") buildings that do not need the energy produced from traditional (hydrocarbon) sources.

In 2010 EU Directive 2010/31/EU on Energy Efficiency of Buildings (EPBD) was adopted. According to this document, starting from 2021 (for administrative buildings—from 2019), all new houses built in the EU should be nearly zero-energy buildings. "Almost zero or very insignificant energy demand of such a building should be covered mainly by renewable energy sources, including such sources located at the location of the building or its surroundings." In 2012 the European Energy Efficiency Directive No. 2012/27/EU (Energy Efficiency Directive) was adopted, which also included a number of measures aimed at saving energy in the real estate segment. In particular, the governments of the EU countries should annually provide reconstruction (energy rehabilitation) of 3% (by area) of the existing stock of state-owned buildings.

Only high energy−efficient buildings, products, and services are allowed to acquire into state ownership. EU countries should develop and adopt long-term strategies for the reconstruction of the existing building stock.

How will the problem of increasing the energy efficiency of buildings be addressed? With the help of well-known practitioners of a complex of measures, the main ones are the mandatory mechanical ventilation with effective heat recovery (heat exchange), reducing the thermal conductivity of walling (walls, roofs, and foundations) using additional insulation, improving the quality of design and construction to achieve standard indicators air permeability of the building and elimination of cold bridges (places of contact of uninsulated building structures with the surrounding food through which high heat loss occurs). It should be noted that this task has been largely solved already now, because, as we noted above, the knowledge and technologies for this are already available. Most of the buildings being built today in Germany, Austria, and Switzerland comply with the requirements of the directive. In a number of regions, leading commitments have been made regarding the construction of energy-efficient buildings. In Brussels, Belgium from

2015 onward, all new buildings have to meet the criteria for passive housing construction, in Luxembourg the deadline was set from 2017, and in Bavaria from 2010 all new administrative buildings have to be built in accordance with these standards. The concept of a passive house developed in the early 1990s by the German professor and engineer Wolfgang Feist who created perhaps the first main provisions for the theoretical foundation of the modern energy-efficient construction (Feist et al., 2005). "Passivity" here means "thermal neutrality," or "thermal inertia" of a building, "indifference" of the internal microclimate to temperature "overboard." Passive house is designed and built in such a way that the need for thermal energy for its heating is negligible. The continuous massive thermal contour of the building and the ventilation system with heat recovery (heat exchange) provide extremely low heat loss combined with a comfortable microclimate (uniform heating of internal surfaces, controlled air exchange and its filtration, etc.).

The relevant standard also regulates the estimated consumption of electrical energy. Regulatory heat consumption, calculated according to a very tough methodology of the Passive House Institute, should not exceed 15 kWh per 1 m^2 of space per year, and the calculated heating capacity should be 10 W/m^2. With these parameters, there is virtually no need for the usual for us heating systems consisting of a heat generator (heating boiler) and radiators installed under the windows. The construction of such buildings does not involve the use of any unique, "innovative," etc. materials and is not a miracle. It is all about competent, qualified design, proper heat engineering calculation (energy modeling), and accurate, high-quality work of builders. A modern house is never built as a "box." Both the heat envelope and engineering systems are designed in a complex. Only in this way high consumer qualities and energy-saving characteristics are achieved.

High energy efficiency of modern buildings is rationally complemented by the use of renewable energy sources. It is primarily about solar electric generation systems (photoelectric) and the use of solar heat to heat water in heating and water supply systems (solar collectors), wood-fired boilers (pellets), as well as heat pumps that use environment energy—low potential heat ground and air. In economically developed countries, as energy-efficient construction develops, traditional methods of heating with the use of hydrocarbons lose their importance.

It should be emphasized that the use of renewable energy sources for electricity and heat supply of buildings is expedient and economically justified precisely in the case of high energy efficiency of the latter. The use of renewable energy sources in "ordinary" buildings, that are not distinguished by high thermal engineering parameters, that is, consume a lot of energy for heating, is inexpedient. German Institute for Buildings predicts that by 2020 the construction of buildings that provide themselves with energy on their own will become a common construction practice (Kohler and Hassler, 2002). One can see that in many EU countries they are not just some futuristic plan adapted from a sci-fi novel. There are roofs, facades, and glass that generate electricity and heat and constitute the usual design used in the European construction market. Almost all buildings in countries like Germany will be "climate-neutral" by 2050, consuming significantly less energy than they are now, while meeting these energy needs with the help of renewable energy sources. This goal constitutes one of the key concepts of the German "Energiewende" energy strategy marking a "big turn" to using renewables and mitigating the climate change.

Zero-energy buildings, plus energy buildings, zero carbon homes—all of these concepts have already become part of the linguistic circulation of Europe and North America. Houses equipped with the smart systems of electric generation and solar heat became an integral part of the landscape of many countries. For example, in Germany and Australia (which are, alongside with California and Scandinavian countries, are the world champions in renewable energy), the number of households equipped with PV modules exceeded 1 million in each country.

Well, one might ask, what are the basic principles of an individual house with a positive energy balance (plus energy building), or, as we sometimes call it, "energy-active building"? As it title suggests, this type of a building generates more energy on average per year than its inhabitants consume. It should be emphasized that we are not talking about self-sufficiency in electricity all year round, energy autonomy, which in the climatic conditions of Central Europe is hardly realizable due to only renewable energy sources with imputed costs or without sacrificing consumer comfort, but about the average annual value (balance) of output/consumption energy.

In the simplest case, it is about replacing the household electricity consumed by electricity generated by a solar generator. According to some available estimates, the average family of four living in an individual house in European conditions consumes 4000—5000 kWh of electricity per year (excluding energy consumption for heating). To generate such a quantity of energy per year in some Central and Eastern European country, a solar power station with an installed capacity of 5—6 kW, located in the southern roof slope and occupying approximately 45—60 m^2. In the summer, such a building produces an excess amount of electricity, which is sold to a local grid company. In winter, on the contrary, the inhabitants are forced to acquire electricity, since in the geographic zones under consideration solar radiation is not enough. At the same time, this decision is half-hearted, since the share of household electricity accounts for only a small part of the energy costs of the average household. This is true both for Central and Northern Europe, as well as for such "cold" country as Russia. About 85% of the energy consumed by the average household per year comes from heating and hot water.

Thus a positive balance on all energy consumed can be achieved by adapting several measures. First of all, this can be done by significantly increasing the energy efficiency of the building to reduce heat demand, as is done in passive houses. Second, this can be achieved using additional engineering equipment that employs renewable energy sources. The most common solution looks like this: a geothermal heat pump that takes on the basic functions of providing the building with heat and hot water, plus solar collectors to support hot water and heating. With this combination, a properly calculated solar power plant, including the appropriate battery capacity, can indeed provide an annual positive energy balance. And, as already mentioned above, in a number of Western European countries the construction of such houses is becoming a common practice already now, many construction companies have included similar houses in their standard product lines. The number of houses built with a positive energy balance is already in the thousands.

In the nonresidential real estate sector, there is also a keen interest in improving energy efficiency and using renewable energy sources. There are outstanding examples of office, retail, and industrial buildings of an energetically (climatically) neutral level, such as the headquarters of the Danish energy company Syd Energy, certified by the Passive House

Institute. The estimated specific energy consumption for heating here is only $8\,kWh/m^2$ per year, geothermal heat pumps are used for heating and hot water (total length of probes is around 10 km), as well as computer servers heat.

Investments in the energy efficiency of buildings from the point of view of the national economy are a win-win strategy. Such investments create new high-tech production and jobs, and at the same time they are financially justified by saving natural resources, reducing imports, and contribute to preserving the environment. Energy efficiency is even called the "key resource of economic and social development" in all countries. Of course, the multiplicative effect for national economies from measures aimed at improving the energy efficiency of buildings is most achieved if they rely on their own scientific developments, technologies, and production (wall and heat-insulating materials, engineering equipment, window structures, etc.). Global energy efficiency investments already exceed $300 billion annually. According to the calculations of the International Energy Agency, economically viable investments in energy efficiency will contribute to a more productive resource allocation within the global economy with the potential to increase total production by $18 trillion by 2035—more than the current size of the economies of North America combined (namely Canada and Mexico) (IEA, 2014). The deliberate rejection of the use of hydrocarbons for heating buildings is becoming a visible European trend. For example, in Denmark since 2013 there is a ban on the installation of gas and liquid fuel (diesel) heating and water boilers in new buildings (in accordance with the Danish Energy Agreement dating back to 2012). This Scandinavian country that has an extensive and efficient network of central heating, the owners of new buildings are left with a simple and straightforward choice: either to connect to the central network or install their own heat pumps. All in all, European countries have taken a steady course of reducing energy consumption and the use of renewable energy in real estate, which is supported by the necessary knowledge, technology, and enthusiasm of the masses. This trend has already been picked up by North America, Japan, and China, which plans to bring the number of square meters of green buildings to 1 billion by the end of 2025. In the long run, this development will obviously have a significant impact on the commodity and energy markets and will lead to a reduction in demand for hydrocarbon fuels in the real estate segment.

The concept of "smart" energy-efficient buildings is closely related to the concept of "smart homes." Smart houses (often called "smart homes") are also perceived as one of the coolest inventions of the 21st century. Who would not dream about living in a house that would take care of the inside temperature, heat the water, open and close the windows to let the fresh air in or even take care of shopping by ordering the food online?

Modern technologies make all that a reality and the plethora of sensors, Bluetooth-connected light bulbs, thermostats steerable by smartphone apps, and watering devices controlled wirelessly, are already finding their way into many households worldwide.

Kagan, a global market intelligence research group and research company, reports that at the end of 2016, the number of smart homes in the United States rose to more than 15 million (S&P Global, 2017).

Fig. 7.2 depicts a cluster of smart houses that have their own home batteries, thermostats, and can communicate with each other and with their owners via Internet and smartphone apps.

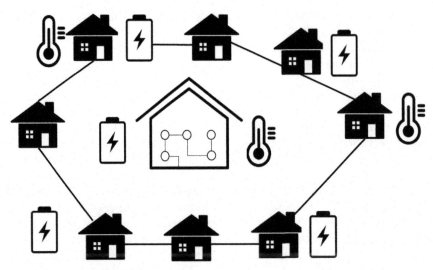

FIGURE 7.2   A cluster of smart houses. *Own results.*

Multiple smart home devices in one home form the basis of an intelligent home ecosystem. High prices, combined with limited consumer demand and long cycles of replacement of appliances, are the main obstacles to the development of the smart home market from the early adoption to the mass market.

Today, there are many networks, standards, and devices that are used to connect an intelligent home, creating interoperability problems and confusing the consumer with the installation and control of several devices.

Some perceive smart homes as a possible breach to their personal security and the possible source of data leak. While many people are actively sharing their personal experiences through social media, they are wary of intelligent devices that have access to personal information about their home and habits. Smart house suppliers should adopt a double stance in order to eliminate this barrier: to ensure that consumers' data is protected, and the intelligent home ecosystem is protected by the right measures for consumers, and to ensure that they are protected.

Smart home architects and builders should take into account the mood of consumers when deciding on the complexity of products, packaging, and other services aimed at consumers, creating a business model that focuses on the customer experience. In particular, 30% of consumers are concerned about privacy issues when considering an investment in smart homes, although 58% believe that smart technology can make their homes safer (Stojkoska and Trivodaliev, 2017).

Consumers' attitudes toward the adoption of smart home appliances are crucial for insurers and their ecosystem partners in developing their Internet of things (IoT) strategies. In fact, the IoT is a concept that is closely linked with the concept of smart homes.

With the increasing intention of buying smart homes, insurers have a significant opportunity to attract and retain customers by offering new and innovative ways to reduce risk and protect their homes. With regard to the above, smart product manufacturers can open a new sales channel and stimulate general adoption through partnerships with insurance companies, which can offer comarketing programs and discounts on products for consumers.

It appears that in order to encourage the adoption of such new technologies, companies need to focus their efforts on "smart home helpers," a central device or system that would allow the customers to control other smart devices in their homes.

Most consumers would rely primarily on technology giants and telecommunication companies to install the new technologies into their homes, while understanding the benefits of connected houses, it remains price-conscious and as such represents the second lowest property among our people, as well as the average willingness to buy or upgrade the smart houses in the future.

Home automation has been around for decades in terms of lighting and simple device control, but recent advances in mobile technology, voice recognition, and data analysis have made it possible for our personal living space to truly enter the connected world, allowing for complete control and self-sufficiency opportunities.

While many smart home security vendors are constantly recording and storing all data in an unsecured cloud at a monthly subscription, others use internal and external HD cameras equipped with monthly or no monthly fee, consumer data protection, and privacy mode when the users are indoors. One way or another, constant recording and monitoring create in many people a feeling of being under constant observation which some individuals describe as paranoid and do not take well.

Although concerns about home security are certainly not new, many of today's security products and services have not been possible without broadband or mobile electronic devices.

Products and services such as IP security cameras, smart locks, and automatic notifications sent directly to the users' smartphones are becoming an integral part of the intelligent home.

Consumer surveys show skepticism among some homeowners about the real value of connected appliances in the home. In addition, customers with higher risk perceptions may be more demanding in terms of the usefulness of smart home applications and may be more critical in system evaluation. The perceived overall risk adversely affects (1) behavioral intentions, (2) power utility (PU), and (3) the source of intelligent home applications. Although the overall risk has a negative impact on the usefulness of an intelligent home, it improves the customer's usability.

In addition, it is important to detect potential problems and offer solutions before consumers even consider giving up their connected products. Since today's IoT consumers are early adopters in the space, they can also promote positive word-of-mouth if brands exceed their expectations.

Table 7.6 provides an overview of the information-based studies describing consumer attitudes toward smart houses.

Overall, it becomes apparent that smart houses are perceived as both a novel and desired trend and as a bit scary piece of technology that constantly monitors its inhabitants, albeit with the noble intentions to ensure their comfort and protection. Since many smart homes are maintained using some computer algorithms and steered using voice

TABLE 7.6   Information-based studies on consumer attitudes toward smart houses.

Research	Type of information	Results	Remarks
Demiris et al. (2004)	Focus group for assessing older adults' perceptions of the technology and ways they believe technology can improve their daily lives. Participants' perceptions of the usefulness of devices and sensors in health-related issues such as preventing or detecting falls, assisting with visual or hearing impairments, improving mobility, reducing isolation, managing medications, and monitoring of physiological parameters elicited	Advanced technologies might help older adult residents in emergency help, prevention and detection of falls, monitoring of physiological parameters, etc.	Concerns expressed about the user-friendliness of the devices, lack of human response and the need for training tailored to older learners. All participants had an overall positive attitude toward devices and sensors that can be installed in their homes in order to enhance their lives
Johnson et al. (2007)	Focus groups of older adults with various impairments. Content analysis was used to identify participants' perceptions of smart home technology	Most participants responded favorably toward the smart door and voice activation than any other smart technology/ application	Results can be used to modify current smart home applications and guide future smart technology/application design
Paetz et al. (2012)	Focus groups for eliciting consumer perceptions of smart home energy management system which optimizes electricity consumption based on different ICT solutions (e.g., variable tariffs, smart metering, smart appliances, and home automation)	Positive group reactions to the smart home environment, many advantages for and the chance to save money	Consumers had issues with giving up high levels of flexibility and adapting everyday routines to fit in with electricity tariffs
Balta-Ozkan et al. (2013)	In-depth deliberative public workshops, expert interviews and the literature review for exploring the social barriers to smart home diffusion (variation by expertise, life stage, and location)	Consumers main concerns are the following: loss of control and apathy, reliability, viewing smart home technology as divisive, exclusive or irrelevant, privacy and data security, cost, and trust	Both experts and the public appeared to agree on some of the more practical social barriers (e.g., reliability, security), deeper, moral concerns about human nature, inequality, and trust are a stronger feature of public discussions
Balta-Ozkan et al. (2014)	Studying differences in perceptions of smart homes in a cross-country comparative context to inform the delivery of smart home services and taking the public perceptions on the role of utilities and government in particular	Policies should leave all paths open due to due to differences in householder preferences and acceptance of smart homes	Implications for service delivery and planning are required for more interdisciplinary research in this area

(Continued)

TABLE 7.6 (Continued)

Research	Type of information	Results	Remarks
Bhati et al. (2017)	Case studies of the perception of Singapore households on smart technology and its usage to save energy including (1) energy consumption in Singapore households, (2) public programs and policies in energy savings, (3) use of technology in energy savings, and (4) household perception of energy savings in smart homes	Behavioral patterns of consumers may not change just to save energy	Even though an individual claimed to be concerned about the environment and energy-saving, it is evident that comfort and security play a bigger role in people's life
Park et al. (2018)	Internet survey conducted in South Korea. The data were analyzed using structural equation modeling and confirmatory factor analysis	Perceived compatibility, connectedness, control, system reliability, and enjoyment of smart home services are positively related to the users' intention to use the services	Negative association between the perceived cost and usage intention. Smart home services have attracted users' interest in the housing context, motivation of users' intention to use these should be studied further
Marikyan et al. (2019)	Literature review and systematic methodology covering 2002–17 from the users' perspective	The notion of becoming isolated and lacking human interaction could pose a challenge for smart home acceptance	Shift from technology-driven research to a consumer-centric approach for exploring all the potential advantages of smart home technology

Own results.

assistants, some consumers mistakenly start associating them with the popular concept of artificial intelligence (AI) that has been recently popularized by the works of science fiction as well as blockbuster movies. In these works of fiction, AI typically becomes self-aware and comes to a conclusion that it does not need humans or even needs to destroy the human race. Many people project these ideas to their own smart houses and began to fear that their dwellings would sooner or later turn against them. Only the bravest (or the technically advanced) ones understand that they have nothing to fear since the technology involving AI and the steering programs for the smart houses used nowadays are fundamentally different.

7.5 Electric vehicles

EVs are becoming the Holy Grail of today's policy-makers who hope they will be able to help to reduce the CO_2 emissions, solve the issues of pollution in big cities as well as assist in mitigating global climate change. Even though the EVs look very appealing and hi-tech, there are many varieties and types of them on the market which makes it quite

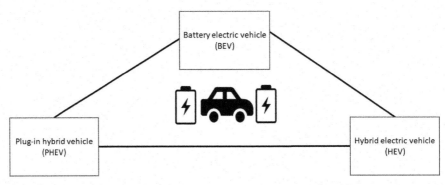

FIGURE 7.3 EV complexity—the electric car "triangle." *Own results.*

cumbersome for a mundane consumer to make herself or himself familiar with the plethora of offers and possibilities.

In fact, EVs are not the high-tech brilliant invention of the 21st century. Over a century ago, Henry Ford's wife Clara Ford preferred an EV to her husband's Ford T model. In Fords' days, electric cars were quite popular and widespread. Thence, Clara Ford used to drive a 1914 Detroit Electric car which could go for about 80 miles on a single charge reaching a top speed of about 20 miles per hour (Bryan, 1997). It is mostly due to Henry Ford's decision to go on with his what we know today as "conventional vehicles" (i.e., using the internal combustion engine which can be fueled by petrol, diesel, or oil derivatives) that made combustion engine cars more popular over the electric ones.

The complexity of the EV types and models can be shown in Fig. 7.3.

In general, there are three main types of EVs: (1) battery-electric vehicle (BEV), (2) plug-in hybrid vehicle (PHEV), and (3) hybrid electric vehicle (HEV).

BEV is an EV that puts in motion via the means of the electric motor. It is powered by electricity and has an installed battery which can be charged from a regular electric socket (e.g., on the wall in a house or an apartment).

PHEV is an EV with an internal combustion engine (either on petrol or diesel) and batteries that can be charged from a normal electric socket at a household. A vehicle like this can operate for several hundreds of kilometers entirely on electricity but when its batteries are finally empty, the vehicle simply automatically switches to the internal combustion engine.

HEV is a vehicle with batteries but without a plug. It has both an internal combustion engine and an electric engine. The combination allows the electric motor and batteries to help the conventional engine operate more efficiently, reducing the consumption of fuel. The internal mechanism switches between the two engines automatically without the driver's intervention. The battery is charged from the energy produced by the combustion engine during driving or while braking. A hybrid car drives several kilometers solely on electricity.

Nowadays, EVs are increasingly more used to prevent local air pollution, particularly in urban areas. Over the last few decades, the use of EVs and PV solar power has grown. Using solar energy to power an electric car can be cheaper for households with PVs since the cost of a solar system at home can also be included in the equation. In addition to understanding what it costs to power an EV, it is also important to know the cost of an important part of the home technology: the EV's equipment and the cost of its installation. This cost depends to a large extent on the size of the system, the quality of the solar panels and the use of the power inverter and the complexity of the installation.

According to the forecasts of the International Energy Agency, by 2020, EVs will occupy a 2% share of the world's passenger car fleet, which in numerical terms will be 20 million units (IEA, 2018). The Centre for Motor Vehicle Research at the Gelsenkirchen Institute (Germany) believes that by 2025 only electric cars and cars with hybrid engines will be sold in Europe in the passenger segment (Welt, 2015). Sweden intends by 2030 to ensure a 100% transition of road transport to noncarbon fuels (EVs and cars on biodiesel are mostly envisaged). By 2040, 75% of the kilometers traveled by passenger transport will be accounted for by EVs (Fridstrom, 2017). The rapidity of the spread of EVs confirms the fact that the number of charging stations for EVs in Japan has already exceeded the number of conventional gasoline filling stations.

In response to consumer dissatisfaction with the long periods of charging of EVs, stations were established and developed fast charging, allowing you to charge batteries 10 times faster than from a household outlet. In December 2014, the ELECTRIC project (European Long-Distance Electric Road Transport Infrastructure Corridor) began to be implemented in Europe, involving the construction of 155 state-of-the-art fast-charging stations on the main roads connecting Sweden, Denmark, Holland, and Germany (Cision, 2015). BMW and Volkswagen automakers unite to build a network of fast-charging stations linking the West and East Coast in the United States.

Further growth of the EV market depends largely on the development of battery technology. Today, the main types of batteries used in EVs are lithium-ion (Li-Ion) batteries and their subspecies. They have long been known, improved, become lighter, more powerful, and cheaper, increasing the availability of EVs for customers. The price of a "fuel device," that is, a battery with which an electric motor is driven, is an important factor in the cost and, accordingly, the competitiveness of an EV.

And there is a promising trend in all that. In particular, it is predicted that the colossal investments in the production of lithium-ion batteries made by Elon Mask of Tesla and other manufacturers will reduce the price of batteries by half by 2020 (Rathi, 2019). In 2012 McKinsey, a consulting company, predicted a price drop from $500−600 to $200 per kilowatt-hour by 2020 (on the US market, an EV becomes an absolute competitor to a traditional gasoline vehicle at about $250 per kilowatt-hour) (Hensley et al., 2012). Meanwhile, manufacturers are approaching this desirable and economically acceptable price level now, and Elon Musk believes that it is possible to reach the price level of $100 before 2020 (Atkin, 2014). It is possible that the use of graphene in the production of batteries will be a breakthrough in the development of battery technology. The Spanish company "Graphenano" that is using graphene for exactly these purposes, allegedly developed batteries that are about 70% cheaper and lighter used today and provide a cruising range of 1000 km and at the same time charge in just 10 minutes (even though no follow-up or reaction from the large automobile producers followed and the company remains unknown to most of the people today) (see Graphene, 2014).

The rapid development of battery technologies for the automotive industry also boosts the growth of the market for household and industrial energy storage systems and, in addition, turns the car into an integral part of the future energy system. By 2030, charging points for EVs and fuel cell vehicles will appear almost everywhere and form a distributed infrastructure for receiving energy from the electrical grid and returning it to the grid.

The Chinese government is making significant efforts to encourage cleaner transportation. This tax breaks, and an additional tax on gasoline, and significant subsidies to citizens for the purchase of an EV (up to $10,000 per one EV). China plans to bring the number of EVs to 500,000 in 2015 and to 5 million by 2020 (Reuters, 2014). By the way, as of today, about 150 million two-wheeled vehicles with electric motors drive the roads and highways of China (Guo et al., 2018). I remember living in Shanghai for a month when I quickly learned to look around carefully before crossing any road due to the fact that e-bikes and scooters make almost no sound when approaching and it is difficult to detect them or hear them as in a case of "regular" motorbikes and cars that emit steady noise and rumble.

Starting from 2016, 30% of public purchases of motor vehicles in China should be accounted for by EVs. The network, which by the end of 2015 should consist of 400,000 stations charging EVs, invested tens of billions of dollars in public investment (Wang et al., 2017). Elon Musk's Tesla is also building its own network of electric stations in China. If one adds here also the administrative capacity of the Chinese leadership to implement its "political will," the global prospects for the internal combustion engine and gasoline do not look that great at all. The latest statistics and forecasts show that the rainbow plans of the raw material giants in part of the Chinese market may not come true and the demand for fuel oil in China will not grow actively and in the long term.

In addition, India, another major Asian player, is also making ambitious plans for the development of renewable energy and alternative transport. Currently, there is a state plan which envisages for the deployment of additional 6−7 million EVs in India until 2020 (Pandit and Kapur, 2015).

When the attitude toward EVs among the consumers is concerned, one can notice that it might differ considerably (similar as in the case with the issue related to smart houses and smart meters). Table 7.7 provides an overview of the information-based studies describing consumer attitudes toward EV.

Over the last few decades, the use of EVs and PV solar power has grown. EVs (plug-in hybrids or all other battery systems) are used to prevent local air pollution, particularly in urban areas.

An energy-efficient household needs to manage its operation using its own generation from the renewable energy sources integrated with EV and renewable energies. Moreover, it is necessary to balance the optimally recharged EVs with grid capacity to increase the predictability of the generation of PV (Mesarić and Krajcar, 2015).

Fattori et al. (2014) describe how a combination of PV power with EVs, intelligent charging and the power of vehicles might optimally function. Amiri et al. (2018) show how the most effective management of EVs through battery swapping stations can be achieved.

When the household PV system produces electricity during the day, any device running in the home—a bulb, a TV, or EV charging station—is powered directly by solar energy. For the EV drivers, the cost calculation for solar panels is shifting from reducing the cost

TABLE 7.7 Information-based studies on electric vehicles.

Research	Type of information	Results	Remarks
Lane and Potter (2007)	Eliciting attitudinal barriers inhibiting the adoption of cleaner vehicles in the United Kingdom using estimating consumer attitudes to low-carbon cars and identifying key "hotspot" factors that influence consumers' adoption of low-carbon products	Electric car buyers have a poor knowledge of cleaner car technologies, the environmental impacts of road transport and car ownership costs	Environmental issues have a very low priority for private and fleet EV buyers
Skippon and Garwood (2011)	Mainstream drivers were given a direct experience of driving a battery-electric vehicle followed by an attitudinal questionnaire, and a vignette exercise to evaluate their attributions of symbolic meaning	Consumers might start to consider electric vehicles as second cars if they had a range of 100 miles, and as main cars if they had a range of 150 miles	Some consumers may be willing to pay modest premiums over conventional vehicles, equivalent to around 3 years' running cost savings
Ziegler (2012)	Discrete choice analysis alternative energy sources and electric vehicles with 598 potential car buyers in Germany	Some population groups have a higher propensity for electric vehicles which can be used by policy and automobile firms	The study highlights the importance of the inclusion of taste persistence across the choice sets
Lebeau et al. (2013)	Large-scale data collection held in Belgium envisaged to measure the perceptions on the advantages and disadvantages of BEVs, the acceptable driving range, the acceptable charging time (both slow and fast), the acceptable maximum speed, the role of the government in the introduction of BEVs, the preferred governmental tools to maximize sales and the consumers' Willingness to Pay (WTP). level of knowledge of the consumers	Knowledge and experience with EV have no impact on the level of acceptance for the driving range. However, consumers with more knowledge want an electric car with a higher maximum speed and desire faster charging durations	The results are relevant for other countries where the potential for electric vehicles is investigated. Future research could include a comparison of our results with similar studies conducted in other countries. Various factors such as taxes, sales prices, and legislation could alter the results
Bahamonde-Birke and Hanappi (2016)	Analysis of the potential of electric vehicles in Austria based on the results of vehicle purchase discrete choice behavioral mixture models with categorized latent variables	Assumptions regarding electromobility hold for the Austrian market (e.g., "green-mindedness" of the young and reluctance of the old), while others are only partially valid (e.g., the power of the engine)	Some policy incentives would have a positive effect for the demand for electric cars, while others—such as an annual Park and Ride subscription or a 1-year-ticket for public transportation—would not increase the willingness to pay for electromobility
Scasny et al. (2018)	Discrete choice experiment to estimate the willingness to pay of a representative sample of consumers intending to buy a car in Poland	Electric vehicles are significantly less preferred than conventional cars, even under public programs	Stimulating the electric vehicle market requires a pricing policy that affects the operating costs and other incentives along with an effective up-front price incentive scheme

Own results.

of utilities to reducing the cost of petrol for rollercoaster. A rule of thumb is that a PV system with a power output of 2.5 kW is usually the right size for a household that adds the load of an EV (Hesse et al., 2017).

In case the household owner is planning to install solar panels and a home electric charger, it is the most cost-effective and logistically advantageous time to install them at the same time. Homeowners of PV will soon be able to make their homes even more sustainable, as PV is also well suited for charging personal electronic vehicles. The surplus electricity obtained from the renewable energy sources could be supplied to the public grid and also to recharge a domestic EV. The power from the PV is first used at home, and the electricity is stored in the battery of the EV. However, when a solar PV system does not produce electricity in winter, the system uses external power. The results of the grid are very high, as the solar PV system produces more electricity in the coming months. However, when a solar PV system does not produce sufficient electricity in winter, it uses external power.

To solve environmental problems, a solar PV hybrid system for charging an environmentally friendly EV and an automatic learning system might also be used. Renewable energy is an important part of the electricity generation, as it has many advantages over conventional sources such as its environmental viability and efficiency.

One of the issues about the EV is the electric charging stations—there are not much of them around recently and the pace of building the new ones is quite slow.

Let us take an example of New York City. New York is installing 50 off-grid solar energy charging stations for Envision's EV range with integrated energy storage. New York believes that the units will generate 650,000 miles of solar-powered EVs per year (PV Magazine, 2018).

There are other new and exciting technologies for the EV, for example, self-loading roads, inductive and wireless recharging, 350 kW recharging, solar panels, connected vehicle solutions, vehicle-to-grid technologies, and new charging technologies using existing roads, garages, and construction infrastructure.

With regard to the above, when one is considering using solar energy to power one's car, the cost of a solar system at home can also be included in the equation. In addition to understanding what it costs to power an EV, it is also important to know the cost of an important part of the home technology: the EV's equipment and the cost of its installation. Thence, solar energy starts to make more sense if you get a system that can supply electricity to the home and EV. This cost depends to a large extent on the size of the system, the regional labor force, the quality of the solar panels, and the use of the power inverter and the complexity of the installation.

As their popularity will grow, EVs will consume more and more electricity which will lead to more generation operations, with a huge increase in the use of generation fuel and air emissions. In particular, EVs connected to the grid may be used instead of or in connection with the storage of electricity in emergency situations or extreme supply shortages.

EV can raise energy prices outside the peak enough to reduce the benefits of certain grid-related storage applications, in particular, the change in energy time and the management of energy costs. In addition, the existing charging model could have some potential for regional power plant development in addition to using distributed PV systems.

With EVs showing explosive growth over the last few years, researchers have found that such distributed mobile storage devices have a high potential for power systems in future networks, especially when they coordinate with renewable energy.

In an EV linked to the household PV system, the EV battery is considered a storage device, as well as a battery that connects the power supply to a distributed PV system.

Most houses can accommodate 2 kW of panels on their roofs, in addition to sufficient panels to compensate or exceed the electricity consumption of the house. However, since electric cars do not allow the battery to be completely lowered to protect it from spoilage, the actual kilometers per kilowatt-hour are a little higher, but it should make little difference, as modern electric cars do not leave much juice in the electric juice pack when they reach zero miles.

Use of EVs for providing grid services also constitutes a frontier battery storage technology. EVs could provide frequency regulation and offer fast response competitive with spinning reserves. Some vehicle technologies could provide peak-shifting services useful for integrating intermittent generation sources (Pierpoint, 2016). It can be shown that EVs can provide indirect storage by interrupting their charging and hence reducing demand to match reduced supply from other sources, as can other demand-side responses.

According to the US Department of Energy (2016), sales of the light EVs constituted about 1.2 million in 2015 (growing at an average 83% p.a. over the past 4 years). If the average battery size is estimated to be 24 kWh, this amounts to just under 30 GWh. Newbery (2016) estimates that the 2012 global car fleet included 773 million, growing since 2000 at 2.9% p.a., which if this continued would give a car fleet of 1.12 billion by 2025. If by then the share of EVs had grown to 10%, there would be 112 million EVs, with 24 kWh/EV with 2.7 TWh, more than the world current pumped supply power (PSP) storage (only a part of that storage would be indirectly accessible). In addition, the EV fleet may not be large enough to cost-effectively provide the ancillary services in the smart grid until 2030 (Bishop et al., 2016). More EVs will be needed to meet the power level and duration requirements that would allow them to participate in an everyday ancillary services market. Moreover, when it comes to the wholescale implementation of EVs, it seems to be a good idea to control the time of charging since most EV owners would most likely charge them during the same time periods (e.g., afterwork hours) which might impose additional pressure on the electrical system.

7.6 Conclusion and discussions

In general, it seems that nowadays smart electricity systems might be irrelevant for the majority of average consumers. This is happening because the costs of installing and running the smart meters to monitor the energy uptakes and to enhance saving on electricity bills might seem higher than to those consumers than the opportunity costs of not caring about the prices of electricity at all. There are many ways how to save on electricity but all of them are quite cumbersome and involve personal commitments and adjustments in habits and behavioral patterns. In order to be attractive for an average consumer, autonomic systems and smart grids of the future need to be more appealing in economic sense. Not only should they help people to save electricity and to appeal to their environment-friendly nature, but they should be able to motivate and to promote ways how to make money on becoming an integral part of smart energy systems by playing the role of proactive prosumers.

Moreover, our findings suggest that electricity consumers are likely to remain rational economic agents in the way they consume the power and pay for energy by maximizing, optimizing, and behaving economically. The acceptance of the smart metering by the general public and the policy-makers is likely to happen if and only if it will mean some economic advantages. In case their further development and implementation would mean additional costs added to the consumers' electricity bills with no real profit or visible advantage for the end users in sight, the majority of people will be against such innovations and will rather support traditional was of delivering and controlling power and energy. Therefore it would be necessary to either reduce the cost of the smart meters, provide smart meters for free or offer subsidies on their installation, or introduce the requirement to install smart meters in households by law.

Overall, this chapter demonstrated how the role and perception of energy through the eyes of the society might differ depending on the situation. All the three domains discussed above in this chapter (smart houses, EVs, and smart meters) yield various possibilities of perception and agreement or disagreement. And yet the overall picture is quite complex.

It is our human nature to like what others like. Hence, we admire the novel technologies that come with information technologies, social networks, or the widespread use of various mobile sensors, meters, trackers, and cameras. In the same time, we are becoming aware that these technologies might strip us of our privacy due to the possible date leakage and hacker attacks. This is especially relevant in the case of energy, in particular, our attitude toward energy and its usage. For decades, we as citizens of the world's developed countries have grown comfortable with an idea that there is always a source of cheap and affordable electrical energy. In any developed Western economy, there are electric plugs everywhere and one would not even think of a problem associated with the inability to charge one's device. This might not be so true for less-developed economies in such regions as Africa or Latin America. Cheap and affordable electricity is not a common thing there. I remember a story one fellow researcher who spent quite a long time in Sierra Leone told me about the access to electricity: in Sierra Leone, smartphones are relatively cheap (one can buy a Chinese smartphone for $100) and ubiquitous. However, one comes to work every morning there is a problem to find a free electric plague since everyone is charging her or his gadget. Smartphones and mobile operator services outstepped the supply of electricity.

These aspects should also be considered when contemplating about the perception of energy through the eyes of the modern society. We live in a globalized but also a diversified world and thence should take it accordingly.

References

Amiri, S.S., Jadid, S., Saboori, H., 2018. Multi-objective optimum charging management of electric vehicles through battery swapping stations. Energy 165, 549–562. Available from: https://doi.org/10.1016/j.energy.2018.09.167.

Angrisani, L., Bonavolontà, F., Liccardo, A., Schiano Lo Moriello, R., Serino, F., 2018. Smart power meters in augmented reality environment for electricity consumption awareness. Energies 11 (9), 2303. Available from: https://doi.org/10.3390/en11092303.

Atkin, E., 2014. We are on the verge of an electric car battery breakthrough. <https://thinkprogress.org/we-are-on-the-verge-of-an-electric-car-battery-breakthrough-6c888ac36001> (accessed 29.04.019.).

Bahamonde-Birke, F.J., Hanappi, T., 2016. The potential of electromobility in Austria: evidence from hybrid choice models under the presence of unreported information. Transport. Res. Part A: Policy Pract. 83, 30–41. Available from: https://doi.org/10.1016/j.tra.2015.11.002.

Baker, E., Chon, H., Keisler, J., 2009. Carbon capture and storage: combining economic analysis with expert elicitations to inform climate policy. Climate Change 96, 379–408. Available from: https://doi.org/10.1007/s10584-009-9634-y.

Balitskiy, S., Bilan, Y., Strielkowski, W., 2014. Energy security and economic growth in the European Union. J. Sec. Sustain. Issue. 4 (2), 125–132. Available from: https://doi.org/10.9770/jssi.2014.4.2(2).

Balta-Ozkan, N., Davidson, R., Bicket, M., Whitmarsh, L., 2013. Social barriers to the adoption of smart homes. Energy Policy 63, 363–374. Available from: https://doi.org/10.1016/j.enpol.2013.08.043.

Balta-Ozkan, N., Amerighi, O., Boteler, B., 2014. A comparison of consumer perceptions towards smart homes in the UK, Germany and Italy: reflections for policy and future research. Technol. Anal. Strat. Manag. 26 (10), 1176–1195. Available from: https://doi.org/10.1080/09537325.2014.975788.

Barbosa, L.D.S.N.S., Bogdanov, D., Vainikka, P., Breyer, C., 2017. Hydro, wind and solar power as a base for a 100% renewable energy supply for South and Central America. PLoS One 12 (3), e0173820. Available from: https://doi.org/10.1371/journal.pone.0173820.

Benders, R., Kok, R., Moll, H.C., Wiersma, G., Noorman, K.J., 2006. New approaches for household energy conservation: in search of personal household energy budgets and energy reduction options. Energy Policy 34, 3612–3622. Available from: https://doi.org/10.1016/j.enpol.2005.08.005.

Bhati, A., Hansen, M., Chan, C.M., 2017. Energy conservation through smart homes in a smart city: a lesson for Singapore households. Energy Policy 104, 230–239. Available from: https://doi.org/10.1016/j.enpol.2017.01.032.

Bhatta, L.K.G., Subramanyam, S., Chengala, M.D., Olivera, S., Venkatesh, K., 2015. Progress in hydrotalcite like compounds and metal-based oxides for CO_2 capture: a review. J. Clean. Prod. 103, 171–196. Available from: https://doi.org/10.1016/j.jclepro.2014.12.059.

Bishop, J., Axon, C., Bonilla, D., Banister, D., 2016. Estimating the grid payments necessary to compensate additional costs to prospective electric vehicle owners who provide vehicle-to-grid ancillary services. Energy 94, 715–727. Available from: https://doi.org/10.1016/j.energy.2015.11.029.

Bowen, F., 2011. Carbon capture and storage as a corporate technology strategy challenge. Energy Policy 39, 2256–2264. Available from: https://doi.org/10.1016/j.enpol.2011.01.016.

Broecks, K.P., van Egmond, S., van Rijnsoever, F.J., Verlinde-van den Berg, M., Hekkert, M.P., 2016. Persuasiveness, importance and novelty of arguments about carbon capture and storage. Environ. Sci. Policy 59, 58–66. Available from: https://doi.org/10.1016/j.envsci.2016.02.004.

Bryan, F.R., 1997. Beyond the Model T: The Other Ventures of Henry Ford. Wayne State University Press.

Buchanan, K., Russo, R., Anderson, B., 2014. Feeding back about eco-feedback: how do consumers use and respond to energy monitors? Energy Policy 73, 138–146. Available from: https://doi.org/10.1016/j.enpol.2014.05.008.

Bulkeley, H.A., Matthews, P.C., Whitaker, G., Bell, S., Wardle, R., Lyon, S., et al., 2015. High level summary of learning: domestic smart meter customers on time of use tariffs. Technical Report. Northern Powergrid (Northeast) Limited, Newcastle Upon Tyne. <http://www.networkrevolution.co.uk/project-library/high-level-summary-learning-domestic-smart-meter-customers-time-of-use-tariffs> (accessed 12.03.19.).

Carroll, J., Lyons, S., Denny, E., 2014. Reducing household electricity demand through smart metering: the role of improved information about energy saving. Energy Econ. 45, 234–243. Available from: https://doi.org/10.1016/j.eneco.2014.07.007.

Chen, J., Cheng, S., Nikic, V., Song, M., 2018. Quo Vadis? Major players in global coal consumption and emissions reduction. Transf. Bus. Econ. 17, 112–133.

Cision, 2015. ELECTRIC project. <http://www.prnewswire.co.uk/news-releases> (accessed 20.05.19.).

Commission for Energy Regulation, 2011. Electricity smart metering technology trials findings report. Information paper CER11080b. <http://www.cer.ie/docs/000340/cer11080(a)(i).pdf> (accessed 12.03.19.).

Coyle, F.J., 2016. 'Best practice' community dialogue: the promise of a small-scale deliberative engagement around the siting of a carbon dioxide capture and storage (CCS) facility. Int. J. Greenhouse Gas Control 45, 233–244. Available from: https://doi.org/10.1016/j.ijggc.2015.12.006.

Darby, S., 2006. The effectiveness of feedback on energy consumption. A review for DEFRA of the literature on metering, billing and direct displays, 486 (2006), 26. <https://www.eci.ox.ac.uk/research/energy/downloads/smart-metering-report.pdf> (accessed 15.02.19.).

Darby, S., 2010. Smart metering: what potential for householder engagement? Build. Res. Inform. 38 (5), 442–457. Available from: https://doi.org/10.1080/09613218.2010.492660.

Demiris, G., Rantz, M.J., Aud, M.A., Marek, K.D., Tyrer, H.W., Skubic, M., et al., 2004. Older adults' attitudes towards and perceptions of 'smart home' technologies: a pilot study. Med. Inform. Internet Med. 29 (2), 87–94. Available from: https://doi.org/10.1080/14639230410001684387.

Dulleck, U., Kaufmann, S., 2004. Do customer information programs reduce household electricity demand – the Irish program. Energy Policy 32, 1025–1032. Available from: https://doi.org/10.1016/S0301-4215(03)00060-0.

Eltigani, D., Masri, S., 2015. Challenges of integrating renewable energy sources to smart grids: a review. Ren. Sustain. Energy Rev. 52, 770–780. Available from: https://doi.org/10.1016/j.rser.2015.07.140.

Faruqui, A., Harris, D., Hledik, R., 2010. Unlocking the €53 billion savings from smart meters in the EU: how increasing the adoption of dynamic tariffs could make or break the EU's smart grid investment. Energy Policy 38 (10), 6222–6231. Available from: https://doi.org/10.1016/j.enpol.2010.06.010.

Fattori, F., Anglani, N., Muliere, G., 2014. Combining photovoltaic energy with electric vehicles, smart charging and vehicle-to-grid. Sol. Energy 110, 438–451. Available from: https://doi.org/10.1016/j.solener.2014.09.034.

Feist, W., Schnieders, J., Dorer, V., Haas, A., 2005. Re-inventing air heating: convenient and comfortable within the frame of the passive house concept. Energy Build. 37 (11), 1186–1203. Available from: https://doi.org/10.1016/j.enbuild.2005.06.020.

Fernández, J., Sotenko, M., Derevschikov, V., Lysikov, A., Rebrov, E.V., 2016. A radiofrequency heated reactor system for post-combustion carbon capture. Chem. Eng. Process. 108, 17–26. Available from: https://doi.org/10.1016/j.cep.2016.07.004.

Fridstrom, L., 2017. From innovation to penetration: calculating the energy transition time lag for motor vehicles. Energy Policy 108, 487–502. Available from: https://doi.org/10.1016/j.enpol.2017.06.026.

Galvin, R., 2014. Estimating broad-brush rebound effects for household energy consumption in the EU 28 countries and Norway: some policy implications of Odyssee data. Energy Policy 73, 323–332. Available from: https://doi.org/10.1016/j.enpol.2014.02.03.

Gerpott, T.J., Paukert, M., 2013. Determinants of willingness to pay for smart meters: an empirical analysis of household customers in Germany. Energy Policy 61, 483–495. Available from: https://doi.org/10.1016/j.enpol.2013.06.012.

Global CCS Institute, 2017. Understanding CCS. <https://www.globalccsinstitute.com/understanding-ccs/industrial-ccs> (accessed 20.01.19.).

Graphene, 2014. Revolutionary graphene polymer batteries for electric cars. <http://www.graphene-info.com/revolutionary-graphene-polymer-batteries-electric-cars> (accessed 19.05.19.).

Guo, Y., Sayed, T., Zaki, M.H., 2018. Evaluating the safety impacts of powered two wheelers on a shared roadway in China using automated video analysis. J. Transp. Saf. Secur. 1–16. Available from: https://doi.org/10.1080/19439962.2018.1447058.

Hall, S., Foxon, T.J., 2014. Values in the smart grid: the co-evolving political economy of smart distribution. Energy Policy 74, 600–609. Available from: https://doi.org/10.1016/j.enpol.2014.08.018.

Hensley, R., Newman, J., Rogers, M., 2012. Battery technology charges ahead. <https://www.mckinsey.com/business-functions/sustainability/our-insights/battery-technology-charges-ahead> (accessed 20.05.19.).

Hesse, H., Schimpe, M., Kucevic, D., Jossen, A., 2017. Lithium-ion battery storage for the grid: a review of stationary battery storage system design tailored for applications in modern power grids. Energies 10 (12), 2107. Available from: https://doi.org/10.3390/en10122107.

IEA, 2014. Energy efficiency: a key tool for boosting economic and social development. <https://www.iea.org/newsroom/news/2014/september/energy-efficiency-a-key-tool-for-boosting-economic-and-social-development.html> (accessed 18.05.19.).

IEA, 2018. Global EV outlook. <https://www.iea.org/gevo2018> (accessed 29.04.19.).

IEA, 2019. Smart grids: tracking clean energy progress. <https://www.iea.org/tcep/energyintegration/smartgrids> (accessed 20.05.019.).

Inderberg, T.H., 2015. Advanced metering policy development and influence structures: the case of Norway. Energy Policy 81, 98−105. Available from: https://doi.org/10.1016/j.enpol.2015.02.027.

Joachain, H., Klopfert, F., 2014. Smarter than metering? Coupling smart meters and complementary currencies to reinforce the motivation of households for energy savings. Ecol. Econ. 105, 89−96. Available from: https://doi.org/10.1016/j.ecolecon.2014.05.017.

Johnson, J.L., Davenport, R., Mann, W.C., 2007. Consumer feedback on smart home applications. Top. Geriatr. Rehab. 23 (1), 60−72. Available from: https://doi.org/10.1097/00013614-200701000-00009.

Kashintseva, V., Strielkowski, W., Streimikis, J., Veynbender, T., 2018. Consumer attitudes towards industrial CO_2 capture and storage products and technologies. Energies 11 (10), 2787. Available from: https://doi.org/10.3390/en11102787.

Kohler, N., Hassler, U., 2002. The building stock as a research object. Build. Res. Inform. 30 (4), 226−236. Available from: https://doi.org/10.1080/09613210110102238.

Koirala, B.P., Araghi, Y., Kroesen, M., Ghorbani, A., Hakvoort, R.A., Herder, P.M., 2018. Trust, awareness, and independence: insights from a socio-psychological factor analysis of citizen knowledge and participation in community energy systems. Energy Res. Soc. Sc. 38, 33−40. Available from: https://doi.org/10.1016/j.erss.2018.01.009.

Kowalska-Pyzalska, A., 2018. What makes consumers adopt to innovative energy services in the energy market? A review of incentives and barriers. Renew. Sustain. Energy Rev. 82, 3570−3581. Available from: https://doi.org/10.1016/j.rser.2017.10.103.

Kraeusel, J., Möst, D., 2012. Carbon Capture and Storage on its way to large-scale deployment: social acceptance and willingness to pay in Germany. Energy Policy 49, 642−651. Available from: https://doi.org/10.1016/j.enpol.2012.07.006.

Krüger, T., 2017. Conflicts over carbon capture and storage in international climate governance. Energy Policy 2017 (100), 58−67. Available from: https://doi.org/10.1016/j.enpol.2016.09.059.

Lane, B., Potter, S., 2007. The adoption of cleaner vehicles in the UK: exploring the consumer attitude−action gap. J. Clean. Prod. 15 (11−12), 1085−1092. Available from: https://doi.org/10.1016/j.jclepro.2006.05.026.

Lebeau, K., Van Mierlo, J., Lebeau, P., Mairesse, O., Macharis, C., 2013. Consumer attitudes towards battery electric vehicles: a large-scale survey. Int. J. Electr. Hybrid Veh. 5 (1), 28−41. Available from: https://doi.org/10.1504/IJEHV.2013.053466.

Leiva, J., Palacios, A., Aguado, J.A., 2016. Smart metering trends, implications and necessities: a policy review. Renew. Sustain. Energy Rev. 55, 227−233. Available from: https://doi.org/10.1016/j.rser.2015.11.002.

Liebe, U., Dobers, G.M., 2019. Decomposing public support for energy policy: what drives acceptance of and intentions to protest against renewable energy expansion in Germany? Energy Res. Soc. Sci. 47, 247−260. Available from: https://doi.org/10.1016/j.erss.2018.09.004.

Liu, Z., Liao, L., Mei, C., 2018. Not-in-my-backyard but let's talk: explaining public opposition to facility siting in urban China. Land Use Policy 77, 471−478. Available from: https://doi.org/10.1016/j.landusepol.2018.06.006.

Lubin, D.A., Esty, D.C., 2010. The sustainability imperative. Harv. Bus. Rev. 88, 42−50.

Marikyan, D., Papagiannidis, S., Alamanos, E., 2019. A systematic review of the smart home literature: a user perspective. Technol. Forecast. Soc. Change 138, 139−154. Available from: https://doi.org/10.1016/j.techfore.2018.08.015.

McNeil, M.A., Feng, W., du Can, S.D.L.R., Khanna, N.Z., Ke, J., Zhou, N., 2016. Energy efficiency outlook in China's urban buildings sector through 2030. Energy Policy 97, 532−539. Available from: https://doi.org/10.1016/j.enpol.2016.07.033.

Mesarić, P., Krajcar, S., 2015. Home demand side management integrated with electric vehicles and renewable energy sources. Energy Build. 108, 1−9. Available from: https://doi.org/10.1016/j.enbuild.2015.09.001.

Milham, S., 2012. Dirty Electricity: Electrification and the Diseases of Civilization, first ed. iUniverse, 128 pp.

Nachreiner, M., Mack, B., Matthies, E., Tampe-Mai, K., 2015. An analysis of smart metering information systems: a psychological model of self-regulated behavioural change. Energy research & social science 9, 85−97. Available from: https://doi.10.1016/j.erss.2015.08.016.

Naus, J., Spaargaren, G., van Vliet, B.J.M., van der Horst, H.M., 2014. Smart grids, information flows and emerging domestic energy practices. Energy Policy 68, 436−446. Available from: https://doi.org/10.1016/j.enpol.2014.01.038.

Newbery, D., 2016. A simple introduction to the economics of storage: shifting demand and supply over time and space. EPRG Working Paper 1626. Cambridge Working Paper in Economics 1661. <http://www.eprg.group.cam.ac.uk/wp-content/uploads/2016/10/1626-Text.pdf> (accessed 19.03.19.).

Obinna, U., Joore, P., Wauben, L., Reinder, A., 2017. Comparison of two residential smart grid pilots in the Netherlands and in the USA, focusing on energy performance and user experiences. Appl. Energy 191 (1), 264–275. Available from: https://doi.org/10.1016/j.apenergy.2017.01.086.

Paetz, A.G., Dütschke, E., Fichtner, W., 2012. Smart homes as a means to sustainable energy consumption: a study of consumer perceptions. J. Consum. Policy 35 (1), 23–41. Available from: https://doi.org/10.1007/s10603-011-9177-2.

Pandit, S., Kapur, D., 2015. Electric vehicles in India: policies, opportunities and current scenario. In ADB Open Innovation Forum. <https://k-learn.adb.org/system/files/materials/2015/05/201505-electric-vehicles-india-policies-opportunities-and-current-scenario.pdf> (accessed 21.05.19.).

Park, E., Kim, S., Kim, Y., Kwon, S.J., 2018. Smart home services as the next mainstream of the ICT industry: determinants of the adoption of smart home services. Univ. Access Inform. Soc. 17 (1), 175–190. Available from: https://doi.org/10.1007/s10209-017-0533-0.

Pepermans, G., 2014. Valuing smart meters. Energy Econ. 45, 280–294. Available from: https://doi.org/10.1016/j.eneco.2014.07.01.

Pierpoint, L., 2016. Harnessing electricity storage for systems with intermittent sources of power: policy and R&D needs. Energy Policy 96, 751–757. Available from: https://doi.org/10.1016/j.enpol.2016.04.032.

Pope, S., 2019. Harvard medical doctor warns against smart meters. https://www.thehealthyhomeeconomist.com/harvard-medical-doctor-warns-against-smart-meters/ (accessed 17.05.19.).

PV Magazine, 2018. Off grid solar powered electric vehicle charging in NYC. <https://pv-magazine-usa.com/2018/12/26/new-york-citys-solar-powered-electric-vehicle-charging-stations> (accessed 18.02.19.).

Rathi, A., 2019. How we get to the next big battery breakthrough. <https://qz.com/1588236/how-we-get-to-the-next-big-battery-breakthrough> (accessed 20.05.19.).

Rausser, G., Strielkowski, W., Štreimikienė, D., 2018. Smart meters and household electricity consumption: a case study in Ireland. Energy Environ. 29 (1), 131–146. Available from: https://doi.org/10.1177/0958305X17741385.

Reiner, D., 2016. Learning through a portfolio of carbon capture and storage demonstration projects. Nat. Energy 1, 15011. Available from: https://doi.org/10.1038/nenergy.2015.11.

Reuters, 2014. China to offer tax breaks on electric cars, limited mostly to local brands. <https://www.reuters.com/article/us-china-autos/china-to-offer-tax-breaks-on-electric-cars-limited-mostly-to-local-brands-idUSKBN0GT0UX20140829> (accessed 20.04.19.).

Rodrigues, C.F.A., Dinis, M.A.P., Lemos de Sousa, M.J., 2016. Gas content derivative data versus diffusion coefficient. Energy Explor. Exploit. 34, 606–620. Available from: https://doi.org/10.1177/0144598716643629.

Roscoe, A.J., Ault, G., 2010. Supporting high penetrations of renewable generation via implementation of real-time electricity pricing and demand response. IET Renew. Power Gener. 4 (4), 369–382. Available from: https://doi.org/10.1049/iet-rpg.2009.0212.

S&P Global, 2017. Smart homes in the U.S. becoming more common, but still face challenges <https://www.spglobal.com/marketintelligence/en/news-insights/blog/smart-homes-in-the-u-s-becoming-more-common-but-still-face-challenges> (accessed 10.03.19.).

Scasny, M., Zverinova, I., Czajkowski, M., 2018. Electric, plug-in hybrid, hybrid, or conventional? Polish consumers' preferences for electric vehicles. Energ. Effic. 11 (8), 2181–2201. Available from: https://doi.org/10.1007/s12053-018-9754-1.

Schultz, P.W., Estrada, M., Schmitt, J., Sokoloski, R., Silva-Send, N., 2015. Using in-home displays to provide smart meter feedback about household electricity consumption: a randomized control trial comparing kilowatts, cost, and social norms. Energy 90 (Part 1), 351–358. Available from: https://doi.org/10.1016/j.energy.2015.06.130.

Simshauser, P., 2016. Distribution network prices and solar PV: resolving rate instability and wealth transfers through demand tariffs. Energy Econ. 54, 108–122. Available from: https://doi.org/10.1016/j.eneco.2015.11.011.

Skippon, S., Garwood, M., 2011. Responses to battery electric vehicles: UK consumer attitudes and attributions of symbolic meaning following direct experience to reduce psychological distance. Transp. Res. Part D: Transp. Environ. 16 (7), 525–531. Available from: https://doi.org/10.1016/j.trd.2011.05.005.

Skjølsvold, T.M., Jørgensen, S., Ryghaug, M., 2017. Users, design and the role of feedback technologies in the Norwegian energy transition: an empirical study and some radical challenges. Energy Res. Soc. Sci. 25, 1–8. Available from: https://doi.org/10.1016/j.erss.2016.11.005.

Stojkoska, B.L.R., Trivodaliev, K.V., 2017. A review of Internet of things for smart home: challenges and solutions. J. Clean. Prod. 140, 1454–1464. Available from: https://doi.org/10.1016/j.jclepro.2016.10.006.

Torriti, J., 2014. Privatisation and cross-border electricity trade: from internal market to European supergrid? Energy 77, 635–640. Available from: https://doi.org/10.1016/j.energy.2014.09.057.

United States Department of Energy, 2012. Carbon capture and storage from industrial sources. <https://energy.gov/fe/science-innovation/carbon-capture-and-storage-research/carbon-capture-and-storage-industrial> (accessed 20.05.19.).

United States Department of Energy, 2016. Fact #918: global plug-in light vehicle sales increased by about 80% in 2015. <http://energy.gov/eere/vehicles/fact-918-march-28-2016-global-plug-light-vehicle-sales-increased-about-80-2015> (accessed 20.03.19.).

Van der Werff, E., Steg, L., 2016. The psychology of participation and interest in smart energy systems: comparing the value-belief-norm theory and the value-identity-personal norm model. Energy Res. Soc. Sci. 22, 107–114. Available from: https://doi.org/10.1016/j.erss.2016.08.022.

Verbong, G.P.J., Beemsterboer, S., Sengers, F., 2013. Smart grids or smart users? Involving users in developing a low carbon electricity economy. Energy Policy 52, 117–125. Available from: https://doi.org/10.1016/j.enpol.2012.05.003.

Wang, Y., Sperling, D., Tal, G., Fang, H., 2017. China's electric car surge. Energy Policy 102, 486–490. Available from: https://doi.org/10.1016/j.enpol.2016.12.034.

Welt, 2015. EV will become a mass phenomenon. <http://www.welt.de/motor/article2148742/Elektroautos-werden-2015-zum-Massenphaenomen.html> (accessed 12.05.19.).

Wennersten, R., Sun, Q., Li, H., 2015. The future potential for carbon capture and storage in climate change mitigation: an overview from perspectives of technology, economy and risk. J. Clean. Prod. 103, 724–736. Available from: https://doi.org/10.1016/j.jclepro.2014.09.023.

Wilson, C., 2014. Evaluating communication to optimise consumer-directed energy efficiency interventions. Energy Policy 74, 300–310. Available from: https://doi.org/10.1016/j.enpol.2014.08.02.

Wood, G., Newborough, M., 2003. Dynamic energy-consumption indicators for domestic appliances: environment, behaviour and design. Energy Build. 35 (8), 821–841. Available from: https://doi.org/10.1016/S0378-7788(02)00241-4.

Ziegler, A., 2012. Individual characteristics and stated preferences for alternative energy sources and propulsion technologies in vehicles: a discrete choice analysis for Germany. Transp. Res. Part A: Policy Pract. 46 (8), 1372–1385. Available from: https://doi.org/10.1016/j.tra.2012.05.016.

Further reading

Bell, M., 2005. Use best practices to design data center facilities. <https://www.it.northwestern.edu/bin/docs/DesignBestPractices_127434.pdf> (accessed 18.05.19.).

Bhardwaj, S., Jain, L., Jain, S., 2010. Cloud computing: a study of infrastructure as a service (IAAS). Int. J. Eng. Inform. Technol. 2 (1), 60–63.

Gifford, G., 2014. Residential storage market to increase to 900 MW by 2018. <https://www.pv-magazine.com/2014/11/12/residential-storage-market-to-increase-to-900-mw-by-2018_100017137> (accessed 18.05.19.).

Gu, C., Fan, L., Wu, W., Huang, H., Jia, X., 2018. Greening cloud data centers in an economical way by energy trading with power grid. Future Gener. Comput. Syst. 78, 89–101. Available from: https://doi.org/10.1016/j.future.2016.12.029.

Koomey, J.G., 2008. Worldwide electricity used in data centers. Environ. Res. Lett. 3 (3), 034008. Available from: https://doi.org/10.1088/1748-9326/3/3/034008.

Liu, J., Terzis, A., 2012. Sensing data centres for energy efficiency. Philos. Trans. R. Soc. A: Math. Phys. Eng. Sci. 370 (1958), 136–157. Available from: https://doi.org/10.1098/rsta.2011.0245.

Porse, E., Derenski, J., Gustafson, H., Elizabeth, Z., Pincetl, S., 2016. Structural, geographic, and social factors in urban building energy use: analysis of aggregated account-level consumption data in a megacity. Energy Policy 96, 179–192. Available from: https://doi.org/10.1016/j.enpol.2016.06.002.

Wang, S., Zhou, A., Hsu, C.H., Xiao, X., Yang, F., 2015. Provision of data-intensive services through energy-and QoS-aware virtual machine placement in national cloud data centers. IEEE Trans. Emerg. Topics Comput. 4 (2), 290–300. Available from: https://doi.org/10.1109/TETC.2015.2508383.

8

Smart grids of tomorrow and the challenges for the future

8.1 Introduction

While the Internet of Things (IoT) is a concept that includes various digital devices (such as smartphones, smart watches, smart meters, digital health monitors, and smoke detectors) that are capable of interacting with each other and conduct a two-way exchange of information and data, the Internet of Energy (IoE) is a subsection of IoT that has a huge impact and a wide use in the field of energy (Reka and Dragicevic, 2018). IoE is going to find wide use in the smart grids that will likely to become ubiquitous in the future replacing the traditional energy regulation and market interaction as we know them today. The IoE is based on the concept of IoT which involves a huge number of devices and appliances connected to Internet or interconnected with one another and smart grids that allow high-speed connections and data and energy transfers (Farhan et al., 2018). Within this framework, there is a large number of largely vertical integrated companies which own about 40% of the electricity generation capacity in the electrical energy sector, while many utilities and electrical cooperatives are "distribution," with 10% and 4% of their electricity generation capacity, respectively (Willis and Philipson, 2018).

Smart grid is the technological paradigm proposed to spread the intelligence of the power distribution and control system from some central hubs to many peripheral nodes, allowing for more precise monitoring of energy losses and more precise regulation and adaptation. This is where it has to rely upon the IoT. Within this context, the IoT mean that well-known features of the Internet applications are going to be available for both machine-to-machine (M2M) and human-to-machine interactions (see, e.g., Gawali and Deshmukh, 2019). By integrating intelligent networks into the IoT, one can take advantage of many benefits—both those already apparent and known and those that no one can even think of today. For example, the system that employs the IoT can take advantage of the widely accepted security and privacy frameworks, wide connectivity and seamless interoperability: the ability to develop cloud systems to facilitate virtualization and distribution of services and the availability of a wide range of widely accepted standards are also key factors.

Social Impacts of Smart Grids
DOI: https://doi.org/10.1016/B978-0-12-817770-9.00008-0

Most recently one important thing occurred thanks to the progress in global connectivity and the wide deployment of high-speed mobile networks, smart meters and sensors together with the unprecedented growth of the computing power and cloud servers. Smart grids and the IoT are now merging into IoT. The use of the IoT communication paradigm for managing intelligent distribution networks would bring many benefits. First, it will be possible to interact with a wide range of information and communication technologies in a uniform way due to the harmonious effect of Internet Protocol (IP) networks. Second, M2M interactions would allow for a decentralization of control procedures, thereby relieving the central network from a high level of communication. Third, high-speed peer-to-peer (P2P) communication is essential to the success of the global open energy market.

Intelligent network technologies, in particular smart metering and sensing, communication and analysis, will play a key role in the definition and management of the value of actors and the assets that create such new values. Energy companies can no longer count on the increase in revenue based on centralized generation. A research carried out by Siemens (2019) shows that demand for energy and supplied electricity is declining due to energy management technologies, energy efficiency, and consumption. Network upgrades offer attractive options for commercial and residential consumers and are capable of allowing at least a certain level of autonomy and can lower their payments to local utilities. With regard to that, it appears logical that state regulatory committees would be well advised to ask utility companies to design network updates for the common benefits of resilience in addition to reliability.

Operational efficiency projects range from automation technologies for substation and distribution to distributed warehouse pilots and smart meter deployments. Any smart grid which extends the information and communications technology (ICT) industry (and involves some degree of information technology) to the mass energy system could provide a new perspective on the integration of technological transformations. In fact, with the maturity of the technology, the smart grid is expected to improve the reliability of the network by eliminating the consumer and producer line, thereby increasing supply and demand, through the promotion of "prosumers" engaged in all types of generation and storage, and increasing the multiway of exchanging information and electrical energy.

Energy storage and its implications are of a particular importance in this debate here. Historically, lead acid batteries are mainly used for home energy storage. With the development of energy storage technologies, lithium-ion (Li-Ion) batteries and their varieties, for example, lithium iron phosphate (LiFePO$_4$), which in addition to the similar technical specifics have better technical characteristics, are gaining an ever-increasing market share. Its price is higher, but gradually, as technology develops and production increases, it decreases. In 2014 the cost of Li-Ion batteries fell by 20%; in 2015 the prices are expected to fall by another 15% (Schmuch et al., 2018). In addition, in many European countries, there are programs of concessional lending to modern energy storage systems that encourage their acquisition. By 2018 a 10-fold increase in the capacity of domestic storage battery systems in the world is expected up to 900 MW compared to the current 90 MW (Few et al., 2018). By the way, Tesla, a company of Elon Musk, known for its futuristic electric vehicles (EVs) and, most recently, launching of space ships targeted at the Earth orbit, the Moon, or even Mars, which is building its giant factory near Reno, in the state of Nevada, intended for the massive production of batteries and energy storage devices, has also already started offering energy storage systems for home use. The widely advertised Powerwall from Tesla is a set of Li-Ion batteries in an elegant case, equipped with an uninterruptible power system. For some reason, it reminds me of iPod that was a mixture of minimalistic curves, white color, and steel and elegant design.

Nowadays, complete system for the conversion and storage of solar electricity is gaining popularity in the European market. Such a system is a single unit—a cabinet the size of a small box which houses Li-Ion batteries, an inverter, an uninterruptible power supply unit, and an electronic power metering system. These products simplify the design and installation of home solar power plants, which also contributes to the growth of their distribution. Nevertheless, they are quite useful, even though stocking up on energy for the future storage is neither possible nor cheap—the tanks used in households in today's development do not allow saving energy "in reserve," "stocking up for later," etc. This technique is designed to smooth out the insolation unevenness during the day and can also be used as a network stabilizer and uninterruptible power supply. The capacity of standard household batteries, as a rule, does not exceed the daily household need for electricity. Creating large tanks for energy storage ("for a week" or more) at home is theoretically possible, but with the available technologies and their cost (more than 1000 EUR/kWh of capacity) it is irrational from an economic point of view. From the point of view of the electric grid economy as a whole, in conditions of a wide distribution of small generation (in a number of countries, as we remember, millions of small solar power plants are already operating), household energy storage systems are very useful because they help stabilize the network and smooth out uneven load.

As a utility company, the intelligent network will make every business chain easier and more flexible. Generation firms will be able to integrate more renewable energies, and transmission companies will have the opportunity to expand their transmission lines through automation, with distribution companies that are trying to develop intelligence on the downstream side of the network. With regard to that, the existing (and very common on the market) investor-owned utility regulators must take into account a completely different landscape, including new and agile competitors, to slow down monopolies, thanks to intelligent network technologies and services.

The adoption of the consumer's lifetime value metrics can help utilities and energy regulators to transform existing activities into consumer-focused activities. Smart grid technologies are capable of creating the potential for individual consumers to produce kilowatts of electricity and become additives rather than subtracting entrants in the energy markets. Thus consumers become potential customers when they install a power-producing device such as a solar panel or a wind turbine (just to name the most popular and notoriously known ones) on their premises. As more households, institutions, and businesses install renewable energy systems, a new order involving the socio-technological distribution system will be emerging.

Smart meters are essential for this order which enables a two-way communication and power transmission between the consumer and the energy supplier (currently an energy buyer). In this kind of arrangement, consumers become proactive consumers or "prosumers." As the electricity grids and the energy markets of the future will evolve and transform, new players will appear on the stage with a plethora of intelligent devices and intelligent services to help assume that their energy bills are minimized.

Three factors considered particularly relevant to smart grid investments should be taken into account: (1) dealing with new risks and uncertainties, (2) dealing with new decision-making areas where investment plans have to be defended, and (3) dealing with the new power positions. For example, some companies based in the Netherlands are investing into becoming important suppliers of flexible gas-fired power plant facilities (in combination with gas storage and liquefied natural gas imports) to bridge the gap between the rigid capacity of the North Sea and coal production in Belgium, Germany, and France, as well as the intermittent supply of renewable energies (Stern, 2002).

All in all, the IoE is coming to alter our everyday lives and the change is going to be unprecedented. There are questions to that: when will the fully developed IoT be operational? Will it be in 5 years or may be in 10? It is clear that there are factors and barriers that might delay the fully functional IoT—among them are technical, economic, and political ones. Let us look at one of the key technical factors, namely the 5G mobile networks that offer the new promising opportunities for the IoT and smart grids of the future.

8.2 Smart grids and 5G networks

The use of mobile data is growing at an enormous speed with about 75% rise in 2015 alone. Today, the use of mobile data worldwide has spiked to about 3.7 exabytes per month (Ericsson, 2019). These figures are predetermined by the growth of video and audio streaming services (just recall such popular ones as YouTube Music, Netflix, or Spotify just to name a few) as well as the use of apps, online games, and social networks.

Nowadays, many people cannot imagine their work and personal life without a high-speed data connection even when not on a readily available Wi-Fi. Currently, the most advanced existing mobile data standard is the 4G global standard. It already allows us to connect to cloud servers and offers much faster speed and more capacity with a more resilient network than the standards available before. However, the 5G mobile standard will offer even more data volume, many more devices that would be available within the same network. In addition, according to some experts, it will also reduce latency and the new level of reliability (Dahlman et al., 2014; Sachs et al., 2018; Parvez et al., 2018).

The 5G mobile standard is a next-generation ultrafast and ultrareliable mobile network boosting range and speed. It will be able to satisfy the growing data requirements for the existing networks using including such novel important technologies as augmented reality, connected vehicles, and IoT. The starting speed of 5G networks will be about 100 Mb/s, which is not that superior from the already-existing Long Term Evolution (LTE) advanced (30–50 Mb/s).

It is envisaged that by 2035 full economic benefits of 5G will be realized around the world. The introduction of the fully operational and functional 5G is highly likely to boost gross domestic product of the world by a great margin and help many fields of our lives—from economy to health care (Latif et al., 2017).

Currently, there are Nokia in Europe and AT&T in United States, which are pursuing 5G. Other big players are Chinese telecommunication companies—since most of the electric and telecommunication hardware are produced and assembled in China nowadays, it makes sense that Chinese companies might benefit from the economy of scale and offer the best conditions and prices.

However, there are some national security issues connected with 5G networks. For example, the producers of hardware required for the deployment of 5G networks often find themselves under fire. When I was writing this book, the big ongoing scandal widely reported in the media was that of the national security risks to 5G networks that a Chinese telecommunication company Huawei was suspected of posing (Inkster, 2019).

Huawei's end-to-end approach to 5G, which provides hardware and software and ongoing operational support, is apparently what makes it as attractive as a solution and helps the company undercut the price of competitors. But it also means identifying vulnerabilities, delivering updates, making corrections, and designing and distributing updates of both hardware and software is in Huawei's hands. It combines the Microsoft or even IBM model which has given us personal computers and operating systems, with the use of phone calls or AT&T's approach to building telecommunication networks. But today we live in a world of virtualization and software-defined hardware. The old method of acquiring data and information is longer valid. Operating systems that make customers dependent on large end-to-end proprietary solutions are not the only way forward. The IoT is also a world of countless sensor and device manufacturers' control systems. Refrigerators, toasters, televisions, security cameras, machines, servers, servers, networks, smartphones, and computers—all of them connected to 5G and its successors, without the owner having a dominant market share—will mean the new giant leap and a digital revolution.

At the same time, there are some fears that being a Chinese company (albeit a private one), Huawei might also pursue a hidden agenda helping the Chinese intelligence. Many researchers claim that China has been the giant of cyber espionage and is believed to be the cause of data breaches in the United States, Great Britain, and Australian government agencies, including the meteorology office, the Commonwealth Scientific and Industrial Research Organization, or the Australian Parliament, just to name a few (Lindsay et al., 2015; Walsh and Miller, 2016). Others claim that China has been involved in cybercrime theft of intellectual property, trade secrets, and commercial trust materials from several companies around the world—as reported in the United States. Combined with the demonstrated intent to conduct widespread cyber espionage, China's intelligence law provides the opportunity to force Huawei to support state intelligence. Section 8.7 of the Chinese Intelligence Act 2017 obliges organizations and citizens to support, assist, and cooperate with the Chinese state intelligence (Parasol, 2018). China's intention to carry out extensive cyber espionage is therefore combined with the country's legal regulations. The obligation for organizations and citizens to provide assistance if necessary, means Chinese companies such as Huawei which increases the risk of supply chains over businesses in countries without any law specifically forcing them to cooperate with intelligence agencies.

In 2018 Le Monde newspaper published an investigation claiming that from January 2012 to January 2017 servers based inside the headquarters of African Union in Addis Ababa were transferring data every single night between 12 midnight and 2 a.m. to unknown servers in China. The media also reported that in addition to what they called a "data theft," there were microphones hidden in desks and walls that enabled to record any conversations at the premises (Maasho, 2018).

With such claims, either supported by the data or not, it is hard to fully concentrate on the effective deployment of faster and more reliable mobile networks that would allow faster data transfers and analysis and thence pave the road to the full-fledged IoT. Huawei offered very plausible technical solutions for a good price and now there are concerns about cybersecurity and national interests that might threaten the development of smart grids of the future. These doubts and controversies, raised by the mass media and social media all around the Internet, have very negative social implications. The claim that the IoT (and the IoE that is a part of it) might be compromised by some national governments or intelligence agencies leading to the personal data theft and cyber espionage, undermine the trust and believes of the ordinary people in smart grids and their usefulness for the good of the society. All these might lead to the pessimistic scenario when the deployment of the smart grids will be slowed down or even shut down at some point due to the security concerns. I will come back to this when discussing the future scenarios for the smart grids.

One can see that the 5G mobile networks are connected with the concept of cloud computing and data centers that enable this technology and ensure its smooth functioning. Nowadays, there is no need for each and every device to have an enormous computing power. Even though every today's smartphone has the computing capacity larger than NASA computes used for sending the mission to the Moon, most of the computing processes are done remotely—that is, using the cloud servers that are located elsewhere. Each time when we search something on Google, a smart meter sends the information on the usage of electricity from some home appliance and it is all transmitted to a data center to be processed and analyzed in seconds.

A data center or data storage center is a specialized building for hosting server and network equipment and connecting subscribers to Internet channels. The data center performs the functions of processing, storing, and distributing information, as a rule, in the interests of corporate clients—it is focused on solving business problems by providing information services. Consolidation of computational resources and data storage facilities in the data center reduces the total cost of ownership of the IT infrastructure due to the possibility of efficient use of hardware, for example, redistribution of workloads, as well as by reducing administrative costs. Most data centers are usually located within or in close proximity to a communication center or point of presence of one or more communication operators. The quality and bandwidth of channels affect the level of services provided, since the main criterion for assessing the quality of any data center is server availability time (uptime). The structure of the typical data center includes (1) information infrastructure, (2) telecommunication infrastructure, and (3) engineering infrastructure.

Information infrastructure includes server hardware and provides basic data center functions—information processing and storage. Telecommunication infrastructure provides interconnection of data center elements, as well as data transfer between data center and users. Engineering infrastructure ensures the normal functioning of the main data center systems. It includes air conditioning, uninterrupted power supply, burglar and fire alarms, and gas fire extinguishing system; remote IP control systems, power management, and access control.

Some data centers offer customers additional services for the use of equipment for automatically avoiding various types of attacks. Teams of qualified specialists around the clock monitor all servers. It should be noted that data center services are very different in price and quantity of services. To ensure the safety of data, backup systems are used. To prevent data theft, data centers use various video surveillance systems to restrict physical access. Corporate (departmental) data centers usually concentrate on servers of the organization. The equipment is mounted in specialized racks and cabinets. As a rule, only rack-mounted equipment, that is standard-sized enclosures adapted for rack mounting, is accepted into a data center. Computers in desktop cases are inconvenient for data centers and are rarely located in them. The main indicator of data center operation is fault tolerance; cost of operation, energy consumption, and temperature control are also important.

The services provided by data centers include: (1) virtual hosting, (2) virtual server, (3) dedicated server, (4) colocation, (5) rent of telecommunication racks, and (6) dedicated area. Virtual hosting means that large data centers usually do not provide a similar mass service due to the need to provide technical and consulting support.

Virtual server ensures the provision of guaranteed and limited part of the server (part of all resources). An important feature of this type of hosting is the separation of the server into several virtual independent servers implemented programmatically.

Dedicated server means that the data center provides the client with a rental server in various configurations. Large data centers mainly specialize in such types of services. Colocation is placing the client's server on the site of the data center for a fee. The cost depends on the power consumption and heat dissipation of the equipment being placed, the bandwidth of the data transmission channel connected to the equipment, as well as the size and weight of the rack.

Rent of telecommunication racks is the transfer of racks to the client for installation of own or client's equipment. Formally, this is a special case of colocation, but with the main difference in that tenants are mostly legal entities.

Dedicated area is a service when, in some cases, the owners of the data center allocate part of the technological areas for special customers, usually financial companies, which have strict internal security standards. In this case, the data center provides a certain dedicated area provided by communication channels, electricity, cooling and security systems, and the client creates his own data center within this space.

The network infrastructure of the data center is as follows: data center communications are most often based on networks using the IP. The data center contains several routers and switches that control traffic between servers and the "outside world." For reliability, the data center is sometimes connected to the Internet using a variety of different external channels from different providers. Some servers in the data center are used to run basic Internet and Internet services that are used internally: mail servers, proxy servers, DNS servers, etc. Network security level is supported by firewalls, VPN gateways, IDS systems, etc. In addition, traffic monitoring systems and some applications are used.

Optimization of data centers was one of the most discussed topics before the onset of the crisis, which brought it to the fore. First, such projects are high initial costs. The cost of building a data center in 2007 exceeded $11,000 per m^2, whereas in 2004 this figure was at the level of about $5000 per 1 km^2. According to a number of expert assessments, the construction and further operation of a large data center (for an enterprise level enterprise) can take up to $15 million or more over 5 years. Second, the content of a modern data center costs "a pretty penny." For example, 1 W of the useful power of a data center costs an average of 3 EUR.

Looking at the structure of the budgets of companies allocated for the maintenance of a modern data center, it becomes clear that there is something to optimize. Thence, in the first place are traditionally the costs of support, that is, in fact, to pay for the work of staff more than 40%. The cost of electricity payments usually occupies the second line. According to various estimates, this is usually 20%–25% (although, of course, much depends on the degree of congestion of the data center, e.g., with 100% filling of the racks, this figure can grow 1.5–2 times). No less significant are the costs of renting space for a data center—most often it is more than about 20%. Surely the crisis will force to adjust both these ratios and the costs themselves for maintaining data centers.

Certainly, the crisis will force to more actively resort to modern means of increasing the efficiency of data centers and reducing the costs of them. Moreover, the highest costs—the costs of building and running the data centers—should also be reduced.

A study was conducted by a specialist service provider in the field of construction of server farms Mortenson. The designers, operators, and owners of corporate data centers, as well as service providers in the field of information technologies operating in the markets of several countries, were surveyed. Mortenson experts agree that today electricity bills are one of the main items on the list of operating costs for data centers (Tarczynska, 2016).

The study showed that the vast majority of data center operators would like to achieve increased energy efficiency of the infrastructure under their control; 84% of data center owners feel the need to consider renewable energy sources (RESs), such as wind and solar farms, as tools to meet future energy needs.

The intensity of work on the Internet for users from all over the world continues to grow, which requires an increasingly developed infrastructure of data centers and additional generating capacity to meet electricity needs. At the same time, data center operators understand that the future belongs to RESs.

In recent years, there has been a rapid growth of large and geographically widespread data centers that require the Internet to support various services such as cloud computing.

Cloud computing is a method of interaction between a client and a server, in which client information is processed and stored on a remote server; reduces the hardware and software requirements of the client's computer.

The global energy crisis has become a serious problem for data centers, which contributes to lower energy use. For example, some data centers that produce "cloud" services are mega data centers, because they include hundreds of thousands of servers that need tens to hundreds of megawatts during a period of maximum energy load. Thus minimizing the power consumption of large "cloud" data centers has recently received a lot of attention from researchers.

In addition to high electricity bills, the huge energy costs of large cloud-based data centers can also have negative environmental effects (e.g., CO_2 emissions and global warming), due to their large carbon footprint. The reason is that most of the electricity produced worldwide is produced using carbon-intensive methods, such as burning coal. This energy, produced from conventional fossil fuels, is commonly called brown energy.

Therefore in order to reduce the negative environmental consequences caused by the rapid growth of energy consumption, many Internet service operators are beginning to take various measures to supply large-scale data centers for renewable (or green) energy. Unlike brown energy, green (or pure) energy is usually produced from renewable sources of energy, such as wind turbines and solar panels, and is thus more environmentally friendly.

For example, large services of Internet operators, such as Google, Microsoft, Apple, and Yahoo, began using renewable sources to power their data centers to reduce their dependence on brown energy. Because for data centers located in different geographic locations for which different RESs are available, depending on local weather conditions, it is important to quickly distribute service requests between data centers.

Unfortunately, due to the unstable nature of RESs such as wind turbines and solar panels, currently renewable sources are much more expensive than brown energy produced from conventional fossil-based fuels. While some data centers are trying to build their own wind or solar photovoltaic power plants, due to problems such as expensive investments and management, many services of Internet operators prefer to work with professional renewable energy producers, integrating green energy into the grid.

For example, the technology giant Google is investing billions of dollars in RESs to make its operations completely green. The company hopes to use solar, wind, and other forms of alternative energy to supply data centers around the world. In 2013 Rick Needham, director of Google Energy and Sustainable Development, told reporters that the company has invested over a billion dollars into 15 projects that possess the potential to produce 2 GW around the world. Needham also added that it was an economic necessity for Google to invest in renewable energy (Stromberg, 2013). In addition to that, it is well known that Google has more than 500,000 servers in data centers that are placed in different geographical locations and consume at least 6.3×10^5 MW per year (Chen et al., 2016).

Google has long been surprised no one by the powerful dynamics of increasing the use of renewable sources for supplying electricity to their data centers. Since Google is an active fighter for "green energy," it also has many patents in various fields, and shows openness and friendliness toward users. No less impressive is the company's data processing centers—this is a whole network of data centers located around the world. Given the fact that investments in the data center in the last year alone amounted to $2.5 billion, it is clear that the company considers this area as strategically important and promising.

Google does not build data centers. The company has a department that deals with development and maintenance, and it cooperates with local integrators who perform the implementation. The main concentration of data centers is in the United States, Western Europe, and Southeast Asia. In addition, some of the equipment is located in rented premises of commercial data centers that have good communication channels. By doing this, Google is strategically focused on moving exclusively to the use of its own data centers—the corporation explains this by increasing demands for the confidentiality of information of users who trust the company and the inevitability of information leaks in commercial data centers. Following the latest trends, it was announced recently that Google is going to lease a "cloud" infrastructure for third-party companies and developers, corresponding to the Compute Engine IaaS service, which will provide the computing power of fixed configurations with hourly pay for use. The main feature of the network of data centers is not so much in the high reliability of a single data center as in geoclustering. Each data center has many high-capacity communication channels with the outside world and replicates its data to several other data centers, geographically distributed throughout the world. Thus even force majeure circumstances like the fall of a meteorite will not significantly affect the safety of data.

Actually, it is clear that by building such huge data centers, a company like Google does not choose their location by chance. Clearly, there are some key criteria that the company's specialists first of all take into account such as the following ones:

1. Enough cheap electricity, the possibility of its supply and its environmentally friendly origin. Adhering to the policy of preserving the environment, the company uses RESs, because one large Google data center consumes about 50−60 MW—enough to be the sole client of the entire power plant. Moreover, renewable sources make it possible to be independent of energy prices. Currently used hydroelectric power stations and wind farms.
2. The presence of a large amount of water that can be used for the cooling system. It can be both a canal and a natural reservoir.
3. Availability of buffer zones between roads and settlements to build a protected perimeter and preserve the maximum confidentiality of the object. At the same time, the presence of highways is required for normal transport communication with the data center.
4. The area of land purchased for the construction of data centers should allow its further expansion and construction of auxiliary buildings or its own renewable sources of electricity.
5. Channels of communication. There should be several of them, and they should be reliably protected. This requirement has become especially relevant after the regular problems of loss of communication channels in a data center located in Oregon (United States). Overhead lines passed through power lines, insulators on which became for local hunters something of a target for shooting competitions. Therefore in the hunting season, the connection with the data center was constantly cut off, and it took a lot of time and considerable effort to restore it. As a result, the problem was solved by laying underground communication lines.
6. Tax incentives. The logical requirement is that the used "green technologies" are much more expensive than traditional ones. Accordingly, payback tax benefits should reduce the already high capital costs in the first stage.

Google has a long way to go before all information centers will be supplied with renewable energy. For a long time, only about 34% of the power consumed by Google is obtained from RESs. This was due to the fact that in some parts of the world it is simply impossible to obtain renewable energy. It was Google's long-term goal to get 100% of its energy from renewable sources of energy. Therefore the company was investing in innovative renewable energy projects that had the great potential to transform energy and help supply clean energy to various businesses and homes around the world and was able to reach its goal in 2017 (Google, 2017).

Google's major rival Apple released a report that says the company reached 100%, using only RESs in all of its data centers and at headquarters a year later (Apple, 2018).

The company's goal was to fully switch to renewable sources—solar, wind, hydro, and geothermal energy, investing in their own production on their territory. Apple claims that the investment pays off. RESs are now used in all data centers, enterprises located in Austin, Elk Grove, Cork, in Munich. All other company facilities around the world use only 75% of renewable energy and it is expected that this figure will increase.

Microsoft also plans to convert the power system and supply cloud services using renewable energy. A long-term contract for the purchase of electricity from Pilot Hill will help the company to provide a more confident power supply to data centers. The only problem that many operator services face is the use of RESs to the maximum extent, which does not allow them to meet their monthly operational budget. The use of RESs can put significant pressure on the budgets of Internet service operators, as the cost of electricity for operating data centers has increased, 20% or even more of the monthly expenses of these enterprises.

However, there exists a completely new system that allows for fast processing of requests in order to maximize the percentage of RESs used to power data center networks and data processing, taking into account the set budget of the Internet service operator. This is the first model of intermittent generation of renewable energy, for example, which is estimated in terms of real-world weather, electricity prices, and loads. This system can significantly increase the use of RESs in the supply of data center networks without exceeding a predetermined budget, despite the unstable nature of RESs, changing electricity prices and loads. For example, the available energy from wind turbines is modeled depending on the external wind speed, while solar-free plants are estimated by tracking the maximum power points per unit area of solar panels and temperature. On the basis of this model, it is possible to formulate the main objectives of the system, in which there are limitations regarding the quality of service that customers require—the frequency of RESs in different places, the peak power limit for each data center, and the monthly budget expenditures.

GreenWare is a system of partial use of RESs for energy supply to data centers. GreenWare quickly distributes data processing requests between data centers to maximize the share of RESs used to power the data center network based on time-varying electricity prices and the availability of RESs in accordance with the geographical location of the centers. GreenWare also guarantees the provision of the required quality of services for customers and the effective maintenance of electricity payments within the budget.

It is assumed that data center networks have a total budget of costs, which can be periodically determined by the operator of the Internet service for each budget period (e.g., a month). The local optimizer is supposed to be present in the network of each individual data center in order to quickly change the number of active servers in order to minimize the power consumption of the data center while maintaining the required level of quality of services. It also assumes that short-term weather conditions (e.g., at 1 hour) and configurations of wind turbines and solar panels of each data center are available. It is a centralized system that manages the data center network to maximize the use of RESs within the budget. Although such a centralized structure is typically used for management in data center networks, the system of partial use of renewable energy can operate in a hierarchical structure.

The main actions of the system of partial use of RESs to supply data centers with energy using GreenWare can be described using the following three main steps:

1. GreenWare calculates budget hourly costs based on the monthly budget of service operators and the cost of electricity already consumed in the previous period, as well as statistics on the load at the same hours in the past (e.g., the last 2 weeks).
2. Based on all changes, GreenWare runs an optimization algorithm to compute the required query (e.g., the share of load allocated to each data center) to the total percentage of renewable energy used to power the data center network within budget; the total cost of electricity is lower than the current hour's budget; and at the level of the required quality of the services provided (e.g., the required response time) for customers guaranteed.
3. GreenWare distributes incoming requests to data centers using a routing mechanism already operating in large-scale cloud data center networks. It was noted that fast request routing has already been implemented by many Internet service operators for matching requests with servers.

With the global energy crisis and environmental problems, for example, global warming, greening data centers is becoming an increasingly important issue in the operation of large-scale data centers for Internet service operators. This system assumes the maximum use of RESs for supplying data centers, based on the time-varying prices for brown energy and the instability of RESs.

This system saves the cost of data transfer from the data center to the offices due to partial use of RESs. The trend of using renewable sources to partially supply data centers in order to reduce maintenance costs can provide an additional competitive advantage to any company that is using it in its everyday business.

8.3 Smart cities and smart grids

Reliability and security of energy supply of the large urban centers depend not only from the development of the energy sector but also from the social efficiency and the quality of life of the population of these centers (Neirotti et al., 2014). This requires mass development and introduction of new organizational and high-tech solutions that ensure the quality of life of the population, the socially oriented development of the economy and the requirements of the functions of the capital status of the city.

The metropolitan environment for all subjects—citizens of the city, the social sphere, innovative technological, and financial business, government agencies—should have the qualities of a "smart" city that combines the interests of the individual and society. The concept of "smart city" is the development of socially oriented infrastructure, including "smart" power supply, smart grid system, "smart" environment, "smart" transport, "smart" home, and "smart" management. The principles of the smart city include the following parameters:

- houses, neighborhoods, and he whole areas in urban centers as well as local energy units;

- the autonomy of the levels of the city;
- social, business, and cultural self-sufficiency of the city;
- development of standards for green building;
- using the latest information and communication technology; and
- introduction of innovative energy technologies (the key energy efficiency), transport and construction (see Ramaswami et al., 2016).

Infrastructure device of the smart city involves the connection of individual sectors through information and communication technologies is carried out according to the principle of "system systems," realizing the compliance of the mechanism of integrated management (Lee et al., 2013).

A characteristic feature of "smart" systems is that their "intelligence" is provided by a variety of infrastructure options used, providing a controlled adaptation of the structure to changing external conditions, the active development of ICT systems and the presence of man as a subject and object of functioning and development of the "socially oriented metasystem" (SoM). This "triple SoM" (e.g., infrastructural, informational, and intellectual) is the framework of the "smart" metropolis, and energy as a system of not only life support, but also the life of the city as a "living" organism, should develop as a metasystem (Hefnawy et al., 2016).

As a matter of fact, the concept of IoE, in which each subject (consumer) not only chooses the information he or she needs, but also makes a decision on her or his own or collective energy supply, taking into account the interests of other members of society and the entire socio-natural environment, becomes a very important aspect for the smart cities. In addition, smart grid also becomes a part of the general smart city model even though it does not just unite material infrastructure objects (power supply system, transport, communications) and the upper level Internet information flows into a single intellectual complex, but also forms energy-information "system of systems" as a single man–machine ecosystem (Gubbi et al., 2013).

Therefore in the smart grid system it is impossible to consider the physical levels of power supply separately—the automatic control system as a superstructure over physical objects. There exists a single SoM that integrates individual infrastructure subsystems both by the types of energy used and by the degree of their localization and centralization. But the main idea is that they carry out multiagent (independent) management of individual objects while coordinating these actions at the level of coordinating common goals, strategic plans, and risk assessment of decisions that are made.

The unified energy-information system smart grid provides a compromise solution in the interests of all the actors involved in the general life support system and the life of the metropolis. The principle of autonomy and network-centric unification of both physical objects and their information models is based on the multiagent management of individual blocks and systems of group management of energy supply in the smart grid system.

Smart grid systems are being actively developed in world practice for the power supply of local facilities and territories. Their introduction was caused by the need to harmonize the conditions and modes of distributed generation, mainly with the help of RESs, in the general grid structure. Decentralized systems are not autonomous, and their connection to a common power grid structure requires a sufficiently developed information and control network built according to the intelligent smart grid methodology (Alanne and Saari, 2006).

With regard to the power supply scheme of large world megapolises, taking into account the annexed territories, the problem of using the smart grid becomes relevant both from the point of view of developing decentralized systems and their integration into the general infrastructure SoM, and from the point of view of increasing the social efficiency of energy saving.

The growing role of human beings in creating the IoE requires the development of new information technology and intelligent network-centric control algorithms. Network-centric management is a multiagent management of individual (local) objects while coordinating their operation with the help of the network energy-information infrastructure. At the same time, it is not the control actions themselves that are centralized, but only the general idea of management and risk assessment of the decisions made.

The presence of a single target network-centric architecture is a distinctive feature of most foreign projects in smart city and smart grid in major cities of the world. One has to mention the spectacular large-scale projects of the smart cities that were implemented in Stockholm (Sweden), Amsterdam (the Netherlands), and Yokohama (Japan) (Madu et al., 2017).

Most of these projects focus on the development of "green" energy using RESs, passive (zero-power) homes, the use of incineration, the development of electric lighting systems, EVs, and energy saving.

The suburban district of Stockholm—Hammarby Sjöstad is supposed to be made an innovative "city of the future"—with a complete renovation of the former industrial zone in order to create a "green city" in the megalopolis (Glazebrook and Newman, 2018).

The concept of a new town planning project is focused on creating a district heating and air-cooling system using electrical installations, using solar panels on the roofs of "passive" houses for local electric heating, reducing the number of cars with their replacement with bicycle traffic and electric transport.

The megaproject "Intelligent City Amsterdam" is also focused on the creation of a "green" energy company (a smart energy city). It is assumed that by 2025 up to 30% of energy (in 2012 it was only 6%) consumed by the city will come from environmentally sound energy sources. Over 10 million m^2 of roof area will be equipped with solar panels, and the office buildings of the Zuidas business park will be equipped with autonomous fuel and energy elements. The city will become a separate part of Amsterdam, including with its own energy complex and a developed intellectual network integrating the transport, social, and energy infrastructure of the region. The "Old South" district, which was considered one of the dirtiest and poorest districts of Amsterdam, is being rebuilt: a new "city within a city" with innovative business centers, office apartments and business-class housing is being built on the site of the former factories and slums. It also requires the reconstruction of the local power supply system with the transition mainly to power supply, carried out both by external cables and through its own power plants (Van Winden, 2016).

One of the most ambitious projects of energy-efficient smart city is the reconstruction of the power supply system of the megalopolis of Yokohama. The project includes various home energy management systems, building energy management systems, community energy management systems, and a dispatch control and data collection system (supervisory control and data acquisition).

A distinctive feature of the Yokohama Smart Grid model is the wide use of storage devices as an intermediate link between consumers and the city's power supply network from autonomous and external sources. The implementation of this system is expected at the expense of the funds of electrical companies supplying equipment, including control systems for energy facilities. The city authorities guarantee the companies' widespread implementation of the smart grid system with a positive operating experience of pilot projects.

Innovative development of energy in the cities of the world is on the way of the development of electric transport, fuel cells (FCs) for autonomous power supply, the transition from the "rosette" technology to the use of network and system batteries. And all these innovations are included in the structure of modern smart grid-powered cities.

Suffice it to say that over the past 3 years, 300,000 have been sold in the United States, and 100,000 EVs have been sold each in China and Japan, with most of the sales in 2015. Moreover, half of sales accounted for "clean" electric cars, and the other half on "hybrids." This is facilitated by a fourfold reduction in the cost of Li-Ion batteries (from $1200 per kWh in 2009 to $300 in 2015), with a further decrease to $80 per kWh by 2020. Due to this, the total cost of Nissan Leaf will be reduced from $35,000 to $15,000, and the total number of sold EVs will increase by 2020 to 17 million units.

In Moscow, at the initiative local government, the infrastructure (charging network) for EVs is being actively developed, which makes it possible to count on a fairly rapid growth in their number, first and foremost, public electric transport.

Progress in the development of Li-Ion batteries stimulates their widespread use as household and network energy storage. In the world, the production and use of FCs for autonomous power supply, especially in the fixed version, is actively growing. Their efficiency is 3.5 times higher than that of the internal combustion engine and two times higher than that of the best gas turbines. Therefore the production of FCs in the world in 5 years increased from 7000 to 35,000 units and their capacity from 60 to 170 MW. These FCs are supposed to be actively used as backup power sources for the most responsible consumers (in hospitals, business centers, banking structures, etc.). It is possible that they will find application for the power supply of individual objects in the zones of decentralized power supply.

All innovative solutions lead to the need to revise the infrastructure of the urban energy sector. In the world, this work is accompanied by the active development of a smart grid-based system. The cost of implementing intelligent networks has reached $20 billion, including the growth in the EU countries estimated to be at a rate of about 27%—25%, about 16% in the United States, around 12% in China, and about 30% in other Asia-Pacific countries. It is assumed that the development of smart grid technologies, among other things, will reduce the needs of energy-deficient regions in fuel and energy resources.

In Russian Federation, the development of a smart city concept began in its largest city and its capital Moscow in 2014 of the concept of an energy-efficient megacity smart city called "New Moscow" and performed following the instructions of the local government in Moscow involving a wide range of specialists and city planners. The concept uses the world experience in creating smart grid systems as the infrastructure foundation of smart cities based on the examples from other smart cities around the world described above. At the same time, special attention is paid to the territorial hierarchy of the smart city model, taking into account the new scheme of the infrastructure development of a megapolis with annexed territories (Drozdova, and Petrov, 2018).

The scheme of territorial development of smart cities contains the steps that determine the volumes and places of public and residential development, the development of the production sector in the associated territories and its transformation within the boundaries of the "old" city, as well as the development of transport, engineering, energy, and social infrastructure. One of the qualitatively new principles of the organization of development of the annexed territories is the cluster approach that helps to integrate the socio-production and infrastructure integration of the settlements.

The development of the cluster principle of forming the territorial-production development of the economy also implies the corresponding development of their energy supply systems. Previously, the connection of all facilities to the unified energy distribution network suggested an increase in the length of the network and caused additional costs for the construction of these network facilities, an increase in losses with a small load, and an increase in the accident rate of the power grid complex (Bünning et al., 2018).

In clusters, due to a combination of appropriate energy consumption and generation, emphasis is placed on the energy self-balance of these entities, which eliminates the need for the extensive development of the distribution grid complex. At the same time, there is still a need for intercluster integration of these energy centers along limited routes and high-voltage network routes with a focus on the future development of the load in new geographic nodes.

For all clusters, its own power supply scheme is foreseen, taking into account the characteristics of local consumers, including with the use of mini heat and power plants, while the centralized heat and power supply from the power plants will operate on the territory of the "old" city (Farzaneh et al., 2016).

In addition to the cluster approach, the interrelation of social—industrial settlements and their energy supply systems is viewed in a new way on the example of key links (objects) of a megapolis such as individual buildings, headquarters, and district formations. At the same time, centralization of external power supply is combined from both the local power complexes and the combination of heat and power (CHP) of the city, and autonomous power supply of a specific load of objects from local sources.

The integration of energy sources and responsible consumers requires the creation of appropriate self-management centers (multiagent management) of energy-information systems, as well as the development of integrating intelligent (multifunctional) smart grid, linking these centers into a common system. For various standard units of the megalopolises, these future systems will be universal (in terms of requirements for reliability and efficiency of energy supply) and qualitatively individual, depending on the structure of the objects themselves and the role of man as a subject of management. At the level of an individual consumer, the smart infrastructure can be represented as an integrated power supply system of a typical smart home.

8.4 Smart homes and smart grids

A "smart" house or an "intelligent building" is a qualitatively new object from the point of view of infrastructure support. It integrates various life support systems (electricity, heat, gas, and water supply) into a common engineering infrastructure complex; focused on large-scale consumer saturation with telecommunications and information services; as a mandatory element of development, it takes into account the availability of individual passenger transport, including the EVs (Wilson et al., 2015).

At the same time, the key features of a "smart" home are in a developed system of monitoring and control over compliance with climatic parameters, lighting, security, and energy-saving and effective (automatic) energy consumption.

"Intellectual building," in addition to these functions, distributes energy and information flows to individual apartments, selects local and external sources of collective energy supply, as well as the possible connection of an object with an external network and issues free (with an excess of its own generation) energy network (Tuominen et al., 2012). The principal feature of such an object is the presence of a storage drive through which the consumer is powered and communicates with the external network.

With the help of these devices one can add power for an individual energy economy during the period of load sinking during peak hours in general distribution networks. Very often, the electric power storage device installed at home or in the country, allows to significantly improve the quality of power supply. In addition, electrical energy storage devices used in individual households constitute the main power sources in emergency situations and centralized shutdowns of electrical networks.

Household energy storage devices are installed mainly in private houses and are constantly in a state of connection. This allows for quite a long time to get enough electricity for lighting and other urgent domestic needs.

However, electric power accumulators perform functions wider than those of a conventional battery. In addition, they create the possibility of direct connection to batteries of consumers with a power supply system at a constant current, bypassing the inverter stage. They represent complex, integrated structures that can not only accumulate energy, but also transform it, making it suitable for further use. These devices occupy one of the leading places in the market of alternative energy devices. They are based on lithium batteries. They consist of a charger or charge controller, a voltage converter (inverter), and a control system. The design of the drives allows you to replace a large number of equipment for emergency systems and alternative power supply. Most models are designed to work not only from the fixed network, but also from solar panels. Their average output power is around 5 kW.

In the future, the building structure itself should take into account their infrastructure requirements and capabilities. There is no need to heat the entire cubic capacity of buildings, to illuminate the entire area of the premises, since it is enough to provide the necessary life support requirements in places where people are constantly located. Moreover, the circulation of energy flows should be ensured by appropriate placement of bedrooms and living rooms, kitchen and auxiliary premises, sources of light and heat, fresh air, and other means of life support.

In connection with the development of new territories, as well as the reconstruction of the residential and industrial sphere in the "old" city borders, the smart city becomes the key object of any smart city. New smarts devices, sensors, and meters will be able to provide high reliability of energy supply of high-rise buildings, saturated with various communications. In order to do this, they themselves should have not only a developed network of external energy supply, but also their own power sources.

These power sources are batteries that are also needed to ensure peak power consumption by objects of continuous use. The high-tech equipment of the business quarters in many European, American, and Asian megapolises with its own energy sources will reduce dependence on external power supply, reduce the number of substation points and supply networks, reduce losses, and, most importantly, increase the reliability of power supply to responsible consumers. Quite often, local cogenerators are connected to an external AC power supply network. The system is supplemented with rectifier converters, which allows the use of modern energy-efficient direct current installations (adjustable motors, FCs, etc.) on its premises. In addition, sometimes inverters are applied in order that own sources of direct current could transfer excess energy to the network. Such projects are also characterized by the presence of electric transport with its own charging facilities (direct or alternating current), as well as the presence of rechargeable batteries (both individual for individual process units and collective ones for the "smart" quarter as a whole). All these objects are connected to a common system using smart grid information management systems and representing an example of IoE.

On the premises of the given smart district in the smart city, the control center performs various functions of infrastructure, social and service provision of the interests of residents and business entities in the district including the following:

- remote monitoring of the state of engineering networks;
- electronic window of access to city services;
- apartment accounting of resources;
- remote control of home devices;
- use of information from a three-dimensional digital model of the city;
- information services based on the user's specific location;
- analysis of the movement of people for adequate generation and distribution of resources (heat, water, and electricity);
- control the level of street lighting;
- control of room lighting based on the schedule and the level of natural lighting;
- management of ventilation of buildings for weather conditions;
- management of climate systems of buildings by occupancy;
- traffic forecasting;
- informing the public about the movements of public transport; and
- remote monitoring and surveillance of potentially dangerous areas of the city.

More functions might be envisaged for local businesses and entrepreneurs residing in smart districts of the smart cities:

- automatic shutdown, the accrual of fines or the regulation of the consumption of defaulters;

- individual regulation of consumption and pricing depending on customer loyalty (level of accounts receivable, advance payments, overdue payments, etc.), as well as taking into account meter calibration;
- automatic informing of all interested services and persons (including the end user, the apartment owner in case of rent, etc.) about possible incidents (accidents, failures, etc.) and about the status of settlements with the consumer;
- predictive repair of equipment to prevent downtime;
- collecting and analyzing data to forecast the generation of resources by generating organizations;
- predictive repair of equipment to prevent downtime;
- forecasting the level of maintenance costs based on the analysis of resource consumption;
- providing, based on analytics, forecast data on consumer preferences and expectations;
- procurement and logistics based on consumer demand analytics; and
- providing asset and real estate management.

The three-level territorial model (buildings, quarters, districts) is integrated into a single whole two-level functional (energy-information) metropolis system, including the collection of consumer data, operational control and risk assessment of multiagent management, and the integration of various energy sources and distribution networks and substations.

The unified energy-information system is able to provide compromise regulation in the interests of all entities involved in the process: the grid company, consumers, distributed generation also create the technological basis of the local markets for regulating regime parameters in the network. The multiagent management of individual energy blocks and group energy management systems in the smart grid-based system is centered around the principle of autonomy and network integration of both physical objects and their information models.

Increased customer requirements for the level of service inevitably lead to the expansion of the range of services provided by energy companies, as well as to ensuring the conditions of reliability and efficiency of the entire system, to the introduction of new administrative, financial and payment mechanisms in the relationship of all economic entities of the energy market.

A comprehensive solution of these tasks in a common "system of systems," which is the energy of the metropolis, requires the development of an intelligent smart grid, which considers both the multiagent management of individual objects and subsystems, and their coordination.

Smart grid is an intelligent human—machine energy-information symbiotic metasystem, which, on the one hand, according to the SoM principle, integrates certain types of power supply at the consumer level into the overall customer-oriented power supply system. On the other hand, it unites the consumers, including the "active" consumers (generating their own energy), infrastructure distribution network, and external supply centers to a common energy complex.

Third, and most importantly, it provides network-centric management of the operation and development of energy. At the same time, multiagent management is combined with the help of infrastructure networks with centralized management in terms of forming common tasks of reliable and efficient energy supply, minimizing functional costs and investment risks for the development of the system. The structure of the smart grid system includes:

- a block of organization (interconnection and coordination) of centralized and decentralized power supply to consumers (at the level of individual facilities, district and cluster formations);
- information block, including the system of organizing information flows from the consumer to the system administrator and to the power supply company (demand for energy services, connection orders, quality requirements), control and management of power flows of energy carriers, financial conditions of energy supplies;
- a situational unit for assessing the state of energy security (resource sufficiency, financial affordability and technological attainability of specified requirements), as well as reliability, including possible damages from interruptions in energy supply, and efficiency (technical, economic, and social) of the city's energy;
- a block of management of perspective development of the fuel and energy complex of a megapolis;
- self-monitoring, self-healing, and self-development unit of the smart grid information management system; and
- a model of organization and interaction of business entities, business structures (including financial ones), and municipal and city government bodies in the energy sector of a megacity.

The integration of the object-based (territorial) smart city energy supply model as well as the functional model of communication between physical and information-controlling objects of the smart grid allows to implement the general principle of researching complex problems: from ideology (shaping the smart city image) to methodology (using the SoM principle defined above) and technology (smart grid design). At the stage of technological design of the smart grid, the key tasks of forming a new image of a "smart" city should be solved—its convenience and attractiveness for residents (social efficiency), ensuring the reliability and safety of infrastructure life support (technological efficiency), and cost reduction (economic efficiency).

The principle of SoM integration in the transition from centralized energy supply systems to stimulating the development of decentralized systems allows changing the conditions and requirements of an efficient energy supply in terms of reducing overall costs, improving the reliability and manageability of the energy infrastructure. Algorithmizing the smart grid design comes down to the fact that for all levels of operation is carried out using the following six principles:

1. Accessibility—providing consumers with electricity, depending on when and where they need it, and depending on the quality paid.
2. Reliability—the ability to withstand physical and informational negative impacts without total outages or high costs of restoration work, the fastest possible recovery (self-healing).
3. Profitability—optimization of electricity tariffs for consumers and reduction of system-wide costs.

4. Efficiency—maximizing the efficiency of using all types of resources and technologies in the production, transmission, distribution, and consumption of electricity.
5. Organic and environment-friendly attitude—reducing negative impacts on the environment.
6. Security—preventing situations in the power industry that are dangerous to people and the environment.

The use of smart grid technologies can help ensure energy security and energy efficiency in the development of the energy system of a megapolis both in the sector of energy production and transmission, and in the distribution sector and energy supply to consumers:

- New materials for power energy and electrical engineering equipment, as well as design and layout solutions will reduce the usable space occupied by the equipment, improve the nominal parameters, as well as the resource and duration of the service interval.
- Powerful voltage converters are capable of controlling the flow of active power in a complexly closed electrical network and reduce the level of short-circuit currents, as well as carry out voltage control at the connection nodes, including to ensure the quality of electricity.
- Effective storage of electrical and thermal energy will smooth out peak consumption in the power system and ensure uninterrupted power supply to end users.
- Sources of distributed generation based on reconstructed boilers will increase the reliability of local power supply systems with the most efficient use of energy resources.
- Automation of the electrical network and new switching equipment will allow for rapid sectioning and reconfiguration of the network, accelerating the restoration of power supply after violations.
- Phasor-based vector measurement systems provide qualitatively new information on transient processes in the power system.
- Digital metering systems for consumers provide ample opportunities to manage consumption.
- Technologies of parallel computing and distributed control allow assessing the risks of disruptions in the operation of the power system and developing adequate response actions in real time.
- Developed information systems for diagnostics and control of equipment condition provide an opportunity for a flexible approach to determining the allowable load and the need for maintenance.

At the same time, the task of introducing new technologies of the smart grid requires a comprehensive solution to the issues of organizational interaction, financing, regulatory support, etc., in the framework of the "smart" city project. The implementation of such a project requires the wide involvement of energy companies, equipment manufacturers, residents, large consumers under the auspices of the city's power structures with a single project management system. In addition, project support requires information support, including a wide range of educational and training programs in the field of energy efficiency and environmental protection, managing a "smart" city throughout its life cycle.

The formation of a new energy image of a megacity and the construction of a "smart" city-wide energy supply system requires a revision of the organizational and legal structure of the city. This model simplifies the old pattern of relationships between economic entities through the development of the information sphere, linking all the participants in the unified energy supply system of the metropolis. The energy efficiency of a megapolis is the state and development of the energy of the smart city with its urgent priorities:

- customer focus (in the interests of the consumer and, above all, the population);
- infrastructure (taking into account both centralized and decentralized power supply); and
- reliability (due to a combination of external and own energy supply of city-forming objects).

Energy efficiency (technical, economic, and social) is the integration of all electricity, heat and gas supply systems into a single system with a focus on the preferential development of electrification, the development of a "new electric world" in the housing and social spheres, in transport and in the manufacturing sector. "The New World of Energy" is an active electric consumer, where electrification is used as the most effective means of ensuring the quality of life and productivity.

Electric energy that is distinguished by its versatility, convenience, and controllability is the main priority of generation (both at large CHP plants and at local energy complexes in social and industrial clusters), distribution network, and consumption within the smart grids of the future.

Manageability in the electric sector of the urban infrastructure is achieved in the most efficient way using the smart grid design technology—a "smart," intellectual (due to the active human participation), and energy-information (physical and information-controlling) system interconnected with each other.

At various levels of the territorial-functional structure of the integrated energy supplying system, there are various types and objects of the smart grid that are considered: a "smart" house, a "smart" quarter, a "smart" district, or a "smart" city. There are some differences in the degree of equipment of these objects with their own energy sources, batteries in household system drives, developed personal and public electric transport, their centers dispatch control and management.

All in all, it becomes clear that the presence of new functions in various economic entities operating in the energy system of the large cities and the need to coordinate their actions in current activities and future development requires reformatting and a general scheme for managing energy saving of world metropolises, combining the interests of individual multiagent management entities and their overall coordination.

8.5 Prosumers and smart grids

With the smart grids firmly occupying our daily routines, the new class of energy users is being forged. This class is called "prosumers" (the word is made up from the words "consumer" and "producer"). However, what does it mean apart from making up the new words in the English vocabulary?

In recent years, the decline in conventional energy sources has led to a change in the energy sector. Today, households and other energy users can produce and consume energy. They can also store the surplus for future use or send it to the grid to share it with other energy users. As a result of the transformation, the smart grid has resulted in the rise of the new class of "prosumers," or the individuals, households, or the communities of both proactively contributing to the energy supply. The term "prosumer" refers to the fact that renewable energy is generated in the home environment and that the surplus energy is stored for future use or that energy consumers who are interested in the smart grid can actively participate in its functioning. Therefore the goal of such prosumers is to produce and use energy, as well as share and distribute excess energy to other users on the grid.

In general, all of these definitions are for energy-producing consumers who consume energy on the grid. Prosumers appear to be quite different from the usual energy consumers: compared to the traditional consumers who consume power from the grid, prosumers generate, consume, and actively transfer or store excess power (it is possible to store excess energy by using the energy storage systems for future use, or to sell excess power to the network or nearby consumers). The existence of prosumers is closely connected with that of smart meters which are used during power generation and integrated with energy management systems, as well as energy storage systems, EVs, and grid systems for efficient integration into the smart grid.

A simple comparison between smart grids and conventional grids show that the latter are improving the efficiency of the energy system in different ways. This includes improving the operation of one's home through smart devices and communication technologies, providing storage space to manage power fluctuations, and balancing local demand and supply. The shift from passive consumers to active consumer operators has maximized a variety of economic, operational, and environmental benefits in microgeneration, demand reduction, response to demand, data management, and energy storage. Such changes include new technologies (smart metering and advanced measurement infrastructure, energy screens, and smart appliances), the implementation of low-cost energy conservation measures, government incentive programs to encourage participation in the energy system and a strong public attitude to mitigate harmful environmental impacts.

It can be shown that every prosumer faces four basic choices: (1) arrange an energy storage system and store the generated energy locally (e.g., within the household or on it premises); (2) rent external energy storage space; (3) sell the generated energy straight away at the local market (e.g., local prosumer community); or (4) sell the generated energy to the central market (e.g., at a special tariff).

Prosumers are truly the new species in the smart energy evolution. In very simple terms, they are the people (or households) who not only consume energy but also create it and perhaps (quite often, in fact) trade it with other prosumers. In the future, when each and every one of us would be able not only to buy energy from the smart grid but also to sell it back to the grid (and not even that—see Chapter 5, Peer-to-peer markets and sharing economy of the smart grids, which explains the principles of P2P electricity markets and trading), new economic incentives will be created.

However, one should not forget that human beings are often opportunistic and selfish. Kashintseva et al. (2018) describe consumer attitudes toward industrial CO_2 capture and storage technologies [called "Industrial Carbon Capture and Storage" (ICCS)] highlighting the "not-in-my-backyard" (NIMBY) attitude most consumers exhibit. Their study shows that the general public does not put matters regarding the environment and global warming as their priorities of interest. The environment is therefore not a pressing concern to the majority of the public, and as a result, even among those few who are concerned about the environment, they are not entirely concerned about the effects of global warming. A vast majority of people in the world have not heard about the environmentally friendly technologies and are also least concerned as to what it might be beneficial or used for. Overall, there are great challenges that are posed to the environment especially when it comes to the realization that many people are less concerned with environmental issues. When people are explained to about the effects of greenhouse gases such as CO_2 to the atmosphere and the general idea of using novel technologies to prevent the damages to the environment, a majority of people seems to support the idea at first. In some countries like China that are quickly adapting the trend of energy-saving technologies, the acceptance of CO_2 emissions and the deployment are perceived by the general public as the contributions to the positive image of the Chinese government. However, when the technical details emerge and people start to understand that saving energy or "going green" is not without a cost (and often a sacrifice on their side of the bargain), their enthusiasm tend to subside.

In a value chain, the end-consumer is the last entity; they are the people or the entities that are the users of the products goods, or services, produced by a firm. In the 21st century, the end-consumer is more knowledgeable and demanding; more educated and has access to diverse information from various sources (Carley et al., 2012). With the increased levels of awareness created about the risks that the increased levels of carbon emissions could pose to the well-being of the human race, and the increased levels of sensitization on the need for environmental management, the end-consumers, have increased their levels of environmental stewardship (Hoffman, 2005) Furthermore, at the individual level, end-consumers are making purchasing decisions based on the reputation of companies in relation to environmental management strategies, implying that early adopters in this field, are poised to enjoy increased business from the modern-day end-consumer (Lubin and Esty, 2010). Finally, the global consumer is increasingly playing an activist role, where they actively engage each other and corporates in the manufacturing sector to agitate for increased awareness and implementation of strategies aimed at managing the carbon emissions from productive activities (Chen et al., 2018). Evidently, from the foregoing, the modern-day end-consumer has a positive attitude and engages in activities that foster environmental stewardship and will support initiatives such as the implementation of ICCS and other related projects that mitigate the growing CO_2 emissions.

Nevertheless, there is also a plethora of issues related to the general public and end-consumers' attitudes toward ICCS technologies and their deployment. There are lots of cases where the clear explanation and communication between the stakeholders and industries and the end-consumers are required. The general public needs to be made aware of all costs and benefits of the ICCS, as well as about the advantages it presents and the outcomes in terms of halting the CO_2 emissions in the short run by the large-scale application of energy-saving technologies. Quite often the communication goes wrong and the pros and cons are not explained correctly. For example, Broecks et al. (2016) demonstrate that people find arguments about climate protection less appealing and persuasive than normative arguments or arguments about benefits of CCS for energy production and economic growth. Furthermore, Kraeusel and Möst (2012) investigate the level and influencing factors of social acceptance of CCS on the example of Germany and find out that the attitude toward CCS is neutral and the level of willingness to pay for CCS technology is much lower than for renewable energy.

Moreover, it might be that the debate on ICCS technologies with the general public and end-consumers should be done on the microlevel: some findings demonstrate that small-scale engagement processes might present a viable alternative to standard community consultation techniques for engagement around the siting of CCS facilities (Coyle, 2016).

According to Wennersten et al. (2015), the major barriers for implementation of ICCS on a large scale are not technical, but economic and social. The key challenge for ICCS is to gain wide public acclaim, which is likely to shape up the future political attitude to it. Such an approach requires a transparent communication about safety aspects early in the planning phase and dealing with potential disasters and hazards such as major leaks of CO_2.

Kashintseva et al. (2018) show that the acceptance of novel solutions for decreasing the amount of CO_2 declines if an individual resides in a region with at least one potential ICCS storage site. The acceptance of these technologies is about three times higher for females than for males. Moreover, it appears that other socio-demographic characteristics seem to have no effect on the willingness-to-accept "green" technologies. An interesting factor is the presence of coal mines in the region or country—when coal mines are presents, the willingness-to-accept sustainable technologies declines. Probably, this is happening due to the fact that when familiar with coal mines and the pollution they cause, consumer associate ICCS facilities with them and extrapolate their perception of coal industry to CCS, which is not always correct.

In addition, adding one more ICCS site to a country or region negatively impacts the overall willingness-to-accept ICCS products and technologies (the reduction of about 20%). In addition, it appears that spillover effects also play a significant role: living next to the region or a country with potential ICCS products and technologies reduces the acceptance level by 40%.

In addition, one needs to explain the effects of ICCS site, as well as the number of ICCS sites, and ICCS neighborhood in greater detail. First of all, it appears that the presence of ICCS sites in the respondent's immediate neighborhood appears to reduce the level of acceptance in a relatively large geographic area. Second, the larger the number of ICCS sites in the country or region, the lower is the willingness-to-accept ICCS in a given location. Finally, living in a country or region bordering another country or region with ICCS sites present, also reduces the consumers' support for accepting similar facilities in their region. NIMBY attitude is prevailing in all of the above cases and in many countries, including the developed economies of the European Union. This social attitude represents a very negative factor that might hamper the development of energy-saving and environmentally friendly technologies and products worldwide. It appears that consumer attitudes toward these technologies might be influenced by a plethora of factors, but overall people remain rather opportunistic when it comes the general good for the society as a whole. Moreover, one should remember that renewable energy schemes offer many monetary incentives and thence represent a potential target for some less law-abiding individuals.

8.6 Some futuristic scenarios and visions

In this section, I will attempt a bit of futuristic predictions for the smart grids of tomorrow and the challenged for the future. My predictions only concern the nearest future (the next few years to come)—it is difficult to envisage anything beyond that due to the short span of the human life and unprecedented changes in technology, economy, and geopolitics that might happen very soon.

It has to be stressed that I am not a fortune-teller of any kind, so the best way to envisage the future will be through the system of good old scenarios that are attempting to reflect all possibilities that might emerge: an optimistic scenario, a realistic scenario (also called "a baseline scenario"), and a pessimistic scenario.

My predictions are based on the available data on the volume of investments into the smart grids in the most dynamic regions of the world: India, Japan, United States, European Union, and China, as opposed to the rest of the world (RoW). It appears one can approximate the path and the speed with which the smart grids of the future are to evolve in the next years to come using the financial data—the amounts of money can be approximated as the pace of growth and evolvement.

According to IEA (2019), investment in smart grid technologies grew by 12% between 2014 and 2016 even though the investments into smart distribution networks are growing slower (about 3% in 2017). The progress in some key elements of the smart grids such as smart meter deployment varies in different countries. It becomes apparent that regulatory change and new business models are required for the smart grids to play their critical integration role in the transition to the carbon-free economy.

Let us look what is going to happen to the smart grids of the future and what it might mean for the human society further down the 21st century. Fig. 8.1 presents three future scenarios for the development of smart grids using the volumes of investments into these grids in various regions and countries across the world.

The methodology applied for building these scenarios is quite simple and straightforward. I took the IEA (2019) data for the investment into smart grid technologies for the several years expressed in billions of USD. From each data pattern, there emerges an obvious trend. For example, the investments into smart grid technologies in India for the past 2 years are growing at a steady but slow pace of 2%. In the case of China and United States this trend was around 34% and 22%, respectively, while in the case of European Union this pace was around 15%. There are differences between countries and regions (one cannot regard European Union as a region, even though it makes the world's largest existing economy). Even in the European Union, the disparities are enormous which is obvious when single Member States are scrutinized. For example, when the deployment of smart meters is concerned, Spain and Italy are used as the examples of the countries deploying them massively on their territories (IEA, 2019).

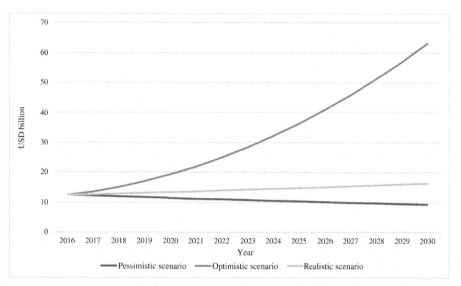

FIGURE 8.1 Future scenarios for the investments in smart grids (2016−30). Source: *Own results based on IEA, 2019. Smart grids: tracking clean energy progress. Available from: <https://www.iea.org/tcep/energyintegration/smartgrids> (accessed 20.05.19.).*

I remember that differences between various EU member states from the time when I worked as a Research Fellow at the University of Nottingham taking care of the EU-funded project called ISAAC (Integrated e-Services for Advanced Access to Heritage in Cultural Tourist Destinations). The project brought together researchers, ICT companies, city authorities and cultural institutions from the five EU countries and Russia and lasted from 2006 until 2009. The main objective of the project was to develop a novel, user-centric ICT platform that would provide tourism e-services sufficient to meet the needs of both tourists and citizens in European cultural destinations through facilitating a virtual access and stimulating learning experience of European cultural heritage assets before, during and after the visit. In the course of the project, we closely worked with the municipalities of the three European cities—Amsterdam, Genoa, and Leipzig. We also conducted a large-scale research that included a questionnaire survey administered in all three cities with an overall number of more than 5000 respondents. Our findings showed that people's views on e-services as well as their familiarity with these services differed tremendously from city to city (Chiabai et al., 2014). It appeared that Amsterdam was the most advanced city in using e-services and its residents and stakeholders were the most advanced Internet users who knew exactly what should be done and how to make the best use of the information and the Internet. Leipzig came the second with slightly less advanced users and fewer tourism e-services and Genova came third. This example clearly shows that in the case of the European Union, the methodology that is looking upon the bloc as if it were a single country, might be quite misleading.

Looking at the trends depicted in Fig. 8.1 one can see some obvious trends and developments. For this analysis, I have deliberately chosen a time horizon until 2030 (further predictions including decades into the future would be quite irrelevant since the number of unpredictable factors that might influence the situation is likely to increase with each year). Now, let us go through them quickly, scenario after scenario, and explain the main outcomes and implications.

1. Optimistic scenario: In this scenario, it is assumed that the development of smart grids in the future will follow an increasing pace. Using the most optimistic estimate of smart grids growth (about 30%−40% annually), the resulting figure would be around $60 billion of total investments into the smart grid technologies. This scenario is, of course, too exaggerated and is unlikely to come true. However, it shows some interesting dynamics and in case new technologies is going to emerge and boost the development of smart grids even further, perhaps the $60 billion figure will be not only reached but also overtaken.
2. Realistic scenario: In this scenario, I assume that the investments into the smart grid technologies are going to increase by about 2% annually (slow and steady pace). With these speed and dynamics, by 2030, the investments into the smart grid technologies would oscillate around $20 billion.
3. Pessimistic scenario: Declining by about 2% annually (slow and steady downturn). With these speed and dynamics, by 2030, the investments into the smart grid technologies would fall below $10 billion.

One might ask: Which of these three scenarios I believe would happen? Or perhaps none of them would? The truth is nobody knows that for sure. Trying to predict the future is a difficult task because the number of variables in the system one is trying to solve is too big. In addition, the links between variables are likely to follow the principles of the chaos theory (similar to the attempts to forecast the weather).

Let me give you an example: imagine a stable thermodynamic system—a hot cup of black tea. Its future state (temperature) can be safely predicted at any time by describing the convection equation whereby the tea will give off heat to the surrounding environment by auto-mixing—the cooled layers of tea will sink along one side of the cup, the hot ones rise along the other (a circular motion will occur inside the liquid). After some time, the tea temperature would be equal the ambient temperature and the dynamic system would be fully stabilized. Knowing this, one can move my cup of tea to any time period by setting the desired state for it, let us say adding milk and cooling it to the right or desired temperature. However, not everything is that simple as it seems. Under certain conditions—for example, the intersection of the parameters of the external (room) and internal environment (cup of tea) convection can cause oscillatory movements in the system or even chaotic processes. By adding milk to my cup of tea, I could involuntarily cause chaos in the system and instead of transferring the cup to the future for 1 minute, it can be transferred to, for example, 30 minutes (the so-called butterfly effect). At the same time, this movement will be absolutely unpredictable but also limited to some time frames (boundaries of the attractor). Moreover, instead of the future, a cup of tea can move into the past (another butterfly effect but now moving to the opposite direction). And at the same time, a cup of tea will steadily move along the past, describing certain spirals, until randomly goes to the future again (the butterfly effect does not reveal itself). Such movements in time from the past to the future and vice versa will look like water rotation on wheels.

Moreover, temporary loops are possible when the same day will be repeating itself over and over for our cup of tea. Moreover, for a cup of tea itself, everything will also be fatal in terms of predicting or changing its future.

Therefore predicting the future is not an easy business. Nevertheless, some general ideas and remarks might be applied to the future of the smart grids. Personally, I believe in the realistic scenario in which the growth of smart grid technologies will be steady but slow. There are so many factors that will likely to slow down the rapid development of such novel technologies as smart meters, EVs, or P2P electricity markets based on the principles of sharing economy. I think one can distinguish political, economic, and social reasons. Let us look at the political reasons first: the growth of Internet and ICTs is going faster than the legislative norms that should be there to regulate it (even though some people, hackers, and political activists alike, proclaim that Internet should be free of any censorship or regulation). With the continuing deployment of smart grids that will be operating on 5G (or perhaps 6G or even 7G) networks, the risks of cyberattacks and hacking are going to increase. While in the case of social networks, one could become a victim to an identity theft or phishing, in the case of energy infrastructure and facilities, the results can be fatal for the large groups of people (just imagine blowing up the nuclear power station and causing something like a Chernobyl disaster that happened in 1986 due to the human error). Therefore as smart grids are going to become ubiquitous, the security concerns are going to grow as well. As a result, new legislation and regulations will be passed with the purpose of restricting the free exchange of information and data. This would mean restricting the smart grids or slowing down their development.

The second reason is economic. As smart grids will be growing in size and in scope and absorbing more and more sensors, storage devices, communication hubs, data centers, solar panels, wind turbines, and the like, they will become more and more expensive to build and to take care of. In some cases, the maintenance cost might become higher than economic benefits. Another case is big data—the amount of data from smart meter measuring or various sensors in the smart energy grid is going to be enormous. Collecting, processing, and analyzing all this data is going to require massive computing power. Some of the work might be done by computers and algorithms (e.g., AI); however, many tasks are going to require humans. It has long been predicted that in the next 10—15 years a new occupation is going to appear—data analyst in the IoT. These people will search for irregularities in the date and pursue new solutions for optimizing the smart grids, so they are going to have high skills and are going to be paid quite well. All these is going to be an additional financial burden for the economy of the future.

Finally, the third reason is social: In spite of all advantages flowing from the smart grid technologies and intelligent energy systems, many people have troubles accepting it. It is more about the thought of a higher power (or a superior intelligence) that is effectively taking decisions about which power line should be shut down during the storm, or which electric appliances should be shut down during the night, than the rational belief in the potential of the smart grids. All in all, these three reasons might mean that smart grids and energy systems of the future are not going to be here as soon as some of us expect.

In addition, there are two key factors that can help the smart grids concept and vision to be accepted by the general community:

1. Benefits for the producers, consumers, and users alike from intelligent and efficient management of energy distribution should be made explained to the general public, made widely available and advertised.
2. A wide set of accessible applications for everyday energy interactions and market scale operations should be created and offered on the market.

There are many benefits that manufacturers, consumers, and users can achieve by managing energy distribution intelligently and efficiently. Comprehensive support for a wide range of applications available for both daily energy interactions and market operations should be provided as open-source application and commercial ones. Most smart grid scenarios include measuring one or more physical amounts that characterize the way energy is generated, transported and consumed in homes and other types of facilities. However, one needs to remember that energy is not necessarily limited to electricity but may also include other sources such as natural gas. Depending on the scenario, the amounts to be measured and the frequency of the measurement can vary considerably. For example, a house can be monitored in many ways—there can be periodic reports on the energy consumption generated periodically and an algorithm can decide what should be done to optimize the use of energy and to increase energy savings.

8.7 Conclusion and discussions

Overall, it seems that the design and implementation of an intelligent grid must include broader social and cultural considerations to ensure the success of intelligent networks. The complexity of future intelligent energy systems is due to the multitude of interactive players who operate as independent decision-makers and stakeholders with stand-alone behavior, goals, as well as attitudes.

While many studies have analyzed the purpose and functionality of smart grid systems, the intelligent grids themselves are only one system in the "smart system" of the IoE that, in its turn, is only a part of the broader system of IoT. Even though many current studies focus on the technical functionality of intelligent networks, they should be considering "a system within systems," with many self-propelled components responding to various economic and environmental problems beyond pure operations.

Investments in maintenance, capacity expansion, and innovation are needed to keep infrastructure systems functional and to keep pace with the changing needs of users and new public priorities. For example, the flexibility required to adapt to the variability and unpredictability of renewable energies can be achieved by investing in the production capacity with a regulated flexibility (especially gas-fired units) or by increasing interconnector capacity with adjacent networks. If, for example, EVs are integrated into the system, the flexibility of the equipment will be further enhanced by the decentralized storage capacity of the electric car batteries.

Moreover, commercial enterprises have also become the players in other segments of the value chain that need to establish sound business arguments to justify their investments, including smart grid development. Thence, it appears that investing in intelligent networks within the market paradigm will differ from investing in the old public monopoly paradigm.

Smart grids are going to bring fundamental changes to the energy markets of tomorrow and there are so many ways we are going to experience that both in our work and in our everyday lives. However, it might be that they it will take us some time to fully get used to their existence and operation, as well as to their interfering with our human lives that are often based on irrationality, laziness, procrastination, and emotions. Big data does not lie, and algorithms do not cry, but the cold machine logic might take some time to sink in.

References

Alanne, K., Saari, A., 2006. Distributed energy generation and sustainable development. Renew. Sustain. Energy Rev. 10 (6), 539–558. Available from: https://doi.org/10.1016/j.rser.2004.11.004.

Apple, 2018. Apple now globally powered by 100 percent renewable energy. Available from: <https://www.apple.com/newsroom/2018/04/apple-now-globally-powered-by-100-percent-renewable-energy> (accessed 12.05.19.).

Broecks, K.P., van Egmond, S., van Rijnsoever, F.J., Verlinde-van den Berg, M., Hekkert, M.P., 2016. Persuasiveness, importance and novelty of arguments about carbon capture and storage. Environ. Sci. Policy 59, 58–66. Available from: https://doi.org/10.1016/j.envsci.2016.02.004.

Bünning, F., Wetter, M., Fuchs, M., Müller, D., 2018. Bidirectional low temperature district energy systems with agent-based control: performance comparison and operation optimization. Appl. Energy 209, 502–515. Available from: https://doi.org/10.1016/j.apenergy.2017.10.072.

Carley, S.R., Krause, R.M., Warren, D.C., Rupp, J.A., Graham, J.D., 2012. Early public impressions of terrestrial carbon capture and storage in a coal-intensive state. Environ. Sci. Technol. 46 (13), 7086–7093. Available from: https://doi.org/10.1021/es300698n.

Chen, T., Gao, X., Chen, G., 2016. The features, hardware, and architectures of data center networks: a survey. J. Parallel Distrib. Comput. 96, 45–74. Available from: https://doi.org/10.1016/j.jpdc.2016.05.009.

Chen, J., Cheng, S., Nikic, V., Song, M., 2018. Quo vadis? Major players in global coal consumption and emissions reduction. Transform. Bus. Econ. 17 (1), 112–133.

Chiabai, A., Platt, S., Strielkowski, W., 2014. Eliciting users' preferences for cultural heritage and tourism-related e-services: a tale of three European cities. Tourism Econ. 20 (2), 263–277. Available from: https://doi.org/10.5367/te.2013.0290.

Coyle, F.J., 2016. 'Best practice' community dialogue: the promise of a small-scale deliberative engagement around the siting of a carbon dioxide capture and storage (CCS) facility. Int. J. Greenhouse Gas Control 45, 233–244. Available from: https://doi.org/10.1016/j.ijggc.2015.12.006.

Dahlman, E., Mildh, G., Parkvall, S., Peisa, J., Sachs, J., Selén, Y., et al., 2014. 5G wireless access: requirements and realization. IEEE Commun. Mag. 52 (12), 42–47. Available from: https://doi.org/10.1109/MCOM.2014.6979985.

Drozdova, I., Petrov, A., 2018. World practice and Russian experience of housing and utilities sector digitization. SHS Web Conf. 44, 00031. Available from: https://doi.org/10.1051/shsconf/20184400031.

Ericsson, 2019. Future mobile data usage and traffic growth. Available from: <https://www.ericsson.com/en/mobility-report/future-mobile-data-usage-and-traffic-growth> (accessed 19.05.19.).

Farhan, L., Kharel, R., Kaiwartya, O., Hammoudeh, M., Adebisi, B., 2018. Towards green computing for Internet of Things: energy oriented path and message scheduling approach. Sustain. Cities Soc. 38, 195–204. Available from: https://doi.org/10.1016/j.scs.2017.12.018.

Farzaneh, H., Doll, C.N., De Oliveira, J.A.P., 2016. An integrated supply-demand model for the optimization of energy flow in the urban system. J. Clean. Prod. 114, 269–285. Available from: https://doi.org/10.1016/j.jclepro.2015.05.098.

Few, S., Schmidt, O., Offer, G.J., Brandon, N., Nelson, J., Gambhir, A., 2018. Prospective improvements in cost and cycle life of off-grid lithium-ion battery packs: an analysis informed by expert elicitations. Energy Policy 114, 578–590. Available from: https://doi.org/10.1016/j.enpol.2017.12.033.

Gawali, S.K., Deshmukh, M.K., 2019. Energy autonomy in IoT technologies. Energy Proc. 156, 222–226. Available from: https://doi.org/10.1016/j.egypro.2018.11.132.

Glazebrook, G., Newman, P., 2018. The city of the future. Urban Plan. 3 (2), 1–20. Available from: https://doi.org/10.17645/up.v3i2.1247.

Google, 2017. Environmental report 2017. Available from: <https://sustainability.google/reports/environmental-report-2017> (accessed 29.04.19.).

Gubbi, J., Buyya, R., Marusic, S., Palaniswami, M., 2013. Internet of Things (IoT): a vision, architectural elements, and future directions. Future Gener. Comput. Syst. 29 (7), 1645–1660. Available from: https://doi.org/10.1016/j.future.2013.01.010.

Hefnawy, A., Bouras, A., Cherifi, C., 2016. IoT for smart city services: lifecycle approach. In: Proceedings of the International Conference on Internet of Things and Cloud Computing, p. 55. Available from: <https://dl.acm.org/citation.cfm?id=2896440> (accessed 27.05.19.).

Hoffman, A.J., 2005. Climate change strategy: the business logic behind voluntary greenhouse gas reductions. Calif. Manage. Rev. 47 (3), 21–46. Available from: https://doi.org/10.2307/41166305.

IEA, 2019. Smart grids: tracking clean energy progress. Available from: <https://www.iea.org/tcep/energyintegration/smartgrids> (accessed 20.05.19.).

Inkster, N., 2019. The Huawei Affair and China's Technology Ambitions. Survival 61 (1), 105–111. Available from: https://doi.org/10.1080/00396338.2019.1568041.

Kashintseva, V., Strielkowski, W., Streimikis, J., Veynbender, T., 2018. Consumer attitudes towards industrial CO_2 capture and storage products and technologies. Energies 11 (10), 2787. Available from: https://doi.org/10.3390/en11102787.

Kraeusel, J., Möst, D., 2012. Carbon capture and storage on its way to large-scale deployment: social acceptance and willingness to pay in Germany. Energy Policy 49, 642–651. Available from: https://doi.org/10.1016/j.enpol.2012.07.006.

Latif, S., Qadir, J., Farooq, S., Imran, M., 2017. How 5G wireless (and concomitant technologies) will revolutionize healthcare? Future Internet 9 (4), 93. Available from: https://doi.org/10.3390/fi9040093.

Lee, J.H., Phaal, R., Lee, S.H., 2013. An integrated service-device-technology roadmap for smart city development. Technol. Forecast. Soc. Change 80 (2), 286–306. Available from: https://doi.org/10.1016/j.techfore.2012.09.020.

Lindsay, J.R., Cheung, T.M., Reveron, D.S. (Eds.), 2015. China and Cybersecurity: Espionage, Strategy, and Politics in the Digital Domain. Oxford University Press, Oxford, UK.

Lubin, D.A., Esty, D.C., 2010. The sustainability imperative. Harv. Bus. Rev. 88 (5), 42–50.

Maasho, A., 2018. China denies report it hacked African Union headquarters, Reuters. Available from: <https://www.reuters.com/article/us-africanunion-summit-china/china-denies-report-it-hacked-africanunion-headquarters-idUSKBN1FI2I5> (accessed 18.05.19.).

Madu, C.N., Kuei, C.H., Lee, P., 2017. Urban sustainability management: a deep learning perspective. Sust. Cities Soc. 30, 1–17. Available from: https://doi.org/10.1016/j.scs.2016.12.012.

Neirotti, P., De Marco, A., Cagliano, A.C., Mangano, G., Scorrano, F., 2014. Current trends in smart city initiatives: some stylised facts. Cities 38, 25–36. Available from: https://doi.org/10.1016/j.cities.2013.12.010.

Parasol, M., 2018. The impact of China's 2016 Cyber Security Law on foreign technology firms, and on China's big data and smart city dreams. Computer Law Secur. Rev. 34 (1), 67–98. Available from: https://doi.org/10.1016/j.clsr.2017.05.022.

Parvez, I., Rahmati, A., Guvenc, I., Sarwat, A.I., Dai, H., 2018. A survey on low latency towards 5G: RAN, core network and caching solutions. IEEE Commun. Surveys Tutor. 20 (4), 3098–3130. Available from: https://doi.org/10.1109/COMST.2018.2841349.

Ramaswami, A., Russell, A.G., Culligan, P.J., Sharma, K.R., Kumar, E., 2016. Meta-principles for developing smart, sustainable, and healthy cities. Science 352 (6288), 940–943. Available from: https://doi.org/10.1126/science.aaf7160.

Reka, S.S., Dragicevic, T., 2018. Future effectual role of energy delivery: a comprehensive review of Internet of Things and smart grid. Renew. Sustain. Energy Rev. 91, 90–108. Available from: https://doi.org/10.1016/j.rser.2018.03.089.

Sachs, J., Wikstrom, G., Dudda, T., Baldemair, R., Kittichokechai, K., 2018. 5G radio network design for ultra-reliable low-latency communication. IEEE Netw 32 (2), 24–31. Available from: https://doi.org/10.1109/MNET.2018.1700232.

Schmuch, R., Wagner, R., Hörpel, G., Placke, T., Winter, M., 2018. Performance and cost of materials for lithium-based rechargeable automotive batteries. Nat. Energy 3 (4), 267. Available from: https://doi.org/10.1038/s41560-018-0107-2.

Siemens, 2019. The future of energy. Available from: <https://new.siemens.com/global/en/company/stories/research-technologies/future-of-energy.html> (accessed 28.04.19.).

Stern, J., 2002. Security of European Natural Gas Supplies. The Royal Institute of International Affairs, London. Available from: <http://bgc.bg/upload_files/file/Security_of_Euro_Gas_.pdf> (accessed 27.05.19.).

Stromberg, J. (2013). Google's Rick Needham is feeling lucky about the future of sustainable energy. Available from: <https://www.smithsonianmag.com/innovation/googles-rick-needham-is-feeling-lucky-about-the-future-of-sustainable-energy-3415041> (accessed 28.05.19.)

Tarczynska, K. (2016). Money lost to the cloud. How data centres benefit from state and local government subsidies. Available from: <http://www.goodjobsfirst.org/sites/default/files/docs/pdf/datacentres.pdf> (accessed 19.05.19.).

Tuominen, P., Klobut, K., Tolman, A., Adjei, A., de Best-Waldhober, M., 2012. Energy savings potential in buildings and overcoming market barriers in member states of the European Union. Energy Build. 51, 48–55. Available from: https://doi.org/10.1016/j.enbuild.2012.04.015.

Van Winden, W., 2016. Smart city pilot projects, scaling up or fading out? Experiences from Amsterdam. In: Regional Studies Association Annual Conference in Austria, Graz, 3–6 April. Available from: <https://pure.hva.nl/ws/files/811939/RSA_paper_upscaling_RG.pdf> (accessed 29.04.19.).

Walsh, P.F., Miller, S., 2016. Rethinking 'Five Eyes' security intelligence collection policies and practice post Snowden. Intell. Nat. Secur. 31 (3), 345–368. Available from: https://doi.org/10.1080/02684527.2014.998436.

Wennersten, R., Sun, Q., Li, H., 2015. The future potential for carbon capture and storage in climate change mitigation—an overview from perspectives of technology, economy and risk. J. Clean. Prod. 103, 724–736. Available from: https://doi.org/10.1016/j.jclepro.2014.09.023.

Willis, H.L., Philipson, L., 2018. Understanding Electric Utilities and De-regulation, first ed. CRC Press.

Wilson, C., Hargreaves, T., Hauxwell-Baldwin, R., 2015. Smart homes and their users: a systematic analysis and key challenges. Pers. Ubiquit. Comput. 19 (2), 463–476. Available from: https://doi.org/10.1007/s00779-014-0813-0.

Further reading

Gharavi, H., Ghafurian, R., 2011. Smart grid: the electric energy system of the future [scanning the issue]. Proc. IEEE 99 (6), 917–921. Available from: https://doi.org/10.1109/JPROC.2011.2124210.

Harper, C., Harper, C.L., Snowden, M., 2017. Environment and Society: Human Perspectives on Environmental Issues. Routledge. Available from: https://www.taylorfrancis.com/books/9781315463247 (accessed 15.05.19.).

Liserre, M., Sauter, T., Hung, J.Y., 2010. Future energy systems: integrating renewable energy sources into the smart power grid through industrial electronics. IEEE Ind. Electron. Mag. 4 (1), 18–37. Available from: https://doi.org/10.1109/MIE.2010.935861.

Mengolini, A., Vasiljevska, J., 2013. The Social Dimension of Smart Grids. Joint Research Centre of the European Commission, Luxembourg. Available from: <https://doi.org/10.2790/94972> (accessed 20.04.19.).

Wade, N.S., Taylor, P.C., Lang, P.D., Jones, P.R., 2010. Evaluating the benefits of an electrical energy storage system in a future smart grid. Energy Policy 38 (11), 7180–7188. Available from: https://doi.org/10.1016/j.enpol.2010.07.045.

C H A P T E R

9

Conclusions

9.1 Introduction

This is a closing chapter of my book. I hope you have enjoyed it despite its length and broad spectrum of ideas, concepts, and implications presented on its pages. Social dimensions of the smart grids represent a very novel topic and I realize that some of what I envisaged might not materialize. There might appear an additional factor that would change the world in many ways. Who could predict the rise of Internet back in the 1960s or 1970s? Who would believe in the popularity of smartphones in the 1980s and 1990s?

This book included some predictions about the future of the smart grids and the energy markets. Predicting the future of the smart grids (and the future in general) is a very complex and ungrateful task. First of all, not all the information required for making the predictions right is available. However, even if it was readily available, it does not mean that a person attempting to see into the future would be able to locate it and make a good use of all of its relevant factors.

Christopher Sims, a Nobel Prize laureate in Economics, is known (among his other numerous achievements) for designing and studying a theoretical framework called "rational inattention." The framework provides a general approach to modeling decision-making of economic agents with limited abilities to obtain and process information. The concept of the rational inattention assumes that all pieces of information are freely available, but agents process only the most important of them by using channels of limited information capacity. Agents simply do not pay full attention even to information that is easily attainable with a negligible cost (especially today, in the era of Internet, Wi-Fi networks, and smartphones).

Predicting the future is a heterogeneous process based in nature. To describe such processes, higher mathematics should be involved, and with this functional analysis should be employed. A particularly important feature of this process is the concept of resistance to external influences. Take the gust of wind as an example of such an external effect and consider two physical processes: the movement of a train and the flight of a kite. The movement of the train going from point A to point B is a steady process. It is clear that wherever direction the wind is blowing, it would need a storm or an almost impossible set of events that would stop the moving train. But a kite launched into the air without a leash is going to fly under the first gust of wind to the direction no one would be able to predict or calculate. Here, launching a kite represents an unstable process.

There are a lot of stable physical processes and the prediction of future events that obey the laws of physics is based on a well-developed mathematical apparatus. The idea is to get a functional description of these processes. For this, statistical methods (multidimensional regression analysis) and heuristic techniques (e.g., the exponential smoothing method) are commonly used.

The main problem in predicting events that obey sustainable processes is the possible lack of data describing important events and process parameters. The accuracy of the forecast depends on and is estimated by probabilistic statistical methods. The higher the probability of occurrence of events, the more reliable the forecast. The probability of events is determined on the basis of the available dynamic range, which is characterized by points in time and levels of a characteristic of a random variable. The higher the frequency of the feature level, the more likely is the occurrence of a certain event and it is then possible to predict the future result with greater accuracy. For example, if one is doing business with fraudsters, she or he will most likely be deceived, and there are no prerequisites for becoming an exception.

The situation with unstable processes is exactly the opposite. The higher the degree of instability of the process, the more difficult is to produce a good forecast. At the same time, the common methods used in forecasting do not imply the analysis of the process for sustainability, although this characteristic is undoubtedly important for assessing the reliability and accuracy of the forecast.

A system with a high degree of instability, unlike systems with a low degree of instability and stable systems, is hypersensitive to any deviations. Under these conditions, the forecast of the future state of the system is a very difficult task. It is possible to describe the system in the form of three models. The first one assumes the determinism of the Universe, the second one is nondeterministic, and the third one is the nonclosure of the Universe as a system.

The first model assumes that there is absolutely complete and full information about the state of the Universe. This assumption allows one to predict its state at any time in the future, and there is absolutely no uncertainty.

The second model assumes that the available information on the state of the Universe does not allow one to predict its state at any time in the future. There is uncertainty which can be reduced to a random value and the probability theory is then applied for description and driving out the implications.

The third model considers the Universe to be an open-loop system. In this case, even a complete and absolute knowledge of the state of the Universe at any time does not allow for any external influences to be taken into account. For example, even if we have complete and absolute information about the ball, it is impossible to understand where it will fly in the next few seconds if there is a football player next to it, since there is no information about the football player. The future of the Universe simply cannot be predicted on the basis of its state.

Predicting the future with absolute accuracy is not realistic because of the impossibility of a complete description of the state of the Universe. In this case, in the case of a collision with a significant instability, an absolutely accurate measurement is necessary. The deterministic Universe remains partially unpredictable. But, at the same time, there are chances to get enough complete information and provide a forecast that is going to be good enough for a certain period of time (a short-term forecast will be more accurate than a long-term one). The same rules and concepts apply to the prediction of the state of a nondeterministic Universe.

Sustainable processes in the past are predictable. Unstable processes remain almost unpredictable in a deterministic and nondeterministic Universe. Indeed, it is enough for a butterfly to randomly flap its wings once to create a whole unregistered tsunami somewhere on the other side of the world. There must be very strict requirements for the initial conditions, volume, and the quality of the statistical aggregate.

Separately, there is a question of the transition from an unstable process to a sustainable one, and vice versa. In theory, this would allow to initiate changes in the future and to manage them. For example, imagine a situation when on a narrow mountain path, a car falls down from a cliff. The system is in a stable position and most likely the driver is going to die or is going to be seriously injured. If it is possible to achieve an unstable position of the system, then this will clearly be a way of progress which is going to allow the driver to get a chance to escape. The imbalance here is a state in which the system can shift under application of effort.

Let us turn to sci-fi and popular culture to explain that. For those who enjoy sci-fi TV series, let me recall an interesting example from the Fox network TV series called "Fringe" created by J.J. Abrams, Alex Kurtzman, and Roberto Orci. In the third episode of the third season of the series (called "The Plateau"), the FBI agents from the paranormal Fringe Division in the parallel Universe face a series of fatal incidents, which were apparently caused by the incredible series of coincidences. In Hoboken, New Jersey, a patient at Bryant Hospital named "Milo" somehow manipulates a chain of seemingly unrelated events. First, he places a pen on the mailbox, which resonates from a passing car on the road nearby. Then, as the pen falls, it attracts the attention of the customer to the cafe who is trying to raise it. The customer encounters a cyclist who does not cope with his bicycle and has to turn sharply, knocking down a tray with fruits on his way. This event draws the attention of a beggar who is trying to take and advantage from the arising situation by trying to carry away several scattered fruits. The fruit shop attendant runs out of the store and takes the fruits from a beggar. This whole incident distracts the driver of a city bus that is just passing by. As a result, the driver does not notice the switching of the traffic lights and knocks down a girl who is crossing the road. Milo leaves the scene, apparently satisfied with what he just accomplished.

As it turns out during the development of the plot, for 2 days in a row the bus following this route passes through a red light without noticing it and always knocks one of the passers-by to death. Moreover, it turns out that the first and second victims worked in the same place—the Bryant Hospital—as co-workers in the same lab pursuing the same project studying the human brain. The FBI Fringe Division agents acknowledge that this is a very interesting and incredibly rare statistical anomaly.

It turns out that Milo has his brain turned into something like a supercomputer as a result of some novel experimental medical treatment and can obtain and process all available information from his surrounding at a glimpse of a millisecond (and then can make the bus driver to hit the doctor who wants to put him back into the mental institution). The story was actually inspired by Daniel Keyese's book "The Flowers for Algernon" written in 1959. Keyese's book, which is both very beautiful and very sad, tells a story of a retarded patient who underwent some experimental treatment which induced his brain capacity. He is capable of learning languages in days and solving difficult mathematical problems, while his social and personal life are also improving. Unfortunately, the treatment regresses and the patient flaps back from his state of enlightenment to his retarders and eventually premature death.

Milo from Fringe TV series is an example of a superrational economic agent. The mathematical model of "Milo's system" must be considered as a simple Markov process occurring in a system with a finite number of states. Many states of the system include nonreturnable states and one absorbing state which means that the process is absorbing with nonessential states. Transition probabilities can both vary in time or remain constant, and therefore it is advisable to consider the homogeneous and inhomogeneous "Milo processes." A process can be both controlled and uncontrolled. In fact, in order for the system to work, it should correspond to the domino principle, when the previous object transfers its energy to the next one and no longer affects the system. In a real situation, it is necessary to state that different Poisson flows of events can act on each object of the system simultaneously which lead to a failure of control of the object.

9.2 Main lessons, ideas, and conclusions

This book addressed the economic and social challenges of smart grids for the contemporary world. It did so in a number of ways, namely through tackling the following issue and points:

- exploration of the power market design for the increasing share of renewables;
- in-depth prediction of the future (including futuristic) visions of the smart grids;
- study of the interconnectedness between the concept of sustainability and the social impacts of the smart grids;
- analysis of the emerging trends in the power markets closely connected to the principle of sharing economy and its implications for the power markets;
- assessment of the available options for market design in power markets from different national and international perspectives (attributed with case studies and examples);
- investigation of the environmental degradation awareness and environmental stewardship associated with the smart grids.

So, what are the main lessons, ideas, and conclusions for the social impacts of smart grids and the future of energy market design?

Well, it is becoming clear that smart grids are going to make humans more efficient in how they do things and how they produce, transport, and use electricity for their everyday needs. They are going to save lots of time, money, and CO_2 emissions and help to re-think how we should deliver services and produce new goods. Most of the people would surely appreciate that, however, there are going to be some who would despise the idea of being guided by the algorithms-driven machines.

We do not have the fully operational smart grids right now, but there are already attempts to create something that vaguely resembles the concept of the Internet of Energy (IoE). Many businesses use smart sensors that help manufacturers to increase their production efficiency by using the data these sensors on their setups. Moreover, there are attempts to model smart grids of the future in the form of autonomic power systems. Many people understand that the quality and scope of the data across the smart grids and the IoE generate an opportunity for much more contextualized and responsive interaction with devices that would help to optimize the use of electricity and save valuable resources.

In spite of all these achievements, most people realize that smart grids of today are still in their infancy because lots of products do not easily connect to each other. There are also many security issues with devices such and unsecured connected cameras, smart meters, electric bulbs, and other devices to be used both in smart homes of the future and in such sectors and transport, heating, or production. These security issues need to be addressed especially in the face of rising concerns about cybersecurity and energy security that are becoming quite debated nowadays.

There is also one fundamental problem in the core of creating vast reliable smart grids based on the IoE that would be the basis for the smart grids of the future. This problem is called compatible standards. Connected sensors, smart meters, electric appliances, and devices need to "speak" one language in order for humans to make sure they "understand" each other and exchange the data they are recording, forwarding or exchanging with each other over the smart grids. If they all run on different standards, they would struggle to communicate and share information and orders, and this might lead to the unforeseen consequences some of which might be disastrous.

The most pressing problem for the creation of the smart grids of the future in one of the ways we envisage them is the existence of hundreds and thousands of different electrical and electronic standards for different applications with myriads of manufacturers following one of even several standards. Therefore, there is an insistent need for standardization and unification which would let more devices and applications to be interconnected in the smart grid. There is a question, however, what standard to choose and whether the all manufacturers and energy providers would suddenly accept this one unifying standard. This is also a question of social impact of smart grids and their acceptance by human beings. Unlike artificial intelligence (AI) that has its strict purpose and mission, human beings are stubborn and egoistic, so it would be very complicated to make everyone agree to one single standard or set of rules—there would always be someone who would object and complain.

All in all, today's smart grids that are based on the IoE constitute a relatively immature market. In addition, we cannot fully grasp full implications and importance of the smart grids of the future except for realizing that they would have an impact on everything where there is a high cost of not intervening. We can predict and daydream about the smart grids of the future, but we do not yet know the exact use cases. What we do know that they have the potential to make some major impacts on human lives and change the energy markets of the future.

9.3 Pathways for further research

This book is about social impacts of smart grids. It is not an electric or chemical engineering book. After all, I am an economist and economics is a social science. Nevertheless, I think that even economists can (and should) study energy-related issues. We have a term for it—we call it "energy economics and policy." In one way or the other, I firmly believe that both social and natural scientists have found at least a couple of interesting ideas and concepts in this work.

The future of the smart grids and energy market design are related to social sciences in the same way they are related to power systems or to the electric engineering. The future is shaped by humans, however irrational and irresponsible beings they tend to be.

When it comes to the pathways for further research, I think it might be interesting to investigate the implications of not only the AI but also the artificial general intelligence (AGI) on the smart grids and energy markets of the future. The study of what the self-aware AI might do when given control and how it would act when operating the smart grids should be very interesting and intriguing.

Moreover, I think that a large-scale study of social acceptance of smart grids should be done using a quantitative approach. I am not talking about a large-scale questionnaire survey carried out in multiple destinations around the globe (like we did in ISAAC project in Amsterdam, Genoa, and Leipzig) but about the comprehensive study based on the Big Data.

Social networks or specially designed smartphone apps can be employed to obtain the amount of data no questionnaire have ever been able to obtain. Do you remember the case of Cambridge Analytica and the U.S. Presidential elections? The origins of the algorithms used by Cambridge Analytica were developed by Michal Kosinski, a young scientist born in Warsaw. He is a psychometrician, a pioneer of the new and exciting field, but his message goes well beyond profiling people using their psychological profiles, Kosinski showed that psychological profiling (and then influence) can be done using individual behavior on the Internet (including Facebook, Twitter, or Instagram). Michal Kosinski worked at the Cambridge Judge Business School from 2008 to 2014 and left for Stanford University 2 years before I started my work at the University of Cambridge. We missed each other by a couple of years, but I interacted with some of his colleagues who helped me to tackle the issues of Big Data in my energy economics and policy projects related to autonomic power system and advanced smart grids. Even though I do not believe that Kosinski and his colleagues helped Donald Trump to win the Presidential elections in United States, it is quite apparent to me that they opened the Pandora's box. From now on, most of the marketing campaigns would be happy to use microtargeting and profiling to increase their revenues and profits. We, consumers, voters, or respondents in scientific studies, have to be alert. Similarly, all quantitative research in social sciences will increasingly incline to using the Big Data obtained from millions of individuals. The good old statistics and econometrics are slowly but gradually becoming obsolete since the new methods are capable of providing more detailed, microprofiled picture of the world and the relationships between many processes that are taking place in it.

9.4 Policy implications

Speaking about the policy implications resulting from the appearance of functional smart grids and the new energy market design imposed by those grids, one has to remember that all of these constitute a very complex system with many variables.

Decision-makers on the power markets of tomorrow would face many issues starting from standardization and unification to regulation of energy tariffs, taxing the sharing economy schemes, and solving the cybersecurity problems. All of these would require more information and more power to process this information and to make the most optimal decisions. Therefore, it feels that all of the above would require powerful AI that would help humans in charge to act rationally. The question is, however, whether the stakeholders and politicians would always rely upon the rational but perhaps unpopular decisions suggested by AI (we all know how politics works).

Decision-makers of the future are going to have one choice: either to believe the algorithmic black box and do what is says, or to use their own human bounded rationality (this is what scares many people when the topic is brought up). However, one should remember even nowadays many processes are automated and humans are fine with relying upon the computing power of the machines. Just remember the high-frequency trading that constitutes more than a half of trades at world's largest stock exchanges nowadays. Algorithms buy and sell in milliseconds spending millions in various currencies and there is no way for human brain to control what they are doing. Yet, stockbrokers, investors, financiers, and even members of the general public can accept this. Perhaps, the smart grids of energy are not going to be very much different from that after all.

9.5 Closing remarks

Well, this book is finished now. I hope it provided you with some food for thought. On the last two hundred-something pages, I faced a number of complex issues related to the "big turn" from fossil fuels to the renewable energy caused by the attempt to mitigate the devastating effects of the global warming and climate changes. I also tried to explain what the role of the power and electricity grids in the carbon-free future would be. Moreover, I tackled the issue of smart grids of the future—digitalized, AI-powered, optimal, and rational—and the ways people are going to perceive them and embrace or reject them.

People have always tended to oppose the new things and be suspicious to innovations of all kinds. Most innovators understand that and try to base their concepts on this assumption. Sometimes it works but sometimes it leads to wrong decision. Smartphones were invented in the early 1990s by IBM, and later Microsoft and Nokia played with an idea but none of them truly believed that consumers would like to buy a device that has large glass screen and no buttons. By the early 2000s, Nokia ruled the market and its cell phones became a legend. Yet, in 2007, Steve Jobs unveiled Apple's first iPhone, and this changed the market forever. In spite of all expectations and predictions, consumer loved their smartphones.

The same might happen to the smart grids. Many people might oppose them or be afraid of them ("Terminator" and "The Matrix" with their "rise of the machines" shaped up our view of smart systems in somewhat twisted and very wrong way), but when they finally are going to be here fully functional and operational, the attitude might change quickly. It always did in most of the cases documented in history.

In general, I have to say that this book was an interesting journey and I hope you have enjoyed it with me all the way through. I would be very curious to see the real smart grids in action and I hope all of us will soon witness their appearance and their success. Come and join me on this exciting journey to the future.

Index